Methods in Molecular Biology

For further volumes:
http://www.springer.com/series/7651

For over 35 years, biological scientists have come to rely on the research protocols and methodologies in the critically acclaimed *Methods in Molecular Biology* series. The series was the first to introduce the step-by-step protocols approach that has become the standard in all biomedical protocol publishing. Each protocol is provided in readily-reproducible step-by-step fashion, opening with an introductory overview, a list of the materials and reagents needed to complete the experiment, and followed by a detailed procedure that is supported with a helpful notes section offering tips and tricks of the trade as well as troubleshooting advice. These hallmark features were introduced by series editor Dr. John Walker and constitute the key ingredient in each and every volume of the *Methods in Molecular Biology* series. Tested and trusted, comprehensive and reliable, all protocols from the series are indexed in PubMed.

Euglenozoa

Methods and Protocols, Volume 1

Edited by

Paul A.M. Michels

School of Biological Sciences, University of Edinburgh, Edinburgh, UK

Michael L. Ginger

School of Applied Sciences, University of Huddersfield, Huddersfield, West Yorkshire, UK

Anna Karnkowska

Institute of Evolutionary Biology, Faculty of Biology, University of Warsaw, Warsaw, Poland

Laura-Isobel McCall

Department of Chemistry and Biochemistry, San Diego State University, San Diego, CA, USA

Ariel M. Silber

Departamento de Parasitologia, Instituto de Ciências Biomédicas – Universidade de São Paulo, Butantã, São Paulo, Brazil

Humana Press

Editors
Paul A.M. Michels
School of Biological Sciences
University of Edinburgh
Edinburgh, UK

Michael L. Ginger
School of Applied Sciences
University of Huddersfield
Huddersfield, West Yorkshire, UK

Anna Karnkowska
Institute of Evolutionary Biology,
Faculty of Biology
University of Warsaw
Warsaw, Poland

Laura-Isobel McCall
Department of Chemistry and Biochemistry
San Diego State University
San Diego, CA, USA

Ariel M. Silber
Departamento de Parasitologia
Instituto de Ciências Biomédicas
– Universidade de São Paulo
Butantã, São Paulo, Brazil

ISSN 1064-3745 ISSN 1940-6029 (electronic)
Methods in Molecular Biology
ISBN 978-1-0716-5141-4 ISBN 978-1-0716-5142-1 (eBook)
https://doi.org/10.1007/978-1-0716-5142-1

This Humana imprint is published by the registered company Springer Science+Business Media, LLC, part of Springer Nature.
The registered company address is: 1 New York Plaza, New York, NY 10004, U.S.A.

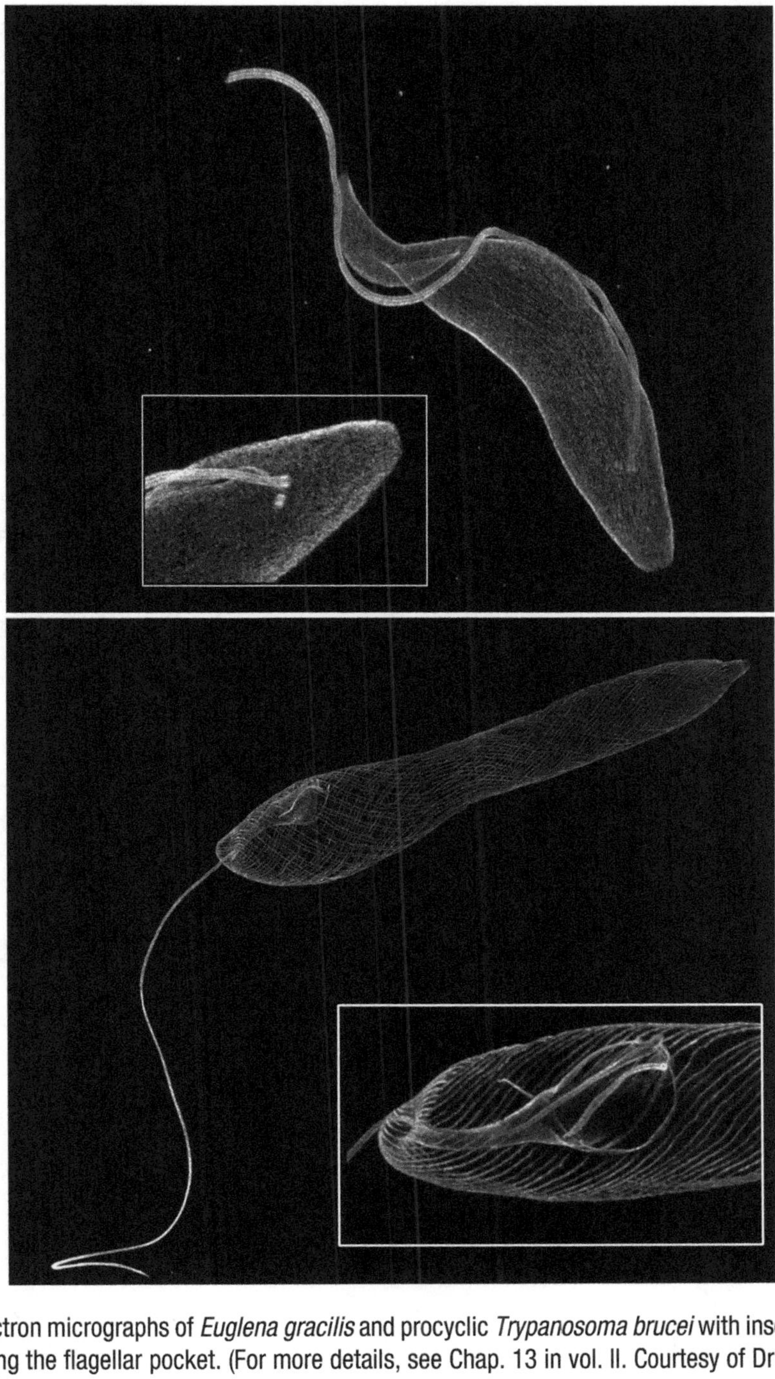

Expansion electron micrographs of *Euglena gracilis* and procyclic *Trypanosoma brucei* with insets detailing the flagellum exiting the flagellar pocket. (For more details, see Chap. 13 in vol. II. Courtesy of Dr Vladimir Varga)

Preface

This volume of Methods in Molecular Biology presents a broad variety of methods used in research on organisms belonging to the protist phylum Euglenozoa. Different groups of euglenozoans have many aspects of extreme biology in common while also displaying interesting differences. The phylum contains the well-known human parasitic trypanosomatids, belonging to the order Kinetoplastea, together with free-living taxa of major ecological and/or translational importance belonging to the orders Diplonemea and Euglenida. Within the Kinetoplastea, there are also many free-living phagotrophic flagellate taxa that are important to the community ecology of many aquatic environments. Occasionally, within the kinetoplastids, the transition from free-living to parasitism is seen, and it was from one free-living kinetoplastid ancestor that the exclusively parasitic trypanosomatid family evolved. Some trypanosomatid species are responsible for serious, mainly neglected topical diseases; other species infect many other animals and plants, and, with rare exception, transmission of trypanosomatids between animal or plant hosts involves a complex life cycle within an insect vector.

For several decades, the human parasitic trypanosomatids have been intensively studied by molecular and cell biologists and biochemists. In contrast, diplonemids have only recently begun to receive wider attention; the realization of their global distribution and the discovery that they are the most abundant eukaryotic microbial heterotrophs in the ocean occurred only in the last decade. In contrast, some euglenids species have received more attention over the years, with the unusual biology of *Euglena gracilis* drawn to the attention of many. More generally, however, the interest in euglenids was, until recently, often limited to botanists and phycologists even though euglenid diversity is dominated by heterotrophic species, rather than photosynthetic algae.

In recent years, a considerable number of researchers who initially worked on parasitic trypanosomatids expanded their investigations to the other, free-living euglenozoans. Their cellular and molecular research showed that all euglenozoans have many aspects of their often-extreme biology in common. Yet, broader comparisons also display interesting differences between the different euglenozoan groups, which are considered likely to inform on the transitions to parasitism in the Kinetoplastea and perhaps the single origin of phototrophy in the Euglenida. The Euglenida, as a group, have also begun to be considered more widely for their potential in different areas of biotechnological industries. For all these reasons, a volume of Methods in Molecular Biology devoted to Euglenozoa seemed timely. Previously, we edited a volume (#2116) specifically devoted to trypanosomatids. We hope that this new book (Volume 1) and the accompanying one (Volume 2), which cover a variety of methods recently developed for experiments on organisms from each of the major euglenozoan groups, will be useful for both colleagues working on the group for which a specific method was developed and for researchers working on organisms of the other groups who may find the method also applicable for their research, if necessary, after some adaptation.

The 44 chapters about Euglenozoa are published in two volumes, arranged in six different sections. The first chapter of this volume presents a review on the history of the research of the different euglenozoans, and their relationship, unique properties, as well as aspects of applied biology. The five subsequent sections each group chapters presenting methods and protocols from a major research discipline applied on different species of euglenozoans.

Finally, we thank all of the authors for the time they have taken to prepare manuscripts for publication in these books, and their patience during the editing process.

Edinburgh, UK *Paul A.M. Michels*
Huddersfield, UK *Michael L. Ginger*
Warsaw, Poland *Anna Karnkowska*
San Diego, CA, USA *Laura-Isobel McCall*
Butantã, São Paulo, Brazil *Ariel M. Silber*

Contents

Contributors

BUNGO AKIYOSHI • *Institute of Cell Biology, School of Biological Sciences, University of Edinburgh, Edinburgh, UK*

MAYKE B. ALENCAR • *Laboratory of Biochemistry of Trypanosomatids—LaBTryps, Department of Parasitology, Institute of Biomedical Science II—ICB II, University of São Paulo—USP, São Paulo, SP, Brazil*

LUCILA ATTALA • *Instituto de Biología Molecular y Celular de Rosario, CONICET-UNR, Rosario, Argentina; Facultad de Ciencias Bioquímicas y Farmacéuticas, Universidad Nacional de Rosario, Rosario, Argentina*

BLYSSALYN V. BIEBER • *Department of Biology, Villanova University, Villanova, PA, USA*

VICTORIA BOSELLI • *Instituto de Biología Molecular y Celular de Rosario, CONICET-UNR, Rosario, Argentina*

GERTRAUD BURGER • *Department of Biochemistry and Robert-Cedergren Centre for Bioinformatics and Genomics, Université de Montréal, Montréal, QC, Canada*

TATIANA M. CÁCERES • *Centro de Investigaciones en Microbiología y Biotecnología-UR (CIMBIUR), School of Sciences and Engineering, Universidad del Rosario, Bogotá, Colombia*

GUILHERME M. P. CARRARA • *Chemistry and Biochemistry, San Diego State University, San Diego, CA, USA*

SERGIO CASTAÑEDA • *Centro de Investigaciones en Microbiología y Biotecnología-UR (CIMBIUR), School of Sciences and Engineering, Universidad del Rosario, Bogotá, Colombia*

DAN CHEN • *Chemistry and Biochemistry, University of Oklahoma, Norman, OK, USA*

AGNIESZKA CZARNOCKA-CIECIURA • *International Institute of Molecular and Cell Biology, Laboratory of RNA Biology, Warsaw, Poland*

JULIA PINHEIRO CHAGAS DA CUNHA • *Cell Cycle Laboratory, and Center of Toxins, Immune Response and Cell Signaling (CeTICS), Butantan Institute, São Paulo, Brazil*

AZUL DE HERNÁNDEZ • *Instituto de Biología Molecular y Celular de Rosario, CONICET-UNR, Rosario, Argentina*

SHANE DENECKE • *Department of Pathobiology, University of Pennsylvania School of Veterinary Medicine, Philadelphia, PA, USA*

FLORENCIA DÍAZ-VIRAQUÉ • *Laboratorio de Interacciones Hospedero-Patógeno—UBM, Institut Pasteur de Montevideo, Montevideo, Uruguay*

RALF ERDMANN • *Department of Systems Biochemistry, Institute of Biochemistry and Pathobiochemistry, Faculty of Medicine, Ruhr University Bochum, Bochum, Germany*

DRAHOMÍRA FAKTOROVÁ • *Institute of Parasitology, Biology Centre, Czech Academy of Sciences, České Budějovice (Budweis), Czech Republic; Faculty of Science, University of South Bohemia, České Budějovice (Budweis), Czech Republic*

MARK C. FIELD • *School of Life Sciences, University of Dundee, Dundee, UK; Biology Centre, Czech Academy of Sciences, Institute of Parasitology, České Budějovice, Czech Republic*

MARÍA ROSA GARCÍA-SILVA • *Laboratorio de Interacciones Hospedero-Patógeno—UBM, Institut Pasteur de Montevideo, Montevideo, Uruguay*

Evgeny S. Gerasimov • *Faculty of Biology, Lomonosov Moscow State University, Moscow, Russia; Institute for Information Transmission Problems, Russian Academy of Sciences, Moscow, Russia*

Michael L. Ginger • *School of Applied Sciences, University of Huddersfield, Huddersfield, West Yorkshire, UK*

Gonzalo Greif • *Laboratorio de Interacciones Hospedero-Patógeno—UBM, Institut Pasteur de Montevideo, Montevideo, Uruguay*

Pranas Grigaitis • *Systems Biology Lab, A-LIFE and AIMMS, Vrije Universiteit Amsterdam, Amsterdam, The Netherlands; Institute of Biological Interfaces, Karlsruhe Institute of Techology, Eggenstein-Leopoldshafen, Germany*

Natalia Gumińska • *International Institute of Molecular and Cell Biology, Laboratory of RNA Biology, Warsaw, Poland*

Jurgen R. Haanstra • *Systems Biology Lab, A-LIFE and AIMMS, Vrije Universiteit Amsterdam, Amsterdam, The Netherlands*

Paweł Hałakuc • *Institute of Evolutionary Biology, Faculty of Biology, University of Warsaw, Warsaw, Poland*

Milou Hoving • *Systems Biology Lab, A-LIFE and AIMMS, Vrije Universiteit Amsterdam, Amsterdam, The Netherlands*

Orsola Iorillo • *Institute of Parasitology, Biology Centre, Czech Academy of Sciences, České Budějovice (Budweis), Czech Republic; Faculty of Science, University of South Bohemia, České Budějovice (Budweis), Czech Republic*

Katarzyna Jankowska • *Faculty of Biology, Biological and Chemical Research Centre, University of Warsaw, Warsaw, Poland*

Vishal C. Kalel • *Department of Systems Biochemistry, Institute of Biochemistry and Pathobiochemistry, Faculty of Medicine, Ruhr University Bochum, Bochum, Germany*

Anna Karnkowska • *Institute of Evolutionary Biology, Faculty of Biology, University of Warsaw, Warsaw, Poland*

Chethan K. Krishna • *Department of Systems Biochemistry, Institute of Biochemistry and Pathobiochemistry, Faculty of Medicine, Ruhr University Bochum, Bochum, Germany*

James A. Letts • *Department of Molecular and Cellular Biology, University of California, Davis, CA, USA*

Julius Lukeš • *Institute of Parasitology, Biology Centre, Czech Academy of Sciences, České Budějovice (Budweis), Czech Republic; Faculty of Science, University of South Bohemia, České Budějovice (Budweis), Czech Republic*

Maja Łukomska-Kowalczyk • *Faculty of Biology, Biological and Chemical Research Centre, University of Warsaw, Warsaw, Poland*

Kacper Maciszewski • *Institute of Parasitology, Biology Centre, Czech Academy of Sciences, České Budějovice, Czechia*

Andrew MacKenzie • *EdinOmics, RRID:SCR_021838, Centre for Engineering Biology, School of Biological Sciences, The University of Edinburgh, Edinburgh, UK*

Rajendra Mandage • *Life Science Research Centre, Faculty of Science, University of Ostrava, Ostrava, Czechia*

Gonzalo Martínez Peralta • *Instituto de Biología Molecular y Celular de Rosario, CONICET-UNR, Rosario, Argentina; Facultad de Ciencias Bioquímicas y Farmacé uticas, Universidad Nacional de Rosario, Rosario, Argentina*

Fraser Massie • *EdinOmics, RRID:SCR_021838, Centre for Engineering Biology, School of Biological Sciences, The University of Edinburgh, Edinburgh, UK*

KEITH MATTHEWS • *Institute for Immunology and Infection Research, Ashworth Laboratories, School of Biological Sciences, University of Edinburgh, Edinburgh, UK*
LAURA-ISOBEL MCCALL • *Department of Chemistry and Biochemistry, San Diego State University, San Diego, CA, USA*
PAUL A.M. MICHELS • *School of Biological Sciences, The University of Edinburgh, Edinburgh, UK*
RAFAŁ MILANOWSKI • *Institute of Evolutionary Biology, Faculty of Biology, University of Warsaw, Warsaw, Poland*
GABRIELA T. MONTANARO • *Laboratory of Biochemistry of Trypanosomatids—LaBTryps, Department of Parasitology, Institute of Biomedical Science II—ICB II, University of São Paulo—USP, São Paulo, SP, Brazil*
TESSA MOSES • *EdinOmics, RRID:SCR_021838, Centre for Engineering Biology, School of Biological Sciences, The University of Edinburgh, Edinburgh, UK*
CARLOS MARIO OSPINA • *Centro de Investigaciones en Microbiología y Biotecnología-UR (CIMBIUR), School of Sciences and Engineering, Universidad del Rosario, Bogotá, Colombia*
LUZ HELENA PATIÑO • *Centro de Investigaciones en Microbiología y Biotecnología-UR (CIMBIUR), School of Sciences and Engineering, Universidad del Rosario, Bogotá, Colombia*
BRUNO R. PINTO • *Laboratory of Biochemistry of Trypanosomatids—LaBTryps, Department of Parasitology, Institute of Biomedical Science II—ICB II, University of São Paulo—USP, São Paulo, SP, Brazil*
CRISTINA POVEDA • *Department of Pediatrics, Division of Tropical Medicine, Baylor College of Medicine, Houston, TX, USA; Texas Children's Hospital Center for Vaccine Development, Houston, TX, USA; Department of Molecular Virology and Microbiology, Baylor College of Medicine, Houston, TX, USA*
MEGAN L. POVELONES • *Department of Biology, Villanova University, Villanova, PA, USA*
MICHAEL POVELONES • *Department of Biology, Villanova University, Villanova, PA, USA*
GALINA PROKOPCHUK • *Institute of Parasitology, Biology Centre, Czech Academy of Sciences, České Budějovice (Budweis), Czech Republic; Faculty of Science, University of South Bohemia, České Budějovice (Budweis), Czech Republic*
JESSE QUATSE • *Department of Pathobiology, University of Pennsylvania School of Veterinary Medicine, Philadelphia, PA, USA*
JUAN DAVID RAMÍREZ • *Centro de Investigaciones en Microbiología y Biotecnología-UR (CIMBIUR), School of Sciences and Engineering, Universidad del Rosario, Bogotá, Colombia; Center for Global Health and Inter-Disciplinary Research, USF Genomics Program, Department of Global, Environmental and Genomic Health Sciences, College of Public Health, University of South Florida, Tampa, FL, USA*
CARLOS ROBELLO • *Laboratorio de Interacciones Hospedero-Patógeno—UBM, Institut Pasteur de Montevideo, Montevideo, Uruguay; Unidad Académica de Bioquímica, Facultad de Medicina, Universidad de la República, Montevideo, Uruguay*
ELVIO RODRÍGUEZ ARAYA • *Instituto de Biología Molecular y Celular de Rosario, CONICET-UNR, Rosario, Argentina; Facultad de Ciencias Bioquímicas y Farmacéuticas, Universidad Nacional de Rosario, Rosario, Argentina*
DANAE SCHULZ • *Department of Biology, Harvey Mudd College, Claremont, CA, USA*
ESTEBAN SERRA • *Instituto de Biología Molecular y Celular de Rosario, CONICET-UNR, Rosario, Argentina; Facultad de Ciencias Bioquímicas y Farmacéuticas, Universidad Nacional de Rosario, Rosario, Argentina*

RUTH SHELTON • *Institute for Immunology and Infection Research, Ashworth Laboratories, School of Biological Sciences, University of Edinburgh, Edinburgh, UK*
DONG-WOO SHIN • *Department of Molecular and Cellular Biology, University of California, Davis, CA, USA*
ARIEL M. SILBER • *Departamento de Parasitologia, Instituto de Ciências Biomédicas – Universidade de São Paulo, São Paulo, Brazil*
INGRID ŠKODOVÁ-SVERÁKOVÁ • *Institute of Parasitology, Biology Centre, Czech Academy of Sciences, České Budějovice (Budweis), Czech Republic; Faculty of Science, University of Ostrava, Ostrava, Czech Republic; Faculty of Natural Sciences, Comenius University, Bratislava, Slovakia*
FAITH A. ST. CLAIR • *Department of Biology, Villanova University, Villanova, PA, USA*
DARIA TASHYREVA • *Institute of Evolutionary Biology, Faculty of Biology, University of Warsaw, Warsaw, Poland*
MATUS VALACH • *Department of Biochemistry and Robert-Cedergren Centre for Bioinformatics and Genomics, Université de Montréal, Montréal, QC, Canada*
ZHIBO YANG • *Chemistry and Biochemistry, University of Oklahoma, Norman, OK, USA; Stephenson Cancer Center, University of Oklahoma Health Sciences Center, Oklahoma City, OK, USA*
VYACHESLAV YURCHENKO • *Life Science Research Centre, Faculty of Science, University of Ostrava, Ostrava, Czechia*
KRISTÍNA ZÁHONOVÁ • *Life Science Research Centre, Faculty of Science, University of Ostrava, Ostrava, Czechia; Institute of Parasitology, Biology Centre, Czech Academy of Sciences, České Budějovice, Czechia; Department of Parasitology, Faculty of Science, Charles University, BIOCEV, Vestec, Czechia; Division of Infectious Diseases, Department of Medicine, Faculty of Medicine and Dentistry, University of Alberta, Edmonton, Canada*
BOŻENA ZAKRYŚ • *Institute of Evolutionary Biology, Faculty of Biology, University of Warsaw, Warsaw, Poland*

Part I

Introduction

Chapter 1

Why Euglenozoans

Michael L. Ginger, Anna Karnkowska, Laura-Isobel McCall, Ariel M. Silber, and Paul A.M. Michels

Abstract

To accompany a new collection of methods and protocols, we discuss the relevance of the microbial eukaryotes belonging to the protist phylum Euglenozoa. For those interested by *Euglena*, applied biology is often relevant: as a starting point for useful natural products including biofuels, nutritional supplements, and metabolites with biomedical potential, or as an environmental agent for bioremediation. Arguably the most widely studied euglenozoans are the parasitic trypanosomatids. Collectively, trypanosomatids cause several serious neglected tropical diseases and economically important diseases of animals and plants; since the early 1900s, drug discovery and disease intervention have been prominent research areas. Yet for those interested in evolution, trypanosomatids and *Euglena* are host to all sorts of extreme biology either not seen or so pronounced in other eukaryotes. Euglenozoans are also relevant in an ecology context: free-living relatives of the trypanosomatids are abundant in freshwater environments. Moreover, the other major euglenozoan group, the diplonemids, are recently recognized as the most abundant heterotrophic protists in the world's oceans, their diversity and abundance at least comparable to major algal groups. Finally, the long history of euglenozoan study illustrates nicely the evolving nature of scientific discovery and reporting since Van Leeuwenhoek first saw *Euglena* in the pioneering days of microscopy.

Key words *Diplonema*, *Euglena gracilis*, Kinetoplast, Kinetoplastid, *Leishmania*, Microbial ecology, Mitochondrial DNA, Neglected tropical diseases, *Trypanosoma*

1 Scene Setting

The Euglenozoa are a broad group of protists rich in biological curios, which belong to an evolutionarily divergent, possibly early-branching eukaryotic lineage called Discoba [1, 2]. They are almost exclusively flagellate; among known euglenozoans only *Perkinsela*, a symbiont of Paramoebae, marine opportunistic pathogens, is known to have lost the ability to build a flagellum or flagella [3]. Morphologically, euglenozoans are highly visual (Fig. 1); cell forms are defined by complex microtubule-based cytoskeletons. In euglenids, elaborate, diverse pellicles adorn the cell surface and underpin a distinctive form of motility based on whole cell

Paul A. M. Michels et al. (eds.), *Euglenozoa: Methods and Protocols, Volume 1*, Methods in Molecular Biology, vol. 3013, https://doi.org/10.1007/978-1-0716-5142-1_1,

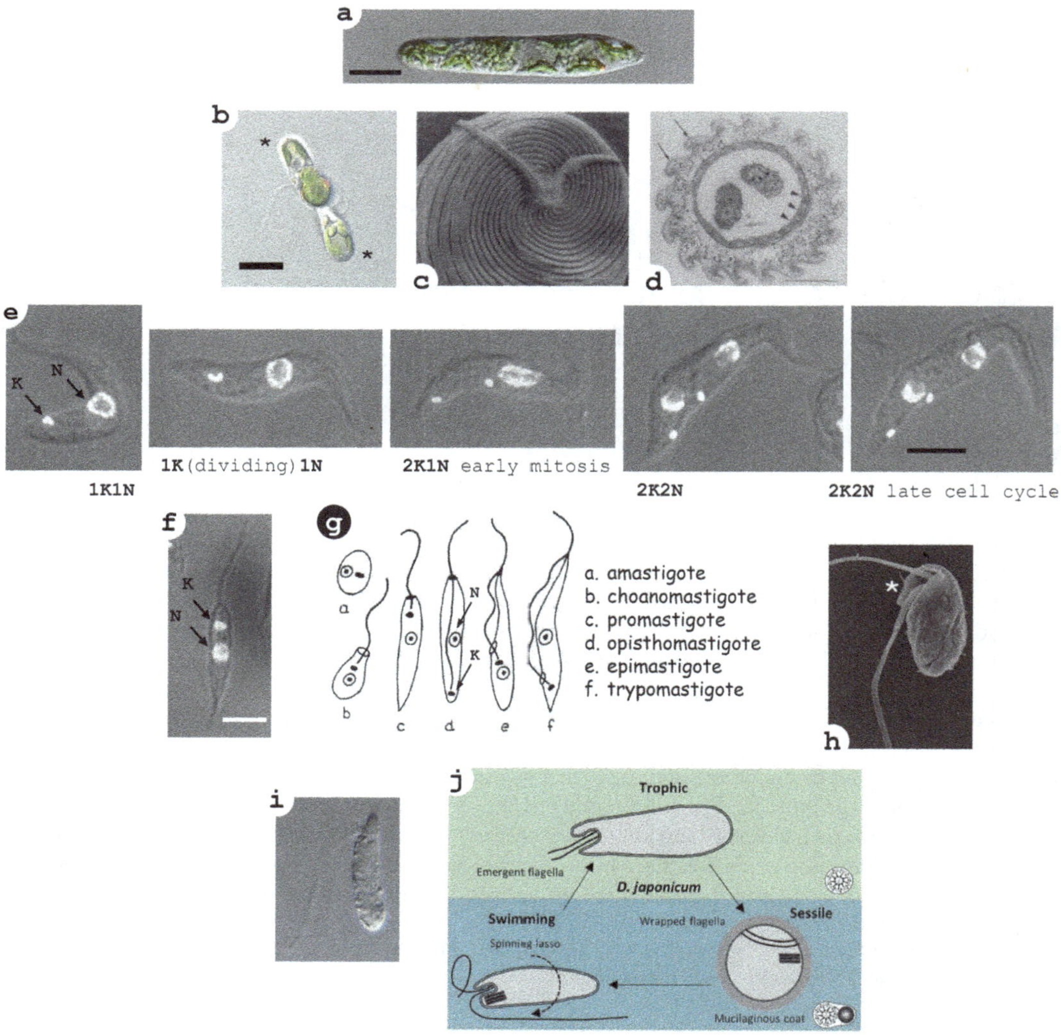

Fig. 1 Morphological diversity within the Euglenozoa. Light micrographs of euglenids *Euglena gracilis* (scale bar, 10 μm) (**a**) (provided by B. Zakryś) and *Rapaza viridis* feeding on prey (**b**) (provided by Y. Kashiyama). In (**b**), two *Rapaza* cells (asterisks) are trying to engulf the prey *Tetraselmis* sp.; the lower cell has lost the competition. (**c, d**) Example of an euglenid pellicle. In (**c**), a scanning electron micrograph shows the anterior cell end of *Distigma proteus* with the two flagella emerging from the canal opening (scale bar, 4 μm). In (**d**), the thin section shows a transverse section through a *D. proteus* canal: 18 peripheral pellicle strips (arrows) terminate leaving 18 sets of microtubules (arrowheads) to form the canal cytoskeleton (scale bar, 0.5 μm); transverse sections through the cell's two flagella are evident in the canal lumen. Images in **c–d** are reproduced with permission from [108]. (**e, f**) Light micrographs of *Trypanosoma brucei* trypomastigotes (**e**) and a *Leishmania tarentolae* promastigote (**f**) (provided by J. Harmer), kinetoplast (K) and nuclear (N) DNA were stained with DAPI. In (**e**), a series of DAPI stained cells illustrate the *T. brucei* procyclic trypomastigote cell cycle: beginning 1K1N, the early cell cycle is characterized by kinetoplast replication and division followed, as a new flagellum elongates, by kinetoplast segregation to yield a 2K1N cell; thereafter mitosis and nuclear segregation occurs prior to cytokinesis and abscission. In trypanosomatids, relative positioning of the kinetoplast and nucleus along the anterior-posterior cell axis combined with the cell body exit point for the flagellum determine cell form (**g**) (cartoons reproduced with permission from [109]). Strikingly, different morphological forms in trypanosomatids can be derived from changes in the expression of one or a few cytoskeletal components. For instance, downregulation of either ClpGM6 or FLAM3, large repetitive proteins in

deformation known as metaboly [4]. Invoking Sonneborn's concept of cytotaxis—the ordering of new cell structures by a parent structure [5]—the flagellum and other cytoskeletal architecture are to the fore in guiding the cell morphogenesis of other euglenozoans [6]. Moreover, many euglenozoans also differentiate through multiple stages, some within complex lifecycles. The visual appeal of euglenozoans thus facilitates ready insight into the relationships between morphotype and behavior, or mutant phenotypes. Yet, far from being curiosities for those with interests in evolution, many extant euglenozoans are also of significant medical, veterinary, ecological, or emerging biotechnological relevance. In the two volumes that follow this introduction, a series of invited methods and protocols are described, which we trust will prove useful for a growing community of euglenozoan researchers and will hopefully also have relevance for researchers interested in other microbial eukaryotes.

Phylum Euglenozoa is composed of four main groups (Fig. 2), one of which, the symbiotids, is largely uncharacterized. It is not considered further in either Volume I or Volume II. The other three groups are the kinetoplastids, which include the relatively well-known parasitic trypanosomatid family, the diplonemids, and the euglenids. Each are represented across the two volumes through a series of molecular methods and protocols. For some of the more recently appreciated euglenozoans, the diplonemids for instance, the protocols focus more on the basics of isolation, cryopreservation, culture, and preliminary biochemical analysis as well as genetic manipulation and genome assembly and annotation. For the trypanosomatids, the advanced protocols described are a culmination of method development over many decades.

2 Euglenids: From Van Leeuwenhoek's "Animalcules" to the Frontiers of Biotechnology Prospecting

Formal descriptions of euglenids began in 1830 with the re-naming of *Cercaria viridis* as *Euglena viridis* by Christian Ehrenberg, but in

Fig. 1 (continued) T. brucei procyclic trypomastigotes, results in kinetoplast-basal body-Golgi-flagellar pocket re-positioning (relative to the nucleus) and an epimastigote-like or, at its extreme, a *Leishmania* promastigote-like appearance [110, 111]. (**h**) A scanning electron micrograph of free-living kinetoplastid, and close trypanosomatid-relative, *Bodo saltans* ingesting two bacterial cells (asterisk) which are caught on the shorter of the cell's two flagella (image, provided by D. Goulding, reproduced with permission from the Wellcome Collection). The cell is ~4 μm long. In contrast to phagotrophic bodonids, the flagellar pocket exit point in osmotrophic trypanosomatids is much more restricted due to the presence of a flagellar pocket collar [112]. (**i**) Light micrograph of diplonemid *Namystinia karyoxenos* (provided by D. Tashyreva). In (**j**), lifecycle biology of another diplonemid, *D. japonicum*, is summarized: when nutritionally stressed (denoted by the blue background) cell morphology changes to sessile and then actively swimming forms. Stage-specific appearance of a PFR and extrusomes (black bars) is shown. The cartoon is reproduced with permission from ref. [86]; the lifecycle biology of *D. japonicum* was described in ref. [88]

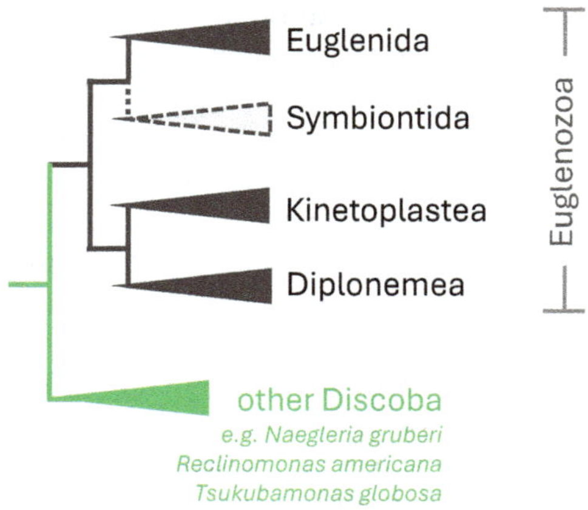

Fig. 2 Relationships within the Euglenozoa. The dotted lines indicate uncertainty with respect to placement of Symbiontida. One multigene phylogeny places Symbiontida as an independent group [16], but other analyses place symbiontids as sister to, or grouped within, the Euglenida

all likelihood *E. viridis* was visible as one of the animalcules described by Antoni van Leeuwenhoek during the 1670s from his examinations of rain-, well-, sea-, and snow-water using early microscopes [7]. Green flagellates resembling *Euglena* species were also evident in environmental samples collected by other pioneers of microscopy and microbiology (e.g. [8]). Large size, eye-catching colorful morphology, distinctive cell motility, plus ubiquity in freshwater undoubtedly explain why euglenids have been seen since the earliest days of cell and microbiology. Yet, only one clade of euglenids, the Euglenophyceae, is (secondarily) photosynthetic owing to acquisition of a plastid of green algal origin. Approximately 3000 taxa are currently classed within Euglenophyceae [9, 10], but curiously, arguably the best known of the Euglenophyceae, and of euglenids generally, *Euglena gracilis*, turned out to be rarer, rather than common, within a recent large-scale study of euglenid population dynamics [11]. Taxonomy of both facultatively phototrophic and the heterotrophic euglenids, initially based on morphological characters, has changed substantially in the last two decades as a consequence of molecular phylogenetics [12–14] and single-cell approaches [15, 16].

Numerous studies have highlighted unusual facets of *E. gracilis* biology: an unusual anaerobic mitochondrial metabolism dependent on wax ester formation [17]; atypical mitochondrial respiratory chain structure [18, 19]; novelty within light-harvesting by the chloroplast photosystems [20]; high tolerance to heavy metal exposure (although perhaps the acidophile *E. mutabilis* is more striking in its response to heavy metal stress (e.g. [21–23]); a capacity for cell surface-mediated environmental sensing, which appears broader than in many other protists [24]; and an unusually large genome (encoding >30,000 protein-coding genes and intron-rich), which still proves fiendishly difficult to assemble or annotate [25]. The origin or evolution of these and other *Euglena* traits are largely unknown, but importantly, the Euglenophyceae cover only a narrow swathe of euglenid diversity [16]. Four groups of heterotrophic euglenids remain largely unstudied—peranemids, anisonemids, ploeotids, and petalomonads [16, 26]. All are in possession of some form of pellicle [4] and a complex microtubule-based cytoskeleton similar to those that are integral to the lifecycle biology of other euglenozoans [27, 28]. Yet, aside from draft data pointing to an intron-rich, high-protein count nuclear genome for heterotroph *Rhabdomonas costata* [29], virtually nothing is known regarding the genome architectures, gene acquisitions and duplications which will have shaped the metabolism, cell biology, or niche adaptations of heterotrophic euglenids. Close to the root of the Euglenophyceae, however, little-studied "green grasper" *Rapaza viridis* has given a glimpse of the range of adaptations that likely await discovery. Classified only in 2012 [30], *R. viridis* recently provided an unexpected sight in euglenids of kleptoplasty—the sequestration of chloroplasts from algal prey by a phagotrophic host cell. It is an observation that may have implications for the establishment of permanent plastids within the Euglenophyceae [31]. Moreover, continued study provided a perhaps rare example among eukaryotes (*see* ref. [32] for further discussion) of a lateral gene transfer shaping adaptation: here, laterally transferred nitrate reductase facilitates inorganic nitrogen use during opportunistic photosynthesis [33]. Despite a long history of study, *Euglena* has lagged behind some protists in the wide availability or applicability of reverse genetics tools. Importantly then, the application in the second study of CRISPR-Cas9 and RNAi to confirm nitrate reductase function points toward the adaptability of contemporary reverse genetics approaches for euglenids currently little-studied in culture or only recently isolated from the field. In Volume II, a proven protocol for genetic engineering of *E. gracilis* is offered (Nagamine et al., Chapter 3 in Vol. II) alongside methods for identification, biodiversity and population dynamics studies in the field (Jankowska et al., Chapter 4 in Vol. I); DNA assembly, annotation, and interrogation (Hałakuc et al., Chapter 7 in Vol. I); sub-cellular fractionation (Tamaki and Ishikawa, Chapter 17 in Vol. II; Tomečková et al., Chapter 16 in Vol. II); and expansion microscopy (Zelená et al., Chapter 13 in Vol. II).

Interesting evolutionary cell biology and biochemistry tell only part of the story for "why euglenids." Translational potential tells the other side of this story. Following indication of extensive secondary metabolism through genome and transcriptome sequencing, secondary metabolites produced by *Euglena* species have begun to be characterized with regards to possible medical applications [34]. As with many algae rich in diverse lipids, euglenophyceans are also considered by some to represent good source material for areas within the biofuel sector although there are technical and economic challenges to overcome, as discussed, for instance, in ref. [35]. Tolerance of a wide range of environmental conditions may be a helpful trait if candidature is to turn into a commercially viable route for biofuel production. Yet, *E. gracilis* is already a protist of industrial significance, grown at scale in North America and South Asia for the preparation of products for healthcare and food industries [26].

In summary, to date, one species only, *E. gracilis*, has been the principal focus for biochemical, molecular, and physiological research of the diverse euglenid group. Yet, these studies, combined with a much smaller volume of reported research from a small handful of other euglenids, offer much intrigue with respect to both evolutionary and translational biology. The stage is set optimistically for a new era of euglenid research.

3 Kinetoplastids: From the "Golden Era of Microbiology" to Twenty-First Century "One Health": A Potted Résumé of 130 Years of Discovery

A unifying characteristic for kinetoplastids is the mass of organellar DNA, readily seen by microscopy using Romanowsky, 4′,6-diamidino-2-phenylindole (DAPI), or other DNA stains, which is proximate to—or in the case of the trypanosomatids, attached to—the flagellar basal bodies. It gives rise to the name for this particular group of euglenozoan flagellates. Originally thought to be involved in the cell's motility, the structure was termed the kinetoplast (or kDNA) derived from the Greek word "kinetikos" meaning "to put in motion." The kinetoplast is now recognized as the mitochondrial DNA present within the single mitochondrion of kinetoplastids. At its extreme, among some free-living kinetoplastids, >60% of cellular DNA is present in the kinetoplast. Moreover, many studies of kDNA and kinetoplastid mitochondrial genome organization have revealed extreme biology: a complex evolutionary picture that, in extant kinetoplastids, yields extensive guide-RNA-mediated U-insertion/deletion editing to transform non-sensical mitochondrial genes into translatable transcripts, alternative editing generating protein diversity, the integration of a complex kDNA S-phase within the overall cell cycle, and the dependence in trypanosomatids of faithful genome segregation upon the physical attachment of kDNA to extra-mitochondrial basal bodies [36–41].

Kinetoplastids, like euglenids, are another taxonomically diverse euglenozoan group, abundant in freshwater environments and with some presence in marine environments too. Three orders are recognized: Neobodonida, Eubodonida, and Parabodonida, alongside the outlying Prokinetoplastida [42]. The vast majority of kinetoplastids are free-living, although parasitism has evolved independently in some lineages, most notably within the parasitic trypanosomatid family—incidentally, a few euglenids have also been considered parasites (discussed further in ref. [43]). Approximately 25 trypanosomatid genera are recognized. The number has risen steadily over the last 15 or so years as a consequence of widespread insect sampling; most trypanosomatids are monogenetic parasites of invertebrates, but occasionally, independently digenetic parasitism of an invertebrate vector and either an animal or plant host has evolved [42].

The study of trypanosomatid biology harks back to the second half of the nineteenth century, the "golden age" of microbiology, when methods advanced and a new discipline of medical microbiology opened with the realization that microorganisms were the aetiological agents of infectious disease. The first pathogenic trypanosomatid parasite to be reliably described was *Trypanosoma evansi* in 1880. The "wonderfully active" parasite with its "undulatory, eel-like" motion was discovered in the Dera Ismael Khan district of Pakistan by veterinarian Griffith Evans through his investigation into the cause of "surra," a fatal disease affecting horses and camels belonging to the Punjab Frontier Force [44]. However, from those early days of trypanosomatid discovery, Bruce is the name more readily known by many. Working alongside rural farmers in KwaZulu-Natal, Surgeon-Major David Bruce and his wife Mary Bruce reported *Trypanosoma brucei* as a cause of nagana (or animal African trypanosomiasis, AAT), a tsetse fly-transmitted cattle-wasting disease common in sub-Saharan Africa [45, 46]. Subsequently, during the early years of the twentieth century, trypanosomatids responsible for three types of serious human diseases in tropical and subtropical countries were identified: *Trypanosoma brucei gambiense* causing sleeping sickness—or human African trypanosomiasis (HAT)—in The Gambia [47], and subsequently HAT caused by *T. b. rhodesiense* in what was then North-East Rhodesia (now Zimbabwe) [48] and elsewhere; *T. cruzi* causing Chagas disease—or American trypanosomiasis—first in Brazil and then in other regions of Latin America [49]; and *Leishmania donovani* and *L. infantum*, causal agents of visceral leishmaniasis, in India and Tunisia, respectively [50]. Curiously, *T. cruzi* was first isolated from reduviid bugs as a protist that looked as though it should be a parasite; the discovery of *T. cruzi* in a young child, Berenice, subsequently confirmed the hypothesis [51]. The 1900s also saw discovery of the two *Trypanosoma* species responsible for the majority of AAT incidence, *T. congolense* and *T. vivax* [52, 53], and the proposition of plant-pathogenic trypanosomatids, too [54], although the notion of *Phytomonas* as a plant pathogen remained contentious until its involvement in hartrot disease of coconut palms was described during the 1970s [55]. With the exception of *T. evansi*, the newly discovered pathogenic trypanosomatids were vector-transmitted; the complexity of the lifecycle and the migratory journeys undertaken by these parasites within their respective invertebrate vectors would become clear during the twentieth and early twenty-first centuries. *T. evansi* and the closely related *T. equiperdum*, in contrast, are mechanically, vertically, or sexually transmitted between the animals in which they cause disease. Consequently, their geographic distribution is very broad [56], but they are extremely closely related [57] to the African trypanosome *T. brucei* discovered by Sir David and Lady Mary Bruce; in recognition of his numerous contributions, not simply the scientific discoveries of three disease agents [58], the former was knighted in 1908, and Mary was herself a recipient of the Royal Red Cross and the Order of the British Empire. The biological "curio" of *T. evansi* and *T. equiperdum* is that they represent another extreme of kinetoplastid mitochondrial biology; they are the parasites, which, because of mutations in a nuclear-encoded ATP synthase subunit [59], grow having reduced (dyskinetoplastic) or lost (akinetoplastic) their kDNA.

Right from the outset, it was clear that different trypanosomatids were related to one another: their flagellate forms were morphologically similar, and each was recognizable from a kinetoplast proximate to the flagellar basal body. Yet, it was only over a 30-year period beginning in the 1960s that an evolutionary relatedness between trypanosomatids and euglenids became clear [60]. Throughout the twentieth century, and indeed into the first three decades of the twenty-first century (e.g. [61–63]), novel strains and species of *Leishmania*, responsible for the very different manifestations of leishmaniasis seen across the tropics and subtropics, have continued to be recognized.

In research, the focus was often the development of medicines to treat the serious, often life-changing or life-ending diseases caused by trypanosomatids—for instance, antimonial compounds introduced in the 1920s to treat leishmaniasis and the repositioning of the antifungal amphotericin B as an antileishmanial therapy from the 1960s onwards; suramin, melarsoprol, and eflornithine introduced in 1922, 1949, and 1990, respectively, to treat HAT. For several reasons, however, the need for new medicines persists. Large populations remain at risk of disease: climate change and shifting patterns of land use, combined with socioeconomic factors, are likely to see the numbers at risk of Chagas disease or leishmaniasis increase further, as geographic ranges for insect vectors permissive for parasite transmission alter [64–66]. There is also still no clinically effective treatment for symptomatic chronic Chagas disease, and toxicity, limited efficacy, and/or drug resistance are seen for many of the medicines currently still in use to treat HAT or leishmaniasis.

Encouragingly, despite limited economic incentive for pharmaceutical companies to take on responsibility for developing medicines to treat trypanosomatid diseases endemic in low-income countries, there has been, from the 1970s to the current day, an increasingly diverse range of funding opportunities for new drug discovery. The first of these funding initiatives was the Special Programme for Research and Training in Tropical Diseases (TDR), created in 1974 by the World Health Organization (WHO) to fund research on a defined group of neglected tropical diseases, with cosponsoring by the United Nations Development Programme (UNDP), the World Bank, and UNICEF. Subsequently, other funders joined the drug discovery venture: the European Economic Community (later the European Union), the National Institutes of Health (NIH) in the USA, and The Wellcome Trust in the UK have all been prominent in supporting collaboration between non-endemic and endemic regions, as have philanthropic and charity organizations, which have come to the fore over the last 25 years or longer. Although new medicines have been hard to come by, there have been interesting diversions—the DNA stain DAPI was discovered from an otherwise unfruitful screen for anti-trypanosome compounds [67, 68]. An older medicine, better known for its relevance to molecular and cell biology, also has an interesting backstory: ethidium bromide, a mainstay reagent of twentieth century molecular biology [69, 70], was the culmination of a trypanosomiasis drug development effort begun by scientists working at the Chemical Research Laboratories in Teddington (UK) and the University of Glasgow. Their effort began to build momentum in the late 1930s with discovery of phenanthrene compounds with trypanocidal activity [71]. A subsequent derivative, ethidium bromide, synthesized and field-tested in the 1950s by scientists of Boots Pure Drug Company Limited of Nottingham (UK) [72], became the successful AAT treatment Homidium. Over 70 years since its efficacy was first seen, there is still some uncertainty why ethidium bromide is effective against AAT, but inhibition of kDNA minicircle replication has appeared as a likely major mode of action [73]. Returning to the present day, as consequences of funding available, improved diagnostics, surveillance, and healthcare infrastructure are contributing to an overall optimistic picture for the possibilities to control diseases caused by trypanosomatids.

It is important to appreciate kinetoplastid, and in particular trypanosomatid, biology has not only been a focus for applied research. The evolution of kinetoplastid mitochondrial genome architectures, replication, and gene expression continues to capture a broad interest [74, 75]. However, many other examples of bizarre or extreme biology have also been seen in trypanosomatids: since the 1960s and the pioneering work of Keith Vickerman [76], *T. brucei* has served as the paradigm for antigenic variation as an immune evasion tactic and research in the area continues to provide surprises [77, 78]; polycistronic transcription, genome-wide *trans*-splicing of mRNA and a near absence of transcription promoter-controlled gene expression; divergent chromosome segregation mechanisms [79]; the novel base "J", which plays roles in transcription termination and initiation [80, 81]; a trypanothione-based system for combatting reactive oxygen species; and extreme peroxisomes known as glycosomes into which many enzymes of carbohydrate metabolism, including glycolysis are imported. In large part, many of these discoveries were facilitated by the amenability of trypanosomatids to in vitro culture or growth in animal models coupled to the development, since the late 1980s of increasingly sophisticated tools for reverse genetics. Methods or protocols describing new or increasingly refined reverse genetics approaches are covered in Part I of Volume II (Arnould et al., Chapter 5 in Vol. II; Hohnen et al., Chapter 6 in Vol. II; Krishna et al., Chapter 7 in Vol. II); others are described in a previous MiMB volume we edited—Trypanosomatids: Methods and Protocols, volume 2116. Amenability of many trypanosomatids to in vitro culture provides biomass from which biochemical studies become feasible. Within Volume II, chapters within Part III document contemporary biochemistry protocols, including those for the study of cell signaling and solute transport. Analysis at genome-, proteome-, or metabolome-wide scales, which are also, at least in part, methods developed from working with parasites readily amenable to laboratory experiments, is covered in Part III of Volume I. Lastly, the chapters in Part II of Volume II, "Cell Biology," provide protocols for high-throughput analysis of parasite morphology (Howell et al., Chapter 10 in Vol. II) and the increasingly widely applied approach of expansion microscopy (Gabiatti et al., Chapter 12 in Vol. II; Zelená et al., Chapter 13 in Vol. II); the latter being an alternative for more resource-heavy, high-resolution imaging by transmission electron, stimulated emission depletion, structured illumination, stochastic optical reconstruction, or Airyscan microscopy.

Finally, while there is concern that the consequences of climate change and shifting patterns of land use will influence the epidemiology of human disease, trypanosomatid pathogenicity has also moved into a newer, broader arena, functional ecology. Here there is renewed interest in the parasitism of bumblebees, honey bees, and solitary bees by some *Crithidia* species, and the contribution this conceivably has on declining pollination globally (discussed further in ref. [82]). This emerging ecosystem relevance is reflected through the inclusion of protocols for culture and genetic manipulation of bumblebee pathogen *C. bombi*, as well as *C. fasciculata*, a mosquito parasite, which has long served as a non-human pathogenic trypanosomatid model (Bieber et al., Chapter 5 in Vol. I).

4 Diplonemids: A Tara Oceans Tale from Invisible Obscurity to Candidate Keystone Status

In contrast to longstanding histories of euglenid and kinetoplastid research, diplonemids, the third major euglenozoan group, can be viewed as a once best kept secret. Until the Tara Oceans expeditions from 2009 to 2013, only a handful of species from two genera were known and diplonemids were considered among the rarer occupants of freshwater and marine environments. Insightfully, even in the absence of molecular data, however, a common evolutionary ancestry between euglenids and diplonemids was suggested following initial descriptions of taxa by Skuja and Griessmann [83]. Now, from 18S rDNA barcoding, diplonemids are recognized as highly abundant, species-rich heterotrophs throughout the world's oceans, found at all depths in the water column. Their diversity and number are at least comparable to other major oceanic protist groups, the stramenopiles and dinoflagellates [84–86]. Such species richness and population abundance point towards keystone microbial species status within the oceans [86], and in that regard, there is evidence of a role for diplonemids in biomineralization [87]. To date, however, only a few species have been brought into culture (reviewed in ref. [86]).

Of the species brought into axenic culture, *Diplonema papillatum* (now also known as *Paradiplonema papillatum*) is currently the model diplonemid for basic, but informative, biochemical studies, and a range of plasmids are available for reverse genetics. Chapters by Valach, Prokopchuk, and Faktorová, alongside their respective colleagues offer further details of how to carry out basic biochemical (Chapter 15 in Vol. II) or cell biology (Chapter 3 in Vol. I) studies and to manipulate *P. papillatum* (Chapter 1 in Vol. II). In culture, *P. papillatum* morphology varies according to the age of batch-fed cultures. Of the small number of other species brought into culture, some provide unequivocal evidence of morphological differentiation in response to nutritional cues (Fig. 1j) [86, 88]. For example, starvation causes *D. japonicum* to transition first into a sessile form and thereafter into an actively swimming form. During this transition or differentiation process, the flagella are elongated and a paraflagellar rod (PFR) assembled alongside each extending axoneme. This is an interesting observation because a PFR has never been evident in *P. papillatum*. Yet, with diplonemids being sister to the kinetoplastids, it would perhaps have been odd if the ornate extra-axonemal structure so distinctive in kinetoplastids and euglenids, and also evident in the enigmatic Symbiontida [89], were not a feature of the third major euglenozoan group, particularly given the importance of the PFR for normal trypanosomatid cell morphogenesis and motility [90–93].

Unsurprisingly, genome and transcriptome sequences in draft or annotated form are becoming available for various diplonemids. Currently, the most completely annotated is a *P. papillatum* draft genome [94], which revealed a genome architecture with a mix of trypanosomatid- and *Euglena*-like characteristics: overall genome size lies intermediate between the expansive *E. gracilis* genome and smaller kinetoplastid genomes; polycistronic transcription occurs, as it does in kinetoplastids; genome-wide *trans*-splicing of mRNA appears a unifying trait within Euglenozoa; and similar to *Euglena*, but in contrast to kinetoplastids, protein-coding genes are intron-rich, albeit *Paradiplonema* genes only contain conventional introns, in contrast to the atypical introns often present in *Euglena*.

From diplonemid genomes and transcriptomes, intermediary metabolism adapted to often nutritionally challenging open ocean environments is evident [94, 95]. It has also been seen that the origin of glycosomes reaches back to a common ancestor of diplonemids and kinetoplastids [96]. In contrast, the evidence available indicates peroxisomal compartmentalization of carbohydrate metabolism does not occur in euglenids or at least does not occur in phototrophic euglenids [24]. Detailed investigation of the glycolytic enzymes present in diplonemids, however, reveals for some (e.g. phosphofructokinase) complex patterns of gene loss, gain, and evolution among different diplonemids [97]. The emerging molecular data also indicate diplonemids, like other euglenozoan groups, present their own variations on extreme biology. Kinetochores, for instance, are different again from the already different mechanism of chromosome segregation seen in kinetoplastids [98], but perhaps their most striking "extreme curio" is mitochondrial genome organization and expression. Here, diplonemids have ripped up the rulebook in a completely different way to kinetoplastids: masses of circular DNA molecules are again present, with a suggestion of structure to the organization of circular DNA within the mitochondrial matrix but the ~18 mitochondrial protein-coding genes present are fragmented and nonsensical across different circles and require *trans*-splicing and various forms of RNA-editing for functional gene expression [99]. Assembly and annotation of peculiar and complex diplonemid and kinetoplastid mitochondrial genomes is the topic covered in the chapter by Gerasimov and colleagues (Chapter 9 in Vol. I). In contrast to kinetoplastids and trypanosomatids, there is little unusual about the *Euglena* mitochondrial genome: it is streamlined in gene content, although not particularly streamlined in comparison to many other eukaryotes, with seven respiratory chain components encoded as intact genes across a series of small (5–8 kb), recombination-prone linear DNA molecules [100].

To close, we return to the broader relevance of diplonemid study; it is surely investigating their role in the microbial and broader ecosystems on the world's oceans that is more critical than their place in the fascinating world of biological curios. We hope methods and protocols offered, or potentially suitable, for diplonemids across Volumes I and II stimulate new research beyond the relatively small number of groups currently interested in these euglenozoans.

5 Closing Discussion: A Nod to the Past, a Look to the Future

That the study of euglenozoan biology dates back to the pioneering days of microscopy, microbiology, and the world's first scientific journal, *Philosophical Transactions* [101], is coincidence. Yet, that early advent is perhaps fitting, given the many applied reasons to study euglenozoans and the deep treasure trove of biological novelties or "curios" offered up by this flagellate protist group.

Intriguingly, there is still an enigmatic fourth euglenozoan group, the Symbiontida, that, like the diplonemids, are also marine dwellers. Known only from environmental DNA sequencing and limited ultrastructural characterization, these euglenozoans have presently only been found in micro-oxic or anaerobic deep-sea habitats [102, 103]. Whether symbiontids are minor or major players in these ecosystems, or are present in other environments, awaits discovery, and perhaps they will assume a greater presence in future volumes of *Methods in Molecular Biology* documenting methods and protocols for euglenozoan protists. There will certainly be relevance for such books as the journeys of discovery, disease intervention, and biotechnological application continue. One could look at any or many euglenozoan taxa to illustrate a point but to take trypanosomiasis by way of continuing relevance: arguably, the economic cost is now greater for AAT than it is for HAT, particularly when considering the spread of animal disease caused by *T. evansi* or *T. equiperdum* across North Africa, Asia, and Latin America. Yet, recent development of new veterinary medicines has been slow while resistance to current medicines and weakening infrastructure reporting incidence and intervention underpin a need for new innovation [104]. In contrast, in the first 50–60 years of the twentieth century, working with less knowledge and resource, a plethora of compounds were developed and used (indeed, continue to be used) to treat AAT [56].

Finally, the Euglenozoa provide a fine illustrative example of how the science of biological discovery and its reporting have evolved since the early days of *Philosophical Transactions* and the diary-style correspondence of the first to discover or chart something novel [7, 8, 101]. Through the twentieth century, into the 1970s, major new reports of discovery from the biologists and the chemists, the latter generally seeking new medicines, were often very concise, sometimes communicated in languages other than English. During this period, collaborations between academic and industry were already in evidence. Through the "golden age" of discovery, the influence and reach of colonial Empire is evident with key discoveries often documented comprehensively in reports to government, rather than through peer-review at a scientific journal. Digitally archived and accessible, the detail, expression, and illustration offered by these pioneers readily grip the reader [44, 45]. Moving into the twenty-first century, collaboration across borders or between stakeholders from different sectors is often easier to facilitate, the author lists describing new data are often longer. Whilst "impactful" discoveries still command attention, with a tide of new journals, data even when negative or without obvious immediate impact have also become an important currency; the "negative" or "without impact" generally being accessible through open access publishing. Within the evolving faces of publishing within the life sciences, the current pioneers of research are adopting "en masse" the sorts of pre-peer review, pre-print practices, which have long been common in some other disciplines. Amid a variety of platforms (plus accompanying platforms for pre-print review) bio$R_x$iv and its younger sibling med$R_x$iv have become the "go-to" for life scientists. Undoubtedly accelerating the rate of scientific discovery, this "new" publishing practice among life scientists [105, 106] can be viewed, however, as a maturation of a life-sciences experiment harking back to the 1960s [107].

Acknowledgments

We thank Yuichiro Kashiyama (Fukui University of Technology), Daria Tashyreva and Bożena Zakryś (University of Warsaw), and Jane Harmer (University of Huddersfield) for use of unpublished images in Fig. 1. AK was funded by the National Science Centre, Poland [grant number 2023/50/E/NZ8/00470]. AMS was supported by Fundação de Amparo à Pesquisa do Estado de São Paulo (FAPESP) [grant numbers 2021/12938-0, 2023/08204-6] and Conselho Nacional de Pesquisas Científicas e Tecnológicas (CNPq) [grant number 307487/2021-0].

References

1. Lukeš J, Leander BS, Keeling PJ (2009) Cascades on convergent evolution: the corresponding evolutionary histories of euglenozoans and dinoflagellates. Proc Natl Acad Sci USA 106:9963–9970. https://doi.org/10.1073/pnas.0901004106
2. Williamson K, Eme L, Baños H et al (2025) A robustly rooted tree of eukaryotes reveals their excavate ancestry. Nature 640:974–981. https://doi.org/10.1038/s41586-025-08709-5
3. Tanifuji G, Cenci U, Moog D et al (2017) Genome sequencing reveals metabolic and cellular interdependence in an amoeba-kinetoplastid symbiosis. Sci Rep 7:11688. https://doi.org/10.1038/s41598-017-11866-x

4. Leander BS, Esson HJ, Breglia SA (2007) Macroevolution of complex cytoskeletal systems in euglenids. BioEssays 29:987–1000. https://doi.org/10.1002/bies.20645
5. Beisson J, Sonneborn TM (1965) Cytoplasmic inheritance of the organization of the cell cortex in *Paramecium aurelia*. Proc Natl Acad Sci USA 53:275–282. https://doi.org/10.1073/pnas.53.2.275
6. Moreira-Leite FF, Sherwin T, Kohl L et al (2001) A trypanosome structure involved in transmitting cytoplasmic information during cell division. Science 294:610–612. https://doi.org/10.1126/science.1063775
7. Lane N (2015) The unseen world: reflections on Leeuwenhoek (1677) 'concerning little animals'. Philos Trans R Soc Lond Ser B Biol Sci 370:20140344. https://doi.org/10.1098/rstb.2014.0344
8. Harris J (1696) Some microscopical observations of vast numbers of animalcula seen in water. Philos Trans 19:254–259. https://doi.org/10.1098/rstl.1695.0036
9. Rackevei AS, Karnkowska A, Wolf M (2023) 18S rDNA sequence-structure phylogeny of the Euglenophyceae (Euglenozoa, Euglenida). J Eukaryot Microbiol 70:e12959. https://doi.org/10.1111/jeu.12959. Epub 2022 Dec 15
10. Zakryś B, Milanowski R, Karnkowska A (2017) Evolutionary origin of *Euglena*. In: Schwartzbach SD, Shigeoka S (eds) Euglena: biochemistry, cell and molcular biology. Springer, Cham
11. Gumińska N, Łukomska-Kowalczyk M, Chaber K et al (2021) Evaluation of V2 18S rDNA barcode marker and assessment of sample collection and DNA extraction methods for metabarcoding of autotrophic euglenids. Environ Microbiol 23:2992–3008. https://doi.org/10.1111/1462-2920.15495
12. Karnkowska A, Bennett MS, Watza D et al (2015) Phylogenetic relationships and morphological character evolution of photosynthetic Euglenids (Excavata) inferred from taxon-rich analyses of five genes. J Eukaryot Microbiol 62:362–373. https://doi.org/10.1111/jeu.12192
13. Kim JI, Linton EW, Shin W (2015) Taxon-rich multigene phylogeny of the photosynthetic euglenoids (Euglenophyceae). Front Ecol Evol 3. https://doi.org/10.3389/fevo.2015.00098
14. Łukomska-Kowalczyk M, Chaber K, Fells A et al (2021) Description of *Flexiglena* gen. nov. and new members of Discoplastis and Euglenaformis (Euglenida). J Phycol 57: 766–779. https://doi.org/10.1111/jpy.13107
15. Chaber K, Łukomska-Kowalczyk M, Fells A et al (2022) Toward the robust resolution of taxonomic ambiguity within Lepocinclis (Euglenida) based on DNA sequencing and morphology. J Phycol 58:105–120. https://doi.org/10.1111/jpy.13220
16. Lax G, Kolisko M, Eglit Y et al (2021) Multigene phylogenies of euglenids based on single-cell transcriptomics of diverse phagotrophs. Mol Phylogenet Evol 159:107088. https://doi.org/10.1016/j.ympev.2021.107088
17. Nakazawa M, Inui H (2023) Understanding wax ester synthesis in *Euglena gracilis*: insights into mitochondrial anaerobic respiration. Protist 174:125996. https://doi.org/10.1016/j.protis.2023.125996
18. He Z, Wu M, Tian H et al (2024) *Euglena*'s atypical respiratory chain adapts to the discoidal cristae and flexible metabolism. Nat Commun 15:1628. https://doi.org/10.1038/s41467-024-46018-z
19. Mühleip AW, Dewar CE, Schnaufer A et al (2017) *In situ* structure of trypanosomal ATP synthase dimer reveals a unique arrangement of catalytic subunits. Proc Natl Acad Sci USA 114:992–997. https://doi.org/10.1073/pnas.1612386114
20. Miranda-Astudillo H, Arshad R, Vega de Luna F et al (2025) A unique LHCE light-harvesting protein family is involved in photosystem I and II far-red absorption in *Euglena gracilis.* bioRxiv 2025.05.07.652572. https://doi.org/10.1101/2025.05.07.652572
21. Halter D, Andres J, Plewniak F et al (2015) Arsenic hypertolerance in the protist *Euglena mutabilis* is mediated by specific transporters and functional integrity maintenance mechanisms. Environ Microbiol 17:1941–1949. https://doi.org/10.1111/1462-2920.12474
22. Kennedy V, Kaszecki E, Donaldson ME et al (2024) The impact of elevated sulfur and nitrogen levels on cadmium tolerance in *Euglena* species. Sci Rep 14:11734. https://doi.org/10.1038/s41598-024-61964-w
23. Mangal V, Stenzler BR, Poulain AJ et al (2019) Aerobic and anaerobic bacterial mercury uptake is driven by algal organic matter composition and molecular weight. Environ

Sci Technol 53:157–165. https://doi.org/10.1021/acs.est.8b04909

24. Ebenezer TE, Zoltner M, Burrell A et al (2019) Transcriptome, proteome, and draft genome of *Euglena gracilis*. BMC Biol 17:11. https://doi.org/10.1186/s12915-019-0626-8
25. Fields O, Hammond MJ, Xu X et al (2025) Advances in euglenoid genomics: unravelling the fascinating biology of a complex clade. Trends Genet 41:251–260. https://doi.org/10.1016/j.tig.2024.07.007
26. Ebenezer TE, Low RS, O'Neill EC et al (2022) *Euglena* International Network (EIN): driving euglenoid biotechnology for the benefit of a challenged world. Biol Open 11:bio059561. https://doi.org/10.1242/bio.059561
27. Yubuki N, Leander BS (2012) Reconciling the bizarre inheritance of microtubules in complex (euglenid) microeukaryotes. Protoplasma 249:859–869. https://doi.org/10.1007/s00709-011-0340-z
28. Wheeler RJ, Gull K, Sunter JD (2019) Coordination of the cell cycle in trypanosomes. Ann Rev Microbiol 73:133–154. https://doi.org/10.1146/annurev-micro-020518-115617
29. Soukal P, Hrdá Š, Karnkowska A et al (2021) Heterotrophic euglenid *Rhabdomonas costata* resembles its phototrophic relatives in many aspects of molecular and cell biology. Sci Rep 11:13070. https://doi.org/10.1038/s41598-021-92174-3
30. Yamaguchi A, Yubuki N, Leander BS (2012) Morphostasis in a novel eukaryote illuminates the evolutionary transition from phagotrophy to phototrophy: description of *Rapaza viridis* n. gen. et sp. (Euglenozoa, Euglenida). BMC Evol Biol 12:29. https://doi.org/10.1186/1471-2148-12-29
31. Karnkowska A, Yubuki N, Maruyama M et al (2023) Euglenozoan kleptoplasty illuminates the early evolution of photoendosymbiosis. Proc Natl Acad Sci USA 120:e2220100120. https://doi.org/10.1073/pnas.2220100120
32. Keeling PJ (2024) Horizontal gene transfer in eukaryotes: aligning theory with data. Nat Rev Genet 25:416–430. https://doi.org/10.1038/s41576-023-00688-5
33. Maruyama M, Kagamoto T, Matsumoto Y et al (2023) Horizontally acquired nitrate reductase realized kleptoplastic photoautotrophy of *Rapaza viridis*. Plant Cell Physiol 64:1082–1090. https://doi.org/10.1093/pcp/pcad044
34. Aldholmi M, Ahmad R, Carretero-Molina D et al (2022) Euglenatides, potent antiproliferative cyclic peptides isolated from the freshwater photosynthetic microalga *Euglena gracilis*. Angew Chem Int Ed Engl 61:e202203175. https://doi.org/10.1002/anie.202203175
35. Kim S, Im H, Yu J et al (2023) Biofuel production from *Euglena*: current status and techno-economic perspectives. Bioresour Technol 371:128582. https://doi.org/10.1016/j.biortech.2023.128582
36. David V, Flegontov P, Gerasimov E et al (2015) Gene loss and error-prone RNA editing in the mitochondrion of *Perkinsela*, an endosymbiotic kinetoplastid. mBio 6(6):e01498. https://doi.org/10.1128/mBio.01498-15
37. Lukeš J, Guilbride DL, Votýpka J et al (2002) Kinetoplast DNA network: evolution of an improbable structure. Eukaryot Cell 1:495–502. https://doi.org/10.1128/EC.1.4.495-502.2002
38. Michieletto D (2025) Kinetoplast DNA: a polymer physicist's topological Olympic dream. Nucleic Acids Res 53:gkae1206. https://doi.org/10.1093/nar/gkae1206
39. Ogbadoyi EO, Robinson DR, Gull K (2003) A higher-order trans-membrane structural linkage is responsible for mitochondrial genome positioning and segregation by fagellar basal bodies in trypanosomes. Mol Biol Cell 14(5):1769–1779. https://doi.org/10.1091/mbc.e02-08-0525
40. Ochsenreiter T, Anderson S, Wood ZA et al (2008) Alternative RNA editing produces a novel protein involved in mitochondrial DNA maintenance in trypanosomes. Mol Cell Biol 28:5595–5604. https://doi.org/10.1128/MCB.00637-08
41. Ochsenreiter T, Cipriano M, Hajduk SL (2008) Alternative mRNA editing in trypanosomes is extensive and may contribute to mitochondrial protein diversity. PLoS One 3:e1566. https://doi.org/10.1371/journal.pone.0001566
42. Kostygov AY, Karnkowska A, Votýpka J et al (2021) Euglenozoa: taxonomy, diversity and ecology, symbioses and viruses. Open Biol 11:200407. https://doi.org/10.1098/rsob.200407
43. Kato K, Yahata K, Nakayama T (2023) Taxonomy of a new parasitic euglenid, *Euglenaformis parasitica* sp. nov. (Eugenales, Euglenaceae) in ostracods and rhabdocoels. Protist 174:125967. https://doi.org/10.1016/j.protis.2023.125967

44. Evans G (1880) Report on 'surra' disease. Part VI – addendum in: Steel JH on his Investigation into an obscure and fatal disease among transport mules in British Burma, published by the Punjab Government, Military Department, No. 439-4467, 3rd Dec. 1880
45. Bruce D (1895) Preliminary report on the tsetse fly disease or nagana in Zululand, available through the Wellcome Collection
46. Skotnes-Brown J (2024) Extracting blood, flies, and ideas: David and Mary Bruce, vernacular experts, and *unakane* in rural Zululand c. 1880s–1900s. In: da Silva MAD, Lynteris C (eds) Rural disease knowledge: anthropological and historical perspectives. Routledge Taylor and Francis Group, Abingdon Oxon
47. Steverding D (2008) The history of African trypanosomiasis. Parasit Vectors 1:3. https://doi.org/10.1186/1756-3305-1-3
48. Stephens JWW, Fantham HB (1910) On the peculiar morphology of a trypanosome from a case of sleeping sickness and the possibility of its being a new species (*T. rhodesiense*). Ann Trop Med Parasitol 4:343–350
49. Steverding D (2014) The history of Chagas disease. Parasit Vectors 7:317. https://doi.org/10.1186/1756-3305-7-317
50. Steverding D (2017) The history of leishmaniasis. Parasit Vectors 10:82. https://doi.org/10.1186/s13071-017-2028-5
51. Chagas C (1909) Nova tripanozomiaze humana. Estudos sobre a morfolojia e o ciclo evolutivo de *Schizotrypanum cruzi n. gen., n. sp.*, ajente etiolojico de nova entidade morbida do homem. Mem Inst Oswaldo Cruz 1: 159–218
52. Broden A (1904) Les infections à trypanosomes au Congo chez l'homme et les animaux (communication préliminaire). Bulletin de la Société d'Etudes Coloniales 11:116–139
53. Ziemann H (1905) Beitrag zur Trypanosomenfrage. Zentralbl Bakteriol Parasitenk Infektionskrankh, Abt I, Orig 38:307–314
54. Lafont A (1909) Sur la présence d'un Leptomonas, parasite de la classe des Flagelles dans le lates de l'Euphorbia pilulifera. Comptes Rendus des Seances la Soc Biol 66:1011–1013
55. Parthasarathy MV, van Slobbe WG, Soudant C (1976) Trypanosomatid flagellate in the Phloem of diseased coconut palms. Science 192:1346–1348. https://doi.org/10.1126/science.192.4246.1346
56. Giordani F, Morrison LJ, Rowan TG et al (2016) The animal trypanosomiases and their chemotherapy: a review. Parasitology 143:1862–1889. https://doi.org/10.1017/S0031182016001268
57. Carnes J, Anupama A, Balmer O et al (2015) Genome and phylogenetic analyses of *Trypanosoma evansi* reveal extensive similarity to *T. brucei* and multiple independent origins for dyskinetoplasty. PLoS Negl Trop Dis 9: e3404. https://doi.org/10.1371/journal.pntd.0003404
58. Bradford JR (1932) Sir David Bruce, 1855-1931. Obit Not Fell R Soc 1:79–85. https://doi.org/10.1098/rsbm.1932.0017
59. Dean S, Gould MK, Dewar CE et al (2013) Single point mutations in ATP synthase compensate for mitochondrial genome loss in trypanosomes. Proc Natl Acad Sci USA 110: 14741–14746. https://doi.org/10.1073/pnas.1305404110
60. Triemer RE, Farmer MA (1991) An ultrastructural comparison of the mitotic apparatus, feeding apparatus, flagellar apparatus and cytoskeleton in euglenoids and kinetoplastids. Protoplasma 164:91–104. https://doi.org/10.1007/BF01320817
61. Desbois N, Pratlong F, Quist D et al (2014) Leishmania (Leishmania) martiniquensis n. sp. (Kinetoplastida: Trypanosomatidae), description of the parasite responsible for cutaneous leishmaniasis in Martinique Island (French West Indies). Parasite 21:12. https://doi.org/10.1051/parasite/2014011
62. Kwakye-Nuako G, Mosore MT, Duplessis C et al (2015) First isolation of a new species of *Leishmania* responsible for human cutaneous leishmaniasis in Ghana and classification of the *Leishmania enriettii* complex. Int J Parasitol 45:679–684. https://doi.org/10.1016/j.ijpara.2015.05.001
63. Lypaczewski P, Matlashewski G (2021) *Leishmania donovani* hydridisation and introgession in nature: a comparative genomic investigation. Lancet Microbe 2:e250–e258. https://doi.org/10.1016/S2666-5247(21)00028-8
64. Brasil LS, Silvério DV, Silva JOA et al (2025) Potential geographic displacement of Chagas disease vectors under climate change. Med Vet Entomol, in press. https://doi.org/10.1111/mve.12810
65. Forsyth C, Agudelo Higuita NI, Hamer SA et al (2024) Climate change and *Trypanosoma cruzi* transmission in north and Central America. Lancet Microbe 5:100946. https://doi.org/10.1016/j.lanmic.2024.07.009
66. Jagadesh S, Combe M, Ginouvès M et al (2021) Spatial variations in leishmaniasis: a

biogeographic approach to mapping the distribution of *Leishmania* species. One Health 13:100307. https://doi.org/10.1016/j.onehlt.2021.100307

67. Dann O, Bergen G, Demant E et al (1971) Trypanocide diamidine des 2-phenyl-benzofurans, 2-phenyl-indens und 2-phenyl-indols. Justus Liebigs Ann Chem 749:68–89. https://doi.org/10.1002/jlac.19717490110
68. Grossgebauer K, Kegel M, Dann O (1977) New fluorescent microscopical technique in diagnostic microbiology. Microsc Acta 80: 71–72
69. Borst P (2005) Ethidium DNA agarose gel electrophoresis: how it started. IUBMB Life 57:745–747. https://doi.org/10.1080/15216540500380855
70. Sharp PA, Sugden B, Sambrook J (1973) Detection of two restriction endonuclease activities in Haemophilus parainfluenzae using analytical agarose—ethidium bromide electrophoresis. Biochemistry 12:3055–3063. https://doi.org/10.1021/bi00740a018
71. Browning CH, Morgan GT, Robb JVM et al (1938) The trypanocidal action of certain phenanthridinium compounds. J Pathol Bacteriol 46:203–204. https://doi.org/10.1002/path.1700460121
72. Watkins TI, Woolfe G (1952) Effect of changing the quaternizing group on the trypanocidal activity of dimidium bromide. Nature 169:506–507. https://doi.org/10.1038/169506a0
73. Roy Chowdhury A, Bakshi R, Wang J et al (2010) The killing of African trypanosomes by ethidium bromide. PLoS Pathog 6: e1001226. https://doi.org/10.1371/journal.ppat.1001226
74. Cadena LR, Hammond M, Tesařová M et al (2024) A novel nabelschnur protein regulates segregation of the kinetoplast DNA in *Trypanosoma brucei*. Curr Biol 34:4803–4812.e3. https://doi.org/10.1016/j.cub.2024.08.044
75. Gerasimov ES, Afonin DA, Škodová-Sveráková I et al (2025) Evolutionary divergent kinetoplast genome structure and RNA editing patterns in the trypanosomatid *Vickermania*. Proc Natl Acad Sci USA 122: e2426887122. https://doi.org/10.1073/pnas.2426887122
76. Vickerman K (1969) On the surface coat and flagellar adhesion in trypanosomes. J Cell Sci 5:163–193. https://doi.org/10.1242/jcs.5.1.163
77. Beaver AK, Keneskhanova Z, Cosentino RO et al (2024) Tissue spaces are reservoirs of antigenic diversity for *Trypanosoma brucei*. Nature 636:430–437. https://doi.org/10.1038/s41586-024-08151-z
78. Keneskhanova Z, McWilliam KR, Cosentino RO et al (2025) Genomic determinants of antigen expression hierarchy in African trypanosomes. Nature 642:182–190. https://doi.org/10.1038/s41586-025-08720-w
79. Akiyoshi B, Gull K (2014) Discovery of unconventional kinetochores in kinetoplastids. Cell 156:1247–1258. https://doi.org/10.1016/j.cell.2014.01.049
80. Kieft R, Cliffe L, Yan H et al (2025) Monoallelic epigenetic regulation of polycistronic transcription initiation by RNA polymerase II in *Trypanosoma brucei*. MBio 16: e0232824. doi: https://doi.org/10.1128/mbio.02328-24
81. Van Luenen HGAM, Farris C, Jan S et al (2012) Glucosylated hydroxymethyluracil, DNA base J, prevents transcriptional readthrough in *Leishmania*. Cell 150:909–921. https://doi.org/10.1016/j.cell.2012.07.030
82. Tiritelli R, Cilia G, Gómez-Moracho T (2025) The trypanosomatid (Kinetoplastida: Trypanosomatidae) parasites in bees: a review on their environmental circulation, impacts and implications. Curr Res Insect Sci 7: 100106. https://doi.org/10.1016/j.cris.2025.100106
83. Roy J, Faktorová D, Benada O et al (2007) Description of *Rhynchopus euleeides* n. sp. (Diplonemea), a free-living marine euglenozoan. J Eukaryot Microbiol 54:137–145. https://doi.org/10.1111/j.1550-7408.2007.00244.x
84. Flegontova O, Flegontov P, Londoño PAC et al (2020) Environmental determinants of the distribution of planktonic diplonemids and kinetoplastids in the oceans. Environ Microbiol 22:4014–4031. https://doi.org/10.1111/1462-2920.15190
85. Flegontova O, Flegontov P, Malviya S et al (2016) Extreme diversity of diplonemid eukaryotes in the ocean. Curr Biol 26:3060–3065. https://doi.org/10.1016/j.cub.2016.09.031
86. Tashyreva D, Simpson AGB, Prokopchuk G et al (2022) Diplonemids - a review on "new" flagellates on the oceanic block. Protist 173: 125868. https://doi.org/10.1016/j.protis.2022.125868
87. Pilátová J, Tashyreva D, Týč J et al (2023) Massive accumulation of strontium and

barium in diplonemid protists. mBio 14: e0327922. https://doi.org/10.1128/mbio.03279-22

88. Tashyreva D, Prokopchuk G, Votýpka J et al (2018) Life cycle, ultrastructure, and phylogeny of new diplonemids and their endosymbiotic bacteria. mBio 9:e02447–17. https://doi.org/10.1128/mBio.02447-17
89. Breglia SA, Yubuki N, Hoppenrath M et al (2010) Ultrastructure and molecular phylogenetic position of a novel euglenozoan with extrusive episymbiotic bacteria: *Bihospites bacati* n. gen. et sp. (Symbiontida). BMC Microbiol 10:145. https://doi.org/10.1186/1471-2180-10-145
90. Bastin P, Sherwin T, Gull K (1998) Paraflagellar rod is vital for trypanosome motility. Nature 391:548. https://doi.org/10.1038/35300
91. Broadhead R, Dawe HR, Farr H et al (2006) Flagellar motility is required for the viability of the bloodstream trypanosome. Nature 440:224–227. https://doi.org/10.1038/nature04541
92. Ginger ML, Collingridge PW, Brown RW et al (2013) Calmodulin is required for paraflagellar rod assembly and flagellum-cell body attachment in trypanosomes. Protist 164: 528–540. https://doi.org/10.1016/j.protis.2013.05.002
93. Lander N, Li ZH, Niyogi S et al (2015) CRISPR/Cas9-induced disruption of Paraflagellar rod protein 1 and 2 genes in *Trypanosoma cruzi* reveals their role in flagellar attachment. mBio 6:e01012. https://doi.org/10.1128/mBio.01012-15
94. Valach M, Moreira S, Petitjean C et al (2023) Recent expansion of metabolic versatility in *Diplonema papillatum*, the model species of a highly speciose group of marine eukaryotes. BMC Biol 21:99. https://doi.org/10.1186/s12915-023-01563-9
95. Škodová-Sveráková I, Záhonová K, Juricová V et al (2021) Highly flexible metabolism of the marine euglenozoan protist *Diplonema papillatum*. BMC Biol 19:251. https://doi.org/10.1186/s12915-021-01186-y
96. Morales J, Hashimoto M, Williams TA et al (2016) Differential remodelling of peroxisome function underpins the environmental and metabolic adaptability of diplonemids and kinetoplastids. Proc Biol Sci 283:20160520. https://doi.org/10.1098/rspb.2016.0520
97. Andrade-Alviárez D, Bonive-Boscan AD, Cáceres AJ et al (2022) Delineating transitions during the evolution of specialised peroxisomes: Glycosome formation in kinetoplastid and diplonemid protists. Front Cell Dev Biol 10:979269. https://doi.org/10.3389/fcell.2022.979269
98. Benz C, Raas MWD, Tripathi P et al (2024) On the possibility of yet a third kinetochore system in the protist phylum Euglenozoa. MBio 15:e0293624. https://doi.org/10.1128/mbio.02936-24
99. Kaur B, Záhonová K, Valach M et al (2020) Gene fragmentation and RNA editing without borders: eccentric mitochondrial genomes of diplonemids. Nucleic Acids Res 48: 2694–2708. https://doi.org/10.1093/nar/gkz1215
100. Dobáková E, Flegontov P, Skalický T et al (2015) Unexpectedly streamlined mitochondrial genome of the euglenozoan *Euglena gracilis*. Genome Biol Evol 7:3358–3367. https://doi.org/10.1093/gbe/evv229
101. Van Leeuwenhoeck A (1677) Observation, communicated to the publisher by Mr. Antony van Leewenhoeck, in a Dutch letter of the 9 Octob. 1676 here English'd: concerning little animals by him observed in rain-well-sea and snow water; as also in water wherein pepper had lain infused. Philos Trans 12:821–831. https://doi.org/10.1098/rstl.1677.0003
102. Monteil CL, Vallenet D, Menguy N et al (2019) Ectosymbiotic bacteria at the origin of magnetoreception in a marine protist. Nat Microbiol 4:1088–1095. https://doi.org/10.1038/s41564-019-0432-7
103. Yubuki N, Leander BS (2018) Diversity and evolutionary history of the Symbiontida (Euglenozoa). Front Ecol Evol 6:100. https://doi.org/10.3389/fevo2018.00100
104. Ungogo MA, de Koning HP (2024) Drug resistance in animal trypanosomiases: epidemiology, mechanisms and control strategies. Int J Parasitol Drugs Drug Resist 25:100533. https://doi.org/10.1016/j.ijpddr.2024.100533
105. Abdill RJ, Blekhman R (2019) Meta-research: tracking the popularity and outcomes of all bioRxiv preprints. eLife 8:e45133. https://doi.org/10.7554/eLife.45133
106. Prosée R, Brown K (2025) Staying ahead of the curve: a decade of preprints in biology.

Biol Open 14:bio062111. https://doi.org/10.1242/bio.062111

107. Cobb M (2017) The prehistory of biology preprints: a forgotten experiments from the 1960s. PLoS Biol 15:e2003995. https://doi.org/10.1371/journal.pbio.2003995

108. Leander BS, Farmer MA (2000) Comparative morphology of the euglenid pellicle. I. Patterns of strip and pores. J Eukaryot Microbiol 47:469–479. https://doi.org/10.1111/j.1550-7408.2000.tb00076.x

109. Hoare C, Wallace F (1966) Developmental stages of trypanosomatid flagellates: a new terminology. Nature 212:1385–1386. https://doi.org/10.1038/2121385a0

110. Hayes P, Varga V, Olego-Fernandez S et al (2014) Modulation of cytoskeletal calpain-like protein induces major transitions in trypanosome morphology. J Cell Biol 206:377–384. https://doi.org/10.1083/jcb.201312067

111. Sunter JD, Benz C, Andre J et al (2015) Modulation of flagellum attachment zone protein FLAM3 and regulation of the cell shape in *Trypaosoma brucei* life cycle transitions. J Cell Sci 128:3117–3130. https://doi.org/10.1242/jcs.171645

112. Gadelha C, Rothery S, Morphew M et al (2009) Membrane domains and flagellar pocket boundaries are influenced by the cytoskeleton in African trypanosomes. Proc Natl Acad Sci USA 106:17425–17430. https://doi.org/10.1073/pnas.0909289106

Part II

Isolation and Culturing

Chapter 2

Isolation and Cultivation of Diplonemids

Daria Tashyreva, Galina Prokopchuk, Orsola Iorillo, Matus Valach, Ingrid Škodová-Sveráková, Dong-Woo Shin, James A. Letts, Gertraud Burger, Julius Lukeš, and Drahomíra Faktorová

Abstract

Marine diplonemids (Diplonemidae and Hemistasiidae), once considered difficult to culture, can now be introduced into stable axenic cultures and grown to high densities. This chapter provides detailed protocols on establishing clonal cultures from environmental cells, their axenization and routine maintenance, as well as instructions on mid- and large-scale cultivation, estimation of cell density, and cryopreservation of selected species.

Key words Diplonemids, Cultivation, Isolation, Axenization, Cryopreservation, Large-scale cultivation

1 Introduction

Diplonemids are a group of predominantly marine flagellates belonging to the phylum Euglenozoa, which also encompasses free-living and parasitic kinetoplastids, photosynthetic and heterotrophic euglenids, and cryptic symbiontids [1]. Diplonemids (or class Diplonemea and order Diplonemida) include four major lineages: classical diplonemids (or family Diplonemidae), hemistasiids (or family Hemistasiidae), eupelagonemids (or family Eupelagonemidae; also known as DSPD I), and a so far unranked lineage termed DSPD II [1, 2].

Unlike euglenids and kinetoplastids that have been known for centuries, diplonemids have remained obscure until recently due to their inconspicuous morphology and difficulties in introducing them into axenic cultures. Indeed, only two Diplonemidae species, *Isonema nigricans* and *Paradiplonema papillatum* (also designated *Diplonema papillatum*), were available in private culture collections until late 1980s, when stable axenic cultures of several diplonemid

Paul A. M. Michels et al. (eds.), *Euglenozoa: Methods and Protocols, Volume 1*, Methods in Molecular Biology, vol. 3013, https://doi.org/10.1007/978-1-0716-5142-1_2,

species were deposited in the American Type Culture Collection (ATCC) [3]; stable culture of the first hemistasiid became available even later [4]. Thanks to focused efforts, the diversity of cultured Diplonemidae and Hemistasiidae has markedly expanded since then [5–7]. However, the other two diplonemid clades, DSPD II and eupelagonemids, currently remain uncultured [2].

The recent publication of the high-quality genome and transcriptome of the model species *P. papillatum* [8], together with a toolbox for its genetic manipulation [9], has made diplonemids an attractive model for studying various cellular and molecular processes. In this chapter, we summarize our experiences in culturing diplonemids and provide detailed protocols for establishing cultures, routine and large-scale cultivation, estimation of cell density, as well as cryopreservation of selected species.

2 Materials

2.1 Isolation from the Environment

1. For collecting samples: plankton nets with mesh size of 20 μm; cell scrapers, scalpels, and tweezers for samples of algal biofilms; 2 or 3 mL disposable plastic Pasteur pipettes to sample sediment; 50 mL Falcon tubes; 1 L glass bottles or jars for water samples; Parafilm; thermometer; salinity meter.
2. For concentrating cells from water samples: reusable Sterifil filtration systems (Merck); 0.8 μm polycarbonate filters (diameter 47 mm, Millipore, Merck); hand pump compatible with filtration system (alternatively: a centrifuge that accommodates 50 mL Falcon tubes).
3. For establishing cultures from single cells: disposable Petri dishes (diameter 50 mm); Parafilm; 1 mL aliquots of molecular grade ethanol in sterile microcentrifuge tubes; glass microscopy slides; 1 mL aliquots of filter-sterilized seawater in sterile microcentrifuge tubes; cleaning tissues for microscope lenses; fine-tip tweezers and scissors; Bunsen burner; 24-well plates; plastic Pasteur pipettes or 10–100 μL pipettors and compatible filter tips; glass Pasteur pipettes; flexible PVC plastic tube (outer diameter 7 mm), 1000 μL sterile pipette tips with filters; inverted microscope with 10×, 20×, and 40× objectives, preferably equipped with left-handed stage controller (if the isolator is right-handed and vice versa), a Petri dish holder, and Plast DIC or phase contrast; Hemi medium of variable salinity (refer to Subheading 2.2.1 for formulation) supplemented with penicillin, streptomycin, and neomycin antibiotic cocktail in 1:100 ratio (P4083, Merck); cell culture incubator with adjustable temperature (10–25 °C).

2.2 Cultivation

2.2.1 Culture Media

Three types of media are currently used for cultivating marine diplonemids: a universal Hemi medium that supports the growth of all currently cultivated diplonemids, and two additional media optimized for high-density growth of *P. papillatum* (referred to as OSS medium and *P. papillatum* medium).

1. Standard Hemi medium: Dissolve 36 g of anhydrous sea salts in 900 mL of deionized water, add 1 mL of Luria-Bertani Broth (LB) prepared according to the manufacturer's protocol and 10 mL of heat-inactivated horse serum (HS), mix well, and adjust the volume to 1 L in a graduated cylinder. To sterilize, filter through a 0.22-µm filter system (connected to an electric pump) into a heat-sterilized glass bottle. Close the bottle aseptically in a laminar flow chamber and store at 4 °C. Optionally, supplement with penicillin, streptomycin, and neomycin antibiotic cocktail at a 1:100 v/v ratio (P4083, Merck) (*see* **Note 1**).
2. *P. papillatum* medium: dissolve 36 g of anhydrous sea salts in 900 mL of deionized water, add 1 g of tryptone powder and 10 mL of heat-inactivated fetal bovine serum (FBS), stir until combined, and adjust the volume to 1 L in a graduated cylinder. Optionally, to prevent bacterial contamination, supplement with 2.5 mL of 40 mg/mL chloramphenicol stock to yield 100 µg/mL final concentration. Prepare the stock by dissolving 2 g of chloramphenicol in 50 mL of ethanol and store at −20 °C. Sterilize and store the medium as above (*see* **Note 1**).
3. OSS medium: dissolve 33 g of Instant Ocean sea salts in 1 L of deionized water and autoclave (or alternatively, use a sterile 0.22 µm filter). If any precipitate forms after autoclaving, decant into a sterile bottle/container before use to remove any insoluble content. Add 10 mL horse serum to 1 L of the autoclaved ocean salt (1× OS) solution just before use, thus creating 1× OSS. If stored for >24 h, keep at 4 °C. Optionally, supplement with ethanol-dissolved chloramphenicol to 40 mg/L to prevent bacterial contamination (*see* **Notes 1** and **2**).

2.2.2 Routine Maintenance of Cultures

Work aseptically in a laminar flow chamber using sterile cell scrapers, pipette tips with filters, and tubes. Disinfect gloves, all surfaces, and items by spraying with 70% ethanol solution. Cultures are maintained in plastic vented cell culture flasks with 25 cm^2 surface area or 12.5 or 25 mL volume. Optimal growth temperature differs among diplonemids species, therefore incubators with adjustable temperature (10–25 °C) may be necessary. An inverted microscope, with 10×, 20×, and 40× objectives, preferably equipped with phase contrast, is required for monitoring growth in flasks.

2.2.3 Mid-Scale Cultivation (0.5–1.5 L)

In addition to materials listed in Subheading 2.2.2, prepare the following: hemocytometer (e.g., Improved Neubauer chamber) with a coverslip, 95% ethanol, a reusable, large surface-area rectangular glass bakeware (ideally at least 15 × 25 cm; preferably Pyrex), fitted covers (use either a plastic lid that comes with the glass bakeware, or tightly cover the container with an aluminum foil), and a rotary platform or a large cultivation chamber with agitation (able to support the weight of the medium-filled container).

2.2.4 Large Scale Cultivation in a Carboy Setup with Active Aeration (>6 L)

1. Medium: the carboy setup has been tested with OSS medium described in Subheading 2.2.1 supplemented with 0.05% (w/v) yeast extract and 1% (v/v) horse serum for *P. papillatum*.
2. Aquarium air pump: aquarium air pumps with additional flow regulators (e.g., Fedour Aquarium Air Pump, Ultra Quiet Aquarium Bubbler for up to ~1100 L with two outlets, 5 W) are recommended to gently aerate the culture and lower the chances of damaging the cells (*see* **Note 3**). Depending on the air pump, the tube diameter of the lines provided by the manufacturer may be smaller than the tubing used in the carboy. In such cases, additional connectors may be needed.
3. Carboy: use of a carboy fitted with a spout is highly recommended to ease sample collection and ensure sterility during the process. Although glass containers can be used, polypropylene Nalgene carboys (e.g., Thermo Scientific Nalgene Polypropylene 23190020 Carboy; cat. no. 02-963-2A) may be safer in case gentle shaking is used to dislodge cells from the surface prior to sampling cultures. A multi-port carboy cap with a minimum of two ports, for inlet and outlet airflow, is required (e.g., ULAB Scientific Plastic Universal Carboy Venting Cap, TPE Gasket and Port Caps, Quick Filling, UCA1004). All parts must be autoclavable.
4. Flexible tubing: tubing that goes into the carboy (e.g., Masterflex® Transfer Tubing, C-Flex®, Clear, 6.35 mm ID × 9.525 mm OD; cat. no. MFLX06422-10) must be autoclavable and should fit snugly through the holes of the multi-port carboy cap. Additional non-autoclavable tubing can be used to connect the carboy to the air pump.
5. Unidirectional vent filters: these are recommended to filter the airflow and minimize contamination and should be autoclaved to ensure sterility (e.g., Cytiva Whatman™ HEPA-Vent Filter, 67235000).
6. Optional parts: autoclavable T-shaped connectors (e.g., 6.35 mm three-way hose barb fitting splicer for liquid/air three-way connector to match the inner diameter of the tubing) to form loops at bottom of the main air input line; flow regulators in the air pump to aerate the culture more gently.

These may come with the aquarium air pump; additional connectors to connect between the carboy and air pump system in case tubing with different sizes are used.

2.3 Cell Counting

A hemocytometer such as a Bürker-Türk or an Improved Neubauer counting chamber with a coverslip, microcentrifuge tubes, sterile 10 mL graduated glass pipettes, pipette gun, deionized water, inverted microscope with a 10× objective. Optionally: cell scrapers.

2.4 Cryopreservation

Sterile 1–2 mL cryovials (e.g., Simport, T311-1); DMSO; 50% (v/v aq.) glycerol solution sterilized by filtering through a 0.22 μm filter; Mr. Frosty cooling unit (Nalgene) or two to three nested styrofoam boxes (a "matryoshka" setup assembled by placing a styrofoam rack into a ~2 cm thick-walled styrofoam box, enclosed into another larger ~3 cm thick-walled styrofoam box).

3 Methods

3.1 Establishing Diplonemid Cultures from Environmental Samples

3.1.1 Collecting Samples

For samples of plankton taken from the ocean, lower the net to a depth of 1–5 m to draw approximately 10 L of seawater, slowly pull upwards, and let water pass through the mesh until the volume is reduced to ~1 L. Swirl the water inside the net and collect the 10× concentrated plankton sample in a 1-L glass bottle (*see* **Note 4**). If a plankton net is not available, refer to Subheading 3.1.2. Store at 15 °C with a cap/lid slightly open until use (typically up to a week), preferably in an incubator with 12:12 h light-dark cycle.

Collect samples of algal biofilms from rocks by pulling them with tweezers or scraping off with a scalpel and use a cell scraper to collect biofilms from aquarium walls. Before proceeding to a new sample, wipe dry with a tissue paper. Draw 2–3 mL of a sediment from coastal shallow water or an aquarium tank with a clean disposable plastic Pasteur pipette. Place the material into a 50 mL Falcon tube and fill it up to 15 mL with seawater from the site (*see* **Note 4**). Never fill to >20 mL to avoid anoxic conditions. For long-distance transportation, seal Falcon tube caps with Parafilm. Store in an incubator as described above with caps slightly open.

3.1.2 Concentrating Samples

In the absence of a plankton net or if a sample requires further concentration, pass the water through a 0.8 μm polycarbonate filter placed inside the Sterifil filtration system. Soft fragile diplonemids are best preserved when the water is allowed to pass by gravity, but the process can be sped up by the occasional use of a handpump. Collect cells concentrated into 20–50 mL with a disposable plastic Pasteur pipette and transfer to a Falcon tube or a small glass bottle. Alternatively, cells can be concentrated by centrifugation in 50 mL Falcon tubes. Diplonemid cultures are typically spun for 10 min at 3000 × g in a refrigerated centrifuge at 15 °C, but lower speed should be tried first for mixed environmental samples.

3.1.3 Establishing Cultures from Single Cells

1. Dispense 1 mL of Hemi medium supplemented with antibiotics (penicillin, streptomycin, and neomycin antibiotic cocktail at a 1:100 v/v ratio) into each well of 24-well microplates. Wash a glass microscopy slide thoroughly with distilled water and disinfect with a cleaning tissue for microscope lenses soaked in 100% ethanol. Regular tissue papers are not recommended as these leave fibers on glass slides.
2. Draw glass Pasteur pipettes manually into fine capillaries: hold the tip of a pipette with tweezers and heat the adjacent part in the flame of a Bunsen burner until the glass becomes soft. Withdraw the pipette from the flame and immediately draw by rapid pulling of the tip with tweezers. If necessary, cut the tip of the capillary with scissors. The inner diameter of the capillary for cell picking should be ~20 μm. Additionally, prepare a capillary with ~100 μm diameter for dispensing water droplets onto glass slides. Optionally: before drawing into microcapillary, bend the thin part of the pipette at a 45° angle (*see* Fig. 1).
3. Pour ~7 mL of sample into a disposable Petri dish, wrap the edges with Parafilm to prevent evaporation, and let it stand for 15 min to 1 h at 15 °C, allowing suspended cells to sink to the bottom of the dish (*see* **Note 5** for an optional enrichment step).
4. Assemble the cell-picking device as shown in Fig. 1: connect sterile 1 mL filter tips to both ends of a ~1 m long flexible PVC tube. Label one side of the tube as a mouthpiece and attach a glass capillary prepared in **step 2** to the other side. Disinfect the capillary by drawing in and pushing out a small amount of molecular grade ethanol.
5. Place the Petri dish in the dish holder attached to the microscope stage. Locate diplonemid cells in the Petri dish under the inverted microscope at 100× or 200× total magnification. For typical diplonemid size and morphology, refer to [1, 5, 10]. Most commonly, diplonemids are found either gliding at the bottom of a Petri dish, or freely swimming in the water column, occasionally stopping at the bottom for a few seconds. Dispense two 3 μL droplets of sterile sea water onto the disinfected glass slide. Draw a small volume (1–2 μL) of sterile seawater into the glass capillary from **step 4** through the mouthpiece and then pick a cell from the Petri dish. Place the slide on the microscope stage and gently push out the cell into one of the droplets, disinfect the capillary with ethanol and withdraw the cell from the droplet. If multiple cells are in the droplet, repeat the procedure by transferring the cell in the next droplet (*see* **Note 6**). Finally, place the cell into a well of 24-well plate containing Hemi medium supplemented with antibiotics (*see* **Note 7**).

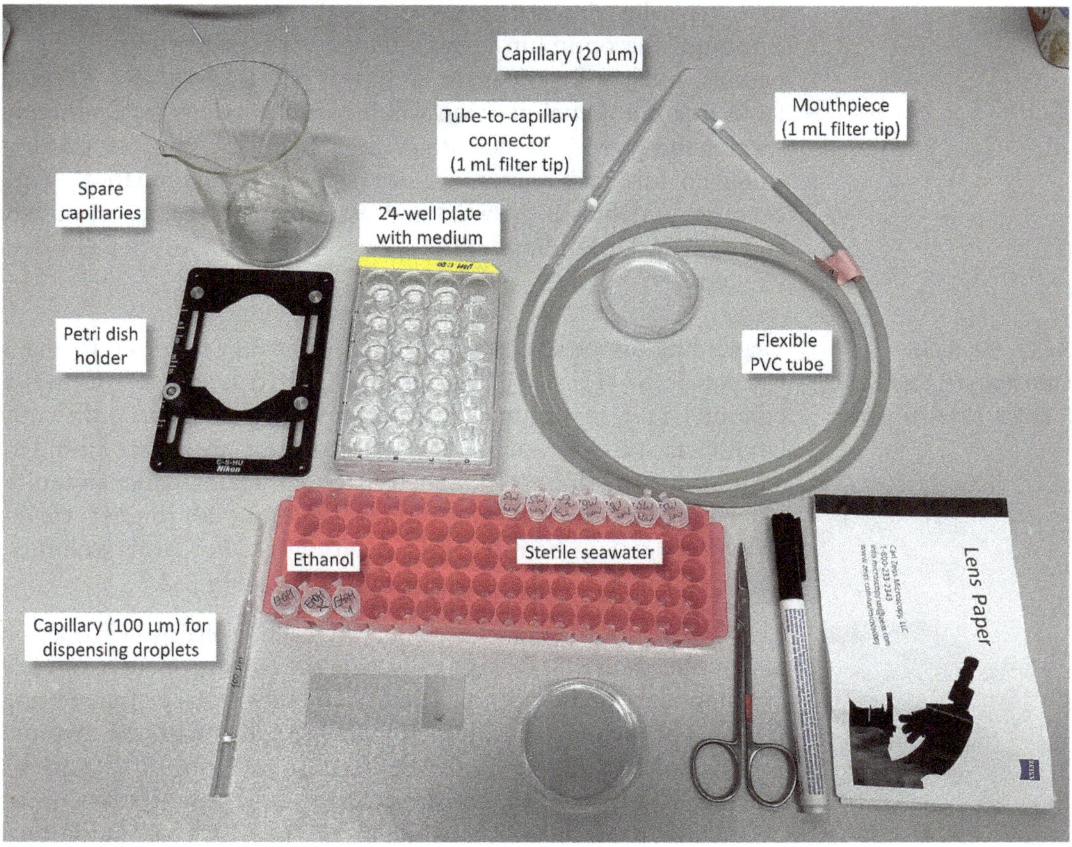

Fig. 1 A set of materials and tools required for isolating environmental diplonemid cells in culture by single-cell picking

6. Wrap the edges of the plate with parafilm to prevent evaporation and incubate at 15 °C in the dark for up to 2 weeks. If isolation is successful, diplonemid cells usually manifest in larger quantities after 1 week. The identification of the isolated organism as a diplonemid should be confirmed by PCR with EukA and EukB primers (5′- AACCTGGTTGATCCTGC CAGT-3′ and 5'-TGATCCTTCTGCAGGTTCACCTAC-3′; expected product length is ~2050 bp) targeting the 18S rRNA gene and sequencing.

3.2 Maintenance and Cultivation

3.2.1 Routine Maintenance

While some diplonemids (hemistasiids, some *Lacrimia* species, *P. papillatum*, *Sulcionema*) form homogenous suspensions or swim freely in the medium, most species glide at the bottom of culture flasks or adhere to surfaces (*Rhynchopus* and *Natarhynchopus* spp., *Diplonema* spp., *Flectonema*, some *Lacrimia* species). Therefore, before passaging, it is recommended to shake flasks vigorously or detach firmly adhered cells with a cell scraper. Transfer a small volume (50–100 μL) of dense culture into a new flask containing 10–15 mL of fresh Hemi medium, or 500 μL for species that typically do not grow to high densities (hemistasiids). Although the optimum growth temperature differs among diplonemids, most will propagate at 13–15 °C (*see* **Note 8**). Keeping flasks in a horizontal position will increase cell density of adhering species and provide sufficient aeration. Regularly monitor cultures using an inverted microscope for signs of contamination (e.g., bacterial and fungal growth) and assessment of cell morphology and density. Depending on the species, subculture every 5–14 days (*see* **Note 9**).

3.2.2 *Mid-Scale Cultivation of* P. papillatum, F. neradi, and R. euleeides

Depending on the size, the medium capacity of a single container is restricted to ~0.5–1.5 L due to the limited height of the water column (<1.5 cm optimally), which is needed to obtain sufficiently large surface area to volume ratio for (mostly) passive aeration. For large-scale growth (>6 L), bench space and maintenance of multiple containers can become rapidly limiting. For higher yields, growth media can be modified (*see* **Note 10**). For larger volumes, employ a carboy system described in Subheading 3.2.3.

1. Prepare an inoculum. Estimate the cell density as described in Subheading 3.3. Passage cells from the late exponential or early stationary phase culture (1 to 5×10^6 cells/mL) by transferring ~5% of the original volume to a culture flask with the medium of choice (e.g., 500 μL to 10 mL) (also *see* **Note 11**). At 20 °C, the wild-type *P. papillatum* strain should enter its late exponential phase in ~72 h, while at 16 °C, it will take additional ~24 h.
2. Prepare the cultivation dish(es). Sterilize a rectangular glass bakeware in an oven at ≥140 °C for at least 2 h, preferably several hours in advance before the inoculation, so that the container has time to cool down to room temperature. To sterilize the container's lid (be it plastic or aluminum), wipe it thoroughly with a paper towel soaked to saturation with 95% ethanol, then immediately put the cover on the still hot bakeware as soon as taken from the oven and wait for ~10 min until ethanol evaporates (*see* **Note 12**). The same procedure can be used to sterilize the container, with any residual ethanol from containers removed by setting it on fire (e.g., with a Bunsen burner); this is usually very brief (1–2 s). While the flame is not necessary, it enhances the sterilization process.
3. Pour sufficient volume of medium into the container to create a layer of 1–1.5 cm, e.g., 500–750 mL to a 30 cm × 18 cm dish. This ensures sufficient aeration of the cells even without continuous shaking (*see* **Note 13**). Add the inoculum prepared in **step 1** to the culture medium at a 1:100 ratio. To reach the late exponential/early stationary phase (i.e., cell density ~ 2.5×10^6 /mL for *P. papillatum* in the regular OSS medium), incubate at

16 °C for ~5–7 days. Alternatively, dilute the inoculum at a 1: 20 ratio to reach similar growth phase in 4–5 days to enter this phase.

4. Monitor the growth of the culture regularly (e.g., once a day) by taking a 10 μL aliquot (from a culture that has been agitated at 30–50 rpm for >10 min to homogenize its content) and screen for the presence of bacteria or fungi under a microscope. Estimate cell density as described in Subheading 3.3. The optimal growth phase to stop the cultivation at depends on the aim of the experiment, so should be determined beforehand in small-scale cultures.
5. Before harvesting the cultures, place the containers on a shaker (or a rotary platform) for 1–3 h (15–20 °C) to detach all surface-adherent cells with gentle, continuous agitation or rocking (20–50 rpm, depending on the container).
6. Harvest the cells by centrifugation (2000 × *g*, 4 °C, 5–10 min). Discard the medium and resuspend the cells in fresh sea salt solution without any carbon source. Spin as before and discard the supernatant. Continue with downstream sample processing as required.

3.2.3 Large Scale Cultivation in a Carboy Setup with Active Aeration (>6 L)

Mid-scale cultivation in dishes is gentler, providing a lower risk of damaging flagella, yet total cell yields may be insufficient for experiments requiring large cell quantities. The following protocol describes a simple set-up for culturing large quantities of *P. papillatum* using a carboy and a fish tank aeration pump (Fig. 2), similar to methods used previously for large-scale cultures of other planktonic protists and algae [11, 12]. Higher densities of cells were obtained by this carboy method than from the Pyrex dish method in the same amount of time (Fig. 3). Although there is still variability in the carboy-grown cultures (Fig. 4), higher total cell yields with significantly less space and labor can be obtained using this setup.

1. Preparation and initial assembly setup: fit the main tubing into the carboy cap, attach in-line vent filters to all lines connected to the carboy cap and connect to the air pump (*see* Fig. 2 for sample set up). Make small perforations in the region of the tubing that will rest near the bottom of the carboy in contact with the medium. Alternatively, perforations can be made on a separate, smaller piece of tubing. This tubing fragment can then be attached as a loop to the main air input tubing using a T-shaped connector (Fig. 2). Check that the vent line is functional and there are no unwanted leaks in the system: when the vent line is placed inside a beaker with water, bubbles formed from the vent line should be steady (1–2 bubbles per second), but not vigorous. Sterilize all parts exposed to the

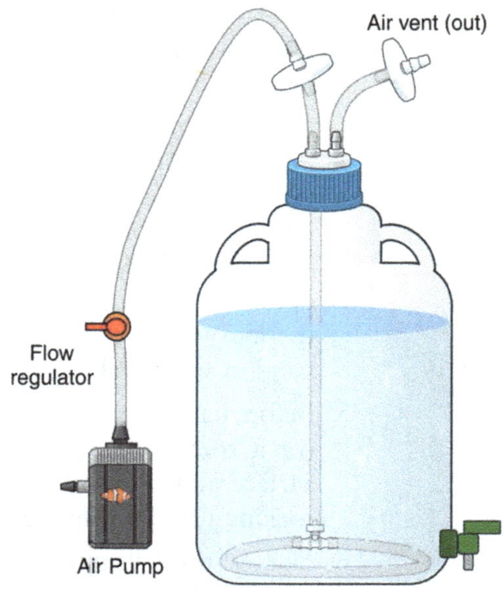

Fig. 2 A schematic of a carboy setup. Parts discussed in the methods are labelled with black text. The airflow of the system is shown in light blue

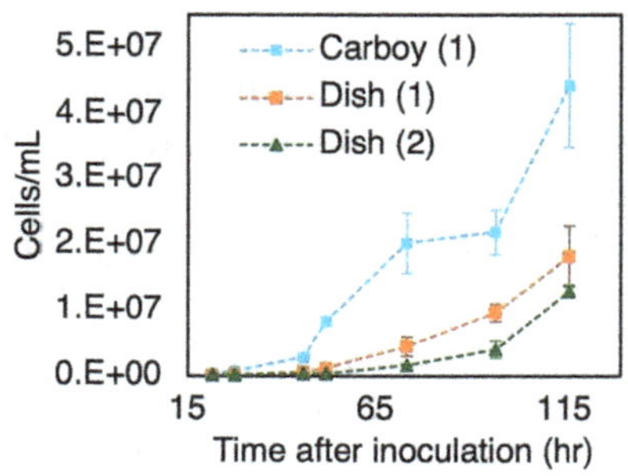

Fig. 3 Growth curves of *P. papillatum* grown in a carboy with 3 L of medium (blue squares) compared to two separate cultures grown in a Pyrex dish with 1 L of medium each (orange square and green triangle). Cells were grown in OS medium with 0.5 g/L yeast extract, 1% (v/v) horse serum, and 40 mg/L chloramphenicol. Cultures were inoculated with 10 mL of the starter culture per 1.5 L of growth medium. These cultures were grown in parallel, at room temperature (22 ± 2 °C). Cells were counted in triplicates using a hemocytometer, and error bars show the average deviation (except the last point in Dish (2), where the error bar is from two counts)

culture medium by autoclaving, except for the components used for the vent line (does not need to be autoclaved due to the positive pressure of the system).

2. Assemble the carboy for culture under a sterile environment (such as a laminar flow chamber) and seal the lines exposed to the outside environment with unidirectional vent filters.

   As foam will form during aeration, fill the carboy up to ~80% of the maximum capacity with cooled, autoclaved

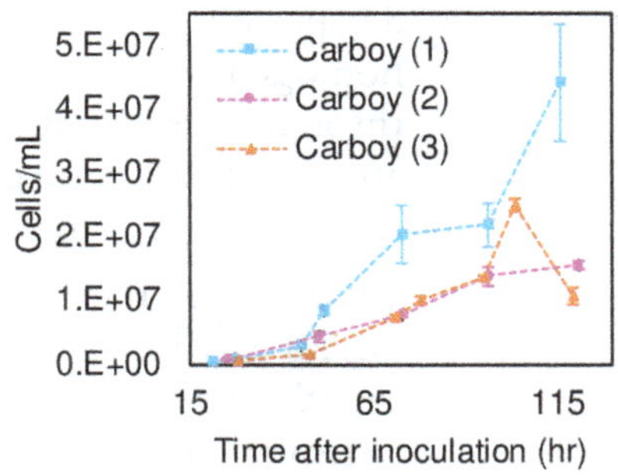

Fig. 4 Growth curves of *P. papillatum* grown in OS medium containing 0.5 g/L yeast extract, 1% (v/v) horse serum, and 40 mg/L chloramphenicol. Shown are growth curves from three independent cultures (3 L blue, 6 L mauve, 6 L orange). These cultures were grown at room temperature (22 ± 2 °C). Cells were counted in triplicates using a hemocytometer, and error bars show the average deviation

medium with desired additives (e.g., OSS medium with 1% (v/v) horse serum and 0.05% yeast extract) mixed with appropriate antibiotics (e.g., 40 mg/L chloramphenicol when culturing wild-type *P. papillatum*).

3. Inoculate the media with ~10–15 mL of starter culture grown to the late exponential phase per liter of media. Cap all exposed tubes by fitting unidirectional vent filters and make sure that the direction of the filters follows the direction of the airflow. This ensures that the medium is protected from potential contamination when the carboy is taken out of the sterile environment. Place the carboy in a cool location away from sunlight. Sufficient growth was observed with carboys placed on a lab bench inside a well-ventilated temperature-controlled room (22 ± 2 °C). Fit the air input lines to the air pump ensuring that the air pump is facing away from machines that output warm air (e.g., tabletop centrifuges). Establish gentle aeration by adjusting the airflow, preferably using the flow regulator. Check the flow rate by placing the vent line inside a clean beaker filled with water. Always maintain positive pressure in the system to prevent unfiltered air from getting inside the carboy.
4. Sampling during growth: dislodge cells from the carboy using a swirling motion. If the carboy is fitted with a spout, spray the spout with 70% ethanol solution before and after opening the spout to prevent contamination. Alternatively, the culture can be sampled by opening the carboy cap in a sterile environment, such as inside a laminar flow chamber. Check the cells under the microscope to ensure that cells swim freely in solution. For *P. papillatum*, fast, freely swimming cells were observed within 24 h post inoculation. Cells are harvested as described in Subheading 3.2.2.

3.3 Cell Counting

1. Scrape adherent cells off all surfaces of a culture flask and homogenize the culture by repeated passing up and down through a large-volume glass or plastic pipette (e.g., 10 mL) attached to a pipette gun. For non-adherent species, agitate the flask thoroughly to achieve a uniform cell distribution.
2. Transfer 180 μL of culture into a microcentrifuge tube containing 20 μL distilled water and immediately mix well to create a slightly hypotonic solution. This treatment will immobilize any fast-swimming cells. Directly proceed to **step 3** as long-term (>1 h) exposure to hypotonic solution may cause cell lysis in certain species.
3. Place the coverslip over the counting grid. Once positioned, the coverslip should remain stationary, as any movement during or after the sample loading may disrupt even distribution of cells and affect accuracy. Fill both chambers by pushing 10 μL of the cell suspension into the lateral capillary gap between the coverslip and the chamber base with a counting grid. Allow the capillary action to evenly distribute the sample under the coverslip.
4. Position the counting chamber under a microscope equipped with a 10× objective lens and focus on the central grid, which consists of nine large primary squares. Count cells within the four corner squares, following standard counting rules: include cells that intersect the top and left borders, but exclude those on the bottom and right borders. For improved accuracy, count cells from the same sample in three chambers resulting in 12 squares in total (*see* **Note 14**).
5. Calculate the concentration of cells per 1 mL as follows:

$$C(\text{cells/mL}) = -X/0.9 \times 10^4$$

Where C—concentration; X—arithmetic mean of cell counts per square; 10^4—multiplication factor converting a volume of 0.1 mm^3 (corresponding to one large square) into cm^3, or mL; 0.9—dilution factor introduced in **step 2**.

3.4 Cryopreservation

3.4.1 Cryopreservation of Paradiplonema papillatum

1. Grow *P. papillatum* in a vented flask to a density of ~5 × 10^6 (exponential phase) in 10–15 mL of medium of choice. Check the culture under the microscope for bacterial or fungal contamination and the condition of diplonemid cells. In a healthy culture, most cells swim and only few stick to the bottom.
2. Harvest cells by centrifugation (2000 × *g*, 4 °C, 5 min). Discard the medium.
3. Add 1.5 mL fresh medium supplemented with sterile DMSO to 8% (5–10% has been successfully tested). Resuspend the cells and distribute to three cryo-tubes (0.5 mL each).

4. Incubate the cryo-tubes at 4 °C or on ice for ~30 min, then transfer in a tube rack −80 °C for at least 3 h (*see* **Note 15** for an alternative protocol).
5. Store cryopreserved cultures at −80 °C for several months (up to a year, if absolutely necessary). Transfer the cryotubes into liquid nitrogen for long-term storage (up to ~15 years).

3.4.2 Cryopreservation of Diplonema ambulator, Flectonema neradi *and* Rhynchopus euleeides

1. Cultivate in 10–15 mL of medium of choice to the late exponential phase (cell growth curve depends on the species and should be determined beforehand). For the tested species, OSS with 2–3% horse serum appears optimal; use the same concentration of horse serum in all steps throughout the procedure described here.
2. Harvest by centrifugation (1000 × *g*, 4 °C, 5 min). Work on ice. If not possible to discard the supernatant without disturbing the cell pellet by decanting, remove as much as possible using a pipette, then resuspend the pellet in the remaining supernatant. Transfer to a 1.5 mL tube, re-spin in the same conditions, and remove the supernatant (*see* **Note 16**).
3. Resuspend in cold medium (1–4 °C) supplemented with cryoprotectants: (1) for *F. neradi*, use OSS with 15% glycerol; (2) for *R. euleeides*, use OSS with 8% DMSO or 5% glycerol; (3) for *D. ambulator*, use OSS with 10% DMSO or 15% glycerol. Keep on ice (or at 4 °C in the fridge) for ~30 min.
4. Transfer the tubes into a styrofoam rack, place it into the "matryoshka" styrofoam box setup, then transfer the whole nested assembly into the −80 °C freezer for >12 h to allow the gradual freezing of the samples before transferring the cryotubes into liquid nitrogen. Alternatively, use a dedicated freezing container (e.g., Mr. Frosty™) instead.

3.4.3 Starting a Diplonemid Culture from a Cryopreserved Stock

1. Place a frozen cryovial on ice and let it slowly thaw.
2. Once thawed, transfer the entire contents into a culture flask with 10–15 mL of medium pre-cooled down to 4–8 °C, but avoid using ice-cold medium as this can enhance cell autolysis.
3. Incubate at 20 °C (or 16–22 °C); recovered cells will start to swim.

4 Notes

1. The salinity of all media can be adjusted. Certain supplements (LB, yeast extract, antibiotics, horse serum or fetal bovine serum, tryptone) can be aliquoted and stored frozen at −20 °C. For OSS medium, it is recommended to prepare 10% stock solutions of tryptone and yeast extract. To prepare them,

dissolve 5 g of yeast extract/tryptone in 50 mL distilled water, filter-sterilize with a 0.22 μm filter or autoclave.

2. To enhance mitochondrial content in *P. papillatum* cultures, supplement with sterile yeast extract to 0.05%. To enhance cell number in *P. papillatum* cultures, supplement with tryptone to 0.05%. For more details see Chapter 15 in Vol. II.
3. Air stones should not be used due to potential difficulties with cleaning and sterilization.
4. Measure the temperature and salinity of samples immediately after collection.
5. Before single-cell picking, diplonemids can be enriched by incubating 100 μL of an environmental sample in 7–10 mL of Hemi medium with antibiotics for 3–4 days. Elimination of bacteria with antibiotics essentially suppresses the growth of bacterivorous protists, allowing only some free-living osmotrophic, parasitic, and certain predatory protists to propagate. Enrichment cultures should be monitored every day because diplonemids can be rapidly eliminated by predatory ciliates and amoebae, or overgrown by fungi.
6. Small droplets of seawater tend to evaporate fast, which dramatically increases the water salinity, and thus, transferring cells into the droplets should happen within 1–2 min after dispensing. If not possible, leave the picked cell inside the 20 μm capillary and dispense a new droplet onto the glass slide with a 100 μm capillary. Increasing the volume of droplets is not recommended because this will make locating diplonemid cells difficult.
7. In certain cases, penicillin-streptomycin-neomycin cocktail does not completely eliminate bacteria. For such samples, other antibiotics can be additionally supplemented, such as 50 μg/mL ampicillin, 50 μg/mL kanamycin, or 10 μg/mL gentamycin. Axenization of cultures is recommended because in many cases bacteria suppress diplonemid growth.
8. Most cultured diplonemids prefer 13–15 °C; hemistasiids isolated from warm waters prefer higher temperatures for growth (20–25 °C); the optimal temperature for *P. papillatum* is 16–22 °C.
9. Most diplonemids can survive extended periods at 4 °C. Low temperature slows down culture growth, which can be exploited for culture preservation (requires passaging once or twice a month). Old batch cultures of adhered cells can be rescued by replacing most of the culture fluid with fresh medium.
10. Maximum cell density can be increased by adjusting the content of horse serum: up to 2% for *P. papillatum*, but even up to

3% (v/v) for diplonemids that do not grow to high density in the regular OSS medium (e.g., *Flectonema neradi*, *Rhynchopus euleeides*). In the case of *P. papillatum*, it is beneficial to add yeast extract up to 0.05% (w/v) to enhance mitochondrial content, or tryptone up to 0.05% (w/v) to enhance cell number. However, adding more of these supplements or combining them is not recommended as cells will begin to swell up and become more sluggish, which indicates a non-physiological response. Not all diplonemids benefit from these supplements, so their influence needs to be tested prior to large-scale cultivation; e.g., the growth of *R. euleeides* is inhibited by the addition of yeast extract >0.01%.

11. While preparing the inoculum is not absolutely necessary, it is a recommended step to ensure that the cell culture has been pre-adapted to the medium and temperature of choice. *P. papillatum* propagates best in the range of 16–22 °C (*see* **Note 8**). In the upper range (22–27 °C), cells grow and divide but are heat-stressed (>28 °C will lead to growth arrest and cell death). In the lower range (4–16 °C), cells divide more slowly than at higher temperatures. Growth temperature and the composition of the medium will ultimately result in physiological changes, e.g., accumulation of storage lipids, changes in composition of the membranes, rate of the mitochondrial metabolism, or the relative and absolute content of most proteins. Therefore, the optimal growth conditions for a particular experiment should be determined beforehand in a small-scale setup.

12. Flat glass containers with plastic covers are more practical and better protect against evaporation, but in the absence of fitted covers, a large piece of aluminum foil can be used to shield the contents. If sterilizing containers using ethanol, add plastic covers onto the containers only once the flame has extinguished. If the bakeware cannot be taken out of the oven while still hot, sterilize it covered with a temporary aluminum foil to ensure the container's sterility. If using just an aluminum foil to cover the container, the whole assembly can be sterilized in the oven without ethanol wiping.

13. Occasional (i.e., once or twice a day) gentle waving motions at 30–50 rpm improve culture aeration and are strongly recommended. Avoid extensive stirring or shaking, which can lead to flagella loss. If an incubator shaker is not available, occasional shaking can be done by hand or on a rotary platform.

14. The optimum cell concentration range is 2.5×10^5 to 1×10^6 cells/ml that corresponds to 25 to 100 cell per square. It is recommended to dilute denser cultures.

15. Alternative for **steps 3–4**: Mix 200 μL of sterile 50% glycerol solution and 800 μL *P. papillatum* cell culture in labelled 1.2 mL cryovials and mix gently using the pipette. Place the cryovials in a dedicated freezing container (e.g., Mr. Frosty™) and let it freeze at −80 °C.
16. The low-speed spins and slow cooling are particularly critical to ensure cell viability after thawing.

Acknowledgments

This work was supported by Czech Science Foundation grant 23-06479X to J.L.; the Fonds de Recherche du Québec—Nature et Technologies (FRQNT; grant 2018-PR-206806 to G.B.), the Natural Sciences and Engineering Research Council of Canada (NSERC; grants RGPIN-2014–05286 and RGPIN-2019–04024 to G.B.), and the National Institute of Health (NIH grant R35 GM137929 to J.A.L.).

References

1. Kostygov AY, Karnkowska A, Votýpka J et al (2021) Euglenozoa: taxonomy, diversity and ecology, symbioses and viruses. Open Biol 11: 200407. https://doi.org/10.1098/rsob.200407
2. Lax G, Okamoto N, Keeling PJ (2024) Phylogenomic position of eupelagonemids, abundant, and diverse deep-ocean heterotrophs. ISME J 18:wrae040. https://doi.org/10.1093/ismejo/wrae040
3. Nerad T (1990) The life history, cytology and taxonomy of *Isonema* and *Isonema*-like flagellates. Dissertation, University of Maryland
4. Yabuki A, Tame A (1996) Phylogeny and reclassification of *Hemistasia phaeocysticola* (Scherffel) Elbrächter & Schnepf. J Eukaryot Microbiol 62:426–429. https://doi.org/10.1111/jeu.12191
5. Tashyreva D, Simpson AGB, Prokopchuk G et al (2022) Diplonemids—a review on "new" flagellates on the oceanic block. Protist 173: 125868. https://doi.org/10.1016/j.protis.2022.125868
6. Tashyreva D, Týč J, Horák A et al (2023) Ultrastructure and 3D reconstruction of a diplonemid protist and its novel membranous organelle. MBio 14:e01921–e01923. https://doi.org/10.1128/mbio.01921-23
7. Tashyreva D, Votýpka J, Yabuki A et al (2025) Description of new diplonemids (Diplonemea, Euglenozoa) and their endosymbionts: charting the morphological diversity of these poorly known heterotrophic flagellates. Protist 177:126090. https://doi.org/10.1016/j.protis.2025.126090
8. Valach M, Moreira S, Petitjean C et al (2023) Recent expansion of metabolic versatility in *Diplonema papillatum*, the model species of a highly speciose group of marine eukaryotes. BMC Biol 21:99. https://doi.org/10.1186/s12915-023-01563-9
9. Faktorová D, Kaur B, Valach M et al (2020) Targeted integration by homologous recombination enables *in situ* tagging and replacement of genes in the marine microeukaryote *Diplonema papillatum*. Environ Microbiol 22: 3660–3670. https://doi.org/10.1111/1462-2920.15130
10. Gawryluk RMR, del Campo J, Okamoto N et al (2016) Morphological identification and single-cell genomics of marine diplonemids. Curr Biol 26:3053–3059. https://doi.org/10.1016/j.cub.2016.09.013
11. Sauer JS, Simkovsky R, Moore AN et al (2021) Continuous measurements of volatile gases as detection of algae crop health. Proc Natl Acad Sci USA 118:e2106882118. https://doi.org/10.1073/pnas.2106882118
12. Hellenbroich D, Valley U, Ryll T et al (1999) Cultivation of *Tetrahymena thermophila* in a 1.5-m^3 airlift bioreactor. Appl Microbiol Biotechnol 51:447–455. https://doi.org/10.1007/s002530051415

Chapter 3

Preparation of Diplonemid Samples for Microscopy

Galina Prokopchuk, Daria Tashyreva, Orsola Iorillo, Bungo Akiyoshi, Julius Lukeš, and Drahomíra Faktorová

Abstract

This chapter provides a comprehensive set of protocols for preparing diplonemid cells for a range of microscopy applications, including live-cell imaging, fixed-cell fluorescence microscopy, endocytosis assay, immunofluorescence assay (IFA), and fluorescence in situ hybridization (FISH). It presents detailed instructions for slide preparation, cell harvesting, chemical fixation, permeabilization, and fluorescent labelling, with critical steps highlighted throughout. These protocols are designed to be adaptable for different experimental goals, enabling researchers to obtain high-quality imaging data for exploring the cellular biology of diplonemids.

Key words Live-cell imaging, Cell fixation, Cell permeabilization, Fluorescence staining, Fluorescence microscopy, Endocytosis assay, Immunofluorescence assay, Fluorescence in situ hybridization

1 Introduction

Microscopy plays a central role in diplonemid research, offering insights into their distinctive morphological features, physiological processes, and molecular organization. Routine light microscopy at low magnification provides an overall view of these microorganisms in their native state, helping to assess culture fitness and monitor cell motility. A more detailed observation of subcellular features of individual cells can be achieved using phase-contrast or, alternatively, differential interference contrast (DIC) microscopy, both enhancing the contrast of transparent diplonemid cells without the need for staining. However, the resolution limits of optical microscopy constrain our ability to study subcellular components. To overcome these limitations, transmission and scanning electron microscopies (TEM and SEM, respectively) are utilized to visualize ultrastructure and surface morphology at nanometer-scale resolution [1–3]. For instance, serial sectioning has enabled 3D reconstructions of the nucleus [4, 5], whereas serial block face SEM has

Paul A. M. Michels et al. (eds.), *Euglenozoa: Methods and Protocols, Volume 1*, Methods in Molecular Biology, vol. 3013, https://doi.org/10.1007/978-1-0716-5142-1_3,

helped to reveal a novel organelle in diplonemids [6]. While the aforementioned techniques uncover structural details, fluorescence and confocal microscopy allow tracking the spatial distribution of organelles, proteins, or other molecular targets within the cell [7, 8]. However, these microscopy techniques require meticulous sample preparation to minimize artifacts. Given that the only diplonemids currently available in culture are marine species, preparing samples to preserve their structure for the imaging can be challenging, largely due to their motility, sensitivity to osmotic stress, and generally poor adhesion to microscope slides, even when coated with common adhesive agents such as poly-L-lysine.

This chapter is focused on sample preparation for a range of visualization applications in diplonemid research. It presents an extensive set of protocols designed for both live- and fixed-cell imaging, covering techniques such as fluorescence staining, tracking of endocytic activity with fluorescent dextran, fluorescence in situ hybridization (FISH) for the detection of bacterial endosymbionts, and immunofluorescence assay (IFA) for protein localization. Many of these protocols were initially developed for the type species, *Paradiplonema papillatum*. However, unlike other diplonemids, this species has shown to be more resilient and is able to tolerate lower osmotic conditions. Accordingly, the protocols presented here have been adapted to ensure their applicability with other diplonemid species as well.

For live-cell imaging, we describe a strategy to immobilize motile cells without compromising their viability using low-gelling-temperature agarose, enabling the application of vital dyes that report on organelles and physiological states. For fixed-cell applications, we provide a general workflow encompassing fixation and permeabilization steps. Choosing a proper fixative is critical as it has a significant impact on preservation quality and compatibility with different imaging techniques. Formaldehyde and its polymerized form, paraformaldehyde (PFA), are commonly used to crosslink proteins and maintain cellular architecture with minimal autofluorescence, making them suitable for most applications. Glutaraldehyde provides superior ultrastructural preservation and is often used in electron microscopy, but its strong autofluorescence makes it unsuitable for fluorescence-based methods. In species where aldehyde fixation fails, osmium tetroxide can be used as an alternative. It reacts with lipids and proteins and can be applied either as a post-fixative for electron microscopy or as a standalone fixative. Accessing intracellular targets in fixed cells requires permeabilization, as many fluorescent dyes and antibody-based probes cannot penetrate intact cell membranes. In this chapter, we describe three permeabilization strategies routinely used for diplonemids: treatment with methanol, ethanol series, and a commercial permeabilization buffer.

2 Materials

2.1 Cell Harvesting

1. Laminar flow cabinet (if sterile conditions are required).
2. Falcon tubes (15–50 mL) for larger culture volumes, and/or microcentrifuge tubes (1.5–2 mL) for small samples.
3. Refrigerated centrifuges compatible with selected tubes.
4. Artificial seawater (ASW; 36 g/L) for washing and resuspension: Dissolve 36 g of anhydrous sea salts in 900 mL of deionized water in a 1 L beaker. Adjust the volume to 1 L in a graduated cylinder and filter through a sterile 0.22 μm filtration system into a sterile bottle.

2.2 Live-Cell Fluorescence Imaging

1. Live-cell fluorescent dyes (e.g., LysoTracker™ Green DND-26, NucBlue™ Live Cell Stain).
2. Artificial seawater (ASW; 36 g/L).
3. Microcentrifuge tubes (1.5 mL).
4. Aluminum foil to wrap microcentrifuge tubes.
5. Microcentrifuge.
6. 2% ultra-low gelling agarose: Add 0.2 g of agarose to 10 mL of ASW. Stir the mixture continuously at a rapid pace on a hotplate stirrer set to 70 °C and boil for 5 min to ensure the agarose fully dissolved (*see* **Note 1**). Allow the solution to cool to ~25 °C.
7. Microscope slides and cover slips (24 mm × 60 mm), or alternatively, 35 mm glass-bottom microscope dish.
8. Lightproof sample box.

2.3 Chemical Fixations for Fixed-Cell Imaging

1. Certified chemical fume hood.
2. Personal protective equipment (gloves, goggles, lab coat).
3. Hermetic containers for solid and liquid waste disposal.

2.3.1 Aldehyde Fixation

Formaldehyde and paraformaldehyde (PFA) fixation (for staining with fluorescent dyes and IFA)

1. Artificial seawater (ASW, 36 g/L) or phosphate-buffered saline (PBS): 137 mM NaCl, 2.7 mM KCl, 10 mM Na_2HPO_4, 1.8 mM KH_2PO_4, pH 7.4.
2. Formaldehyde-based fixative. To prepare a 1 mL of 4% formaldehyde solution, mix 250 μL of 16% methanol-free formaldehyde (e.g., Thermo Fisher) with 750 μL of ASW or PBS (for *P. papillatum*). If using a 37% formaldehyde (formalin), dilute 108 μL in 892 μL of ASW or PBS to obtain a final volume of 1 mL.

3. PFA-based fixative. To prepare a 4% PFA solution, add 0.8 g of PFA powder to 20 mL of ASW in a 50 mL Falcon tube. Incubate the tube in a 65 °C water bath, vortex occasionally to promote dissolution. Complete dissolution typically takes 1–1.5 h (*see* **Note 2**). Once dissolved, allow the solution to cool to room temperature (RT) before use. The fixative can be stored at 4 °C in the dark for up to 2 weeks or at −20 °C for several months.

Glutaraldehyde fixation (for electron microscopy of P. papillatum)

1. 2.5% (v/v) glutaraldehyde: Dilute 1 part of 25% aqueous glutaraldehyde solution in 9 parts of 0.2 M sodium cacodylate buffer. Protect from light. Store aliquots at −20 °C for short-term use.
2. Phosphate-buffered saline (PBS).

2.3.2 Osmium-Based Fixation (for FISH, Post-Fixation Staining with Fluorescence Dyes)

1. 4% water-based OsO_4 (commercially available).
2. Graduated disposable plastic Pasteur pipettes.
3. 2× artificial seawater (ASW, 72 g/L).
4. Deionized water.

2.4 Cell Adhesion to Microscopy Slides

1. Adhesive microscope slides (e.g., poly-L-lysine-coated, Sigma). To prepare in-house adhesive gelatin-coated slides, soak standard microscope slides in a cleaning solution (2.5 M NaOH, 105 mL of absolute ethanol, 70 mL of ultrapure or distilled water) for 2 h. Meanwhile, prepare a 0.1% (w/v) gelatin solution by dissolving 0.2 g of gelatin in 200 mL of distilled water. Add 0.02 g of $CrK(SO_4)_2$, which corresponds to a concentration of 0.01% w/v. Heat the mixture to 65 °C on a magnetic hotplate stirrer, stirring continuously until the gelatin is completely dissolved. Afterward, filter the solution through a 0.22 μm filter. Rinse the cleaned slides thoroughly with distilled water, then dip them into the gelatin solution, and air dry. Store dried slides in a microscope slide box until use.
2. Pencil or oil-based resistant marker.
3. Blotting paper.
4. Fume hood.
5. Coplin jar or similar container.
6. *Optional*: liquid blocking PAP pen.

2.5 Permeabilization of Fixed Cells

1. Permeabilization agents:
   (a) 100% methanol (pre-chilled to −20 °C).
   (b) Commercial permeabilization buffer (e.g., eBioscience™ Permeabilization Buffer (10×), Invitrogen 00-8333-56).

To prepare 1× or 2× concentration, dilute 10× buffer with ultrapure water.

(c) 50%, 80%, 100% molecular biology grade ethanol (to prepare 50% and 80% concentrations, dilute 100% ethanol with ultrapure water).

2. 50 mL Falcon tube or similar sealable container.
3. Phosphate-buffered saline (PBS).

2.6 Processing Live Cells with Fluorescent Markers for Post-Fixation Imaging (for Staining with Fixable Fluorescent Dyes and/or Detecting Endogenously Expressed Fluorescence Signals)

1. Fixable fluorescent dyes (e.g., Hoechst 33342).
2. Phosphate-buffered saline (PBS).
3. Antifade DABCO mounting medium: 1% (w/v) 1,4-diazabicyclo[2.2.2]octane, 90% glycerol, 50 mM sodium phosphate, pH 8.0. *Optional*: add 100 ng/mL DAPI for nucleic acid staining.
4. Aluminum foil.
5. Microscope slides and cover slips (24 mm × 60 mm).
6. Lightproof storage box.

2.7 Fluorescence Staining of Fixed Cells

1. Fluorescent dyes (e.g., RedDot1, DAPI, SYBR Green, acridine orange).
2. Phosphate-buffered saline (PBS).
3. Antifade mounting medium: homemade DABCO mounting medium (*Alternative*: 50–90% (v/v) glycerol, 0.5% (w/v) N-propyl gallate, 20 mM Tris, pH 8.0) or commercial options (e.g., ProLong™ Gold Antifade Mountant, ProLong™ Diamond Antifade Mountant). Antifade mounting medium can be supplemented with nucleic acid stain, such as DAPI or Hoechst.
4. Aluminum foil.
5. Adhesive microscope slides and cover slips (24 mm × 60 mm).
6. Lightproof storage box.

2.8 Vital Staining of P. papillatum with Fluorescent Dextran (to Monitor Endocytosis)

1. Fluorescein isothiocyanate (FITC)-dextran (e.g., Sigma Aldrich). Prepare a 50 mg/mL stock solution.
2. Sterile artificial seawater (ASW, 36 g/L).
3. Phosphate-buffered saline (PBS).
4. Antifade mounting medium with nucleic acid stain.

2.9 Immuno-fluorescence Assay

1. 4% PFA solution, 100% methanol (for slide-based workflow), 1× commercial permeabilization buffer (for tube-based workflow).
2. Primary antibody.

3. Secondary antibody conjugated to fluorophore.
4. Phosphate-buffered saline (PBS).
5. PBS with 0.05% (v/v) Tween® detergent (PBS-T).
6. 5.5% FBS (v/v) in PBS-T or 5% (w/v) of skim milk powder for blocking and 3% (w/v) of bovine serum albumin (BSA) in PBS-T or 5% (w/v) of skim milk powder for antibody incubation.
7. Humid chamber (homemade option: fill a sealable plastic box with blotting paper, spray it with distilled water, and wrap the container with aluminum foil).
8. Parafilm.
9. Coplin jar or similar container.
10. Aluminum foil.
11. Adhesive microscope slides and cover slips (24 mm × 60 mm).
12. Antifade mounting medium with or without nucleic acid stain.
13. Lightproof storage box.

2.10 Fluorescence In Situ Hybridization (FISH) with Bacterial Eub338 Probe for Endosymbiont Screening

1. Eub338 oligonucleotide probe (5′-labelled with fluorophores, such as FITC, Cy3, Cy5, ATTO, or Alexa Fluor dyes).
2. Hybridization oven with a horizontal shaker.
3. 50 mL Falcon tube or similar sealable container.
4. 3-ply tissue paper.
5. 5 M NaCl stock solution: Dissolve 29.22 g of NaCl in 80 mL of ultrapure water. When dissolved completely, adjust the volume to 100 mL. Store at RT.
6. 1 M Tris–HCl stock solution, pH 7.2: Dissolve 12.1 g of Tris base in 70 mL of deionized water. Adjust the pH to 7.2 by adding 10 N HCl dropwise while monitoring with a pH meter. Once adjusted, bring the volume up to 100 mL with deionized water. Store at 4 °C.
7. Hybridization buffer (900 mM NaCl, 20 mM Tris–HCl, 0.01% SDS) with 35% (v/v) formamide: In a 50 mL Falcon tube, add 10 mL of ultrapure water, 9 mL of 5 M NaCl, 1 mL of 1 M Tris–HCl (pH 7.2), 17.5 mL of deionized formamide, and 50 μL of 10% SDS. Fill up to 50 mL with ultrapure water and mix well. Store at 4 °C.
8. Washing buffer (225 mM NaCl, 20 mM Tris–HCl, 0.01% SDS): In a 0.5 L graduated cylinder, add 400 mL of deionized water, 7 mL of 5 M NaCl, 10 mL of Tris–HCl (pH 7.2), and 0.5 mL of 10% SDS. Fill up to 500 mL with deionized water, mix well and transfer to a 0.5 L bottle. Store at 4 °C. If SDS precipitates, warm the solution to 30–35 °C until it fully re-dissolves.
9. Parafilm.

10. Adhesive microscope slides and cover slips (24 mm × 60 mm).
11. Antifade mounting medium with nucleic acid stain (e.g., DAPI).
12. Lightproof storage box.

3 Methods

3.1 Cell Harvesting

For optimal viability and metabolic activity, collect diplonemids during the exponential growth phase, unless the experimental design specifically requires cells in a different physiological state. If diplonemids are to be cultured or processed under aseptic conditions, perform all sample handling within a laminar flow cabinet to ensure sterility.

1. Based on the cell density and required sample volume, transfer the appropriate amount of culture into a suitable microcentrifuge or Falcon tube and centrifuge at 3000 × *g* for 10 min (*see* **Note 3**) in a refrigerated centrifuge set to +15 °C (*see* **Note 4**).
2. Discard the supernatant without disturbing the pellet.
3. Gently resuspend the cells in ASW. If the culture was harvested in a Falcon tube, you may use a small volume (e.g., ~1 mL) of ASW to resuspend the cells and then transfer the suspension to a microcentrifuge tube.
4. Centrifugate the resuspended cells again under the same conditions to eliminate any residual components of culture medium, which could interfere with staining procedures.
5. Discard the supernatant. At this point, the cells are ready for downstream applications.

3.2 Live-Cell Fluorescence Imaging

1. After completing the standard harvesting procedure described in the previous subheading, resuspend the washed cell pellet in 1 mL of sterile ASW to obtain a uniform cell suspension (*see* **Note 5**).
2. Add the selected fluorescent dye(s) (*see* **Note 6**) directly to the cell suspension to achieve a final concentration recommended by the manufacturer (typically in the range of 0.1–1 μM for most commercial dyes). From this point onward, protect the sample from light to prevent photobleaching.
3. Gently mix the suspension by pipetting or inverting the tube to ensure even distribution of the dye throughout the cell population.

4. Incubate the cell-dye mixture at the recommended temperature (usually culturing temperature or RT) for 15–30 min, following the manufacturer's instructions.
5. After incubation, centrifugate the sample at 3000 × *g* for 10 min.
6. Discard the supernatant without disturbing the pellet.
7. To wash out any unbound dye, resuspend the pellet in 1 mL of sterile ASW and centrifuge again under the same conditions. Repeat the wash if necessary.
8. After the wash, resuspend the cells in 0.5 mL of ASW.
9. To immobilize the cells for microscopy, mix 0.5 mL of 2% ultra-low gelling agarose with an equal volume of the stained cell suspension (*see* **Note 7**). Immediately apply the agarose-cell mixture as a thin layer onto a microscope slide and place a coverslip on top, or alternatively, dispense the mixture into a microscope dish. Keep the slide or dish horizontal in a light-proof sample box to protect it from light until imaging.
10. As soon as the sample solidifies (*see* **Note 8**), proceed with fluorescence microscopy imaging using the appropriate filters for the selected dyes.

3.3 Chemical Fixations for Fixed-Cell Imaging

Warning: Handle chemical fixatives and perform all fixation steps in a certified chemical fume hood while wearing appropriate personal protective equipment, including gloves, goggles, and a lab coat. When working with osmium tetroxide (OsO_4), it is advisable to wear double gloves and to protect surfaces with absorbent paper. Strictly follow safety regulations and dispose solid and liquid waste according to established safety protocols.

3.3.1 Aldehyde Fixation

Formaldehyde and paraformaldehyde (PFA) fixation (for staining with fluorescent dyes and IFA)

1. Harvest cells as described in Subheading 3.1 and, depending on the size of the pellet, resuspend them in 250–500 μL of the 4% fixative. Mix gently until a homogenous suspension is obtained.
2. For slice-based IFA assay (that we recommend for *P. papillatum*), spread the cells immediatelly on the microscopy slides, incubate at RT for 15 min and after fixation rinse the cells with PBS.
3. For tube-based workflow, incubate the cells in the tubes for 15–30 min and after fixation, centrifuge the sample at 1500–2000 × *g* (*see* **Note 9**) for 10 min at RT. Discard the supernatant into a designated waste container, then resuspend the pellet in 200–500 μL of either distilled water (FISH,

fluorescence staining) or PBS (IFA, fluorescence staining). Repeat the wash once more under the same conditions.

Glutaraldehyde fixation (for electron microscopy of P. papillatum)

1. Prepare cell pellet as described in Subheading 3.1 and carefully resuspend it in 2.5% glutaraldehyde. The volume of fixative should be sufficient to completely cover the pellet (typically 250–500 μL, depending on pellet size).
2. Incubate the suspension for 1 h at RT or up to several hours at 4 °C.
3. Following fixation, pellet the cells again at 1500–2000 × *g* for 10 min and carefully discard the supernatant into a designated waste container.
4. Wash the fixed cells twice in sterile PBS (or 0.2 M sodium cacodylate buffer) to remove residual glutaraldehyde. For each wash, gently resuspend the pellet in 500 μL buffer, centrifuge, and discard the supernatant.

3.3.2 Osmium-Based Fixation (for FISH, Post-Fixation Staining with Fluorescent Dyes)

1. Prepare a fresh 2% (v/v) OsO_4 working solution immediately before use (*see* **Note 10**). Using a graduated disposable plastic Pasteur pipette, mix 100 μL of 4% aqueous OsO_4 solution with 100 μL of 2× ASW (7.2% w/v). Mix thoroughly by pipetting.
2. Harvest cells as described in Subheading 3.1, leaving ~20 μL of fluid. Resuspend the cell pellet in this residual volume.
3. Carefully add 200 μL of 2% OsO_4 solution to the resuspended cells using a Pasteur pipette and mix well. Ensure not to spill the OsO_4 and do not allow it to come into contact with the exterior of the microcentrifuge tube.
4. Incubate the cell suspension for 15 min at RT.
5. Pellet the fixed cells by centrifugation at 1500–3000 × *g* for 10 min. Discard the supernatant into a designated osmium waste container, following hazardous waste disposal guidelines.
6. Resuspend the pellet in 1.2 mL of distilled water, centrifuge, and discard the supernatant. Repeat this washing step four times to ensure complete removal of osmium residues. Finally, resuspend the pellet in 200–500 μL of distilled water.

3.4 Cell Adhesion to Microscopy Slides

1. Label an adhesive microscope slide using a pencil or oil-based resistant marker. *Optional*: draw a hydrophobic barrier using a liquid-blocking PAP pen to confine the sample area.
2. Apply a large drop (~200 μL) of the fixed cell suspension to the middle of the slide and spread it evenly using a pipette tip.

3. Allow the cells to settle and adhere for 15 min at RT.
4. Carefully remove the excess liquid by gently touching the top of the droplet with blotting paper, avoiding disturbance to the cells.
5. Transfer the slide to a fume hood and let any residual liquid to evaporate completely. It is crucial not to overdry the cells. Monitor the slide, and as soon as the liquid has evaporated, immediately proceed to **step 6**.
6. Thoroughly rinse the slide by dipping it into a jar with deionized water, then dry it by tilting and gently tapping onto a tissue paper.

3.5 Permeabilization of Fixed Cells

3.5.1 With 100% Methanol for Staining with Fluorescent Dyes and IFA

Place the slide with adhered cells into a 50 mL Falcon tube or a similar container filled with 100% methanol pre-chilled to −20 °C or spread 400μL of 100% methanol directly on the slide (recommended for slice-based IFA assay of *P. papillatum*). Incubate the sample for 15–20 min at RT. After incubation, remove the slide from methanol and rinse it with PBS (*see* **Note 11**). To remove excess PBS, tilt the slide and gently tap it against tissue paper, avoiding direct contact with the cell area to prevent detachment or damage.

3.5.2 With Commercial Buffer for Staining with Fluorescent Dyes and IFA

Permeabilization can be performed either directly in suspension or on slides with affixed cells. To permeabilize in suspension, fix the cells as described in **steps 1** and **2** of Subheading 3.3.1 *Formaldehyde and PFA fixation*. Next, add an equal volume of 2× permeabilization buffer directly to the cell suspension. Centrifugate the sample at 1500–2000 × *g* for 10 min at RT. Discard the supernatant into a designated waste container and resuspend the pellet in 200–500 μL of 1× permeabilization buffer. Continue with downstream application or proceed with cell adhesion as described in Subheading 3.4. To permeabilize cells that have already been fixed and adhered to slides, apply 250 μL of 1× permeabilization buffer directly onto the slide, incubate for 10 min at RT, and then wash thoroughly with PBS or distilled water.

3.5.3 With Ethanol for FISH and DAPI Staining

Sequentially apply 50%, 80%, and 100% ethanol directly onto the horizontally positioned slide, incubating with each concentration for 3 min. Between each step, remove excess ethanol by tilting the slide and tapping it onto tissue paper.

Sample slides not processed immediately for downstream applications can be stored in 100% ethanol or methanol at −20 °C in sealed Falcon tubes or jar for several days.

3.6 Processing Live Cells with Fluorescent Markers for Post-Fixation Imaging (for Staining with Fixable Fluorescent Dyes and/or Detecting Endogenously Expressed Fluorescence Signals)

1. Harvest cells as described in Subheading 3.1. If the experiment involves only fluorescent protein-expressing cells (e.g., YFP-tagged [9]), and no live staining, proceed directly to **step 3**.
2. If using a fixable fluorescent dye(s), add the dye to the suspension of live cells at the manufacturer's recommended working concentration. Incubate in the dark for the specified time (typically 10–30 min) at RT or as instructed. Centrifuge at 1500–2000 × *g* for 10 min, discard the supernatant, and resuspend the pellet in PBS.
3. Fix the cells using a formaldehyde-based fixative as described in Subheading 3.3.1 *Formaldehyde and PFA fixation*.
4. Resuspend the fixed cells in a small volume (~10 μL) of antifade DABCO mounting medium with or without DAPI, depending on the experimental requirements.
5. Mount the suspension onto a microscope slide and place a coverslip on top, avoiding air bubbles. Store the slides at 4 °C in the dark until imaging.

3.7 Fluorescence Staining of Fixed Cells

1. Harvest cells as described in Subheading 3.1 and fix them using an appropriate protocol described in Subheading 3.3.
2. Allow the cells to adhere to slides either before or after staining, depending on whether the staining will be performed in suspension (in tubes) or on the slide surface (*see* Subheading 3.4). Add the fluorescent dye(s) to either the cell suspension or directly onto the cells that have adhered to the slides (*see* **Note 12**) at the recommended working concentration. Incubate in the dark for the specified time (typically 10–30 min) at RT or as instructed by the manufacturer.
3. After staining, rinse the cells by dipping slides in PBS (at least two times) to remove unbound dye. For suspended cells, centrifugate at 1500–2000 × *g* for 10 min, discard the supernatant, and then resuspend the pellet in PBS or distilled water.
4. If not already adhered, allow the stained cells to attach to adhesive microscope slides as described in Subheading 3.4.
5. Apply a drop of antifade mounting medium with or without nucleic acid stain, depending on the experimental requirements, and place a coverslip on top, avoiding air bubbles. Store the slides at 4 °C in the dark until imaging.

3.8 Vital Staining of P. papillatum with Fluorescent Dextran (to Monitor Endocytosis)

1. Collect 1 mL of an exponentially growing *P. papillatum* culture as described in Subheading 3.1. Resuspend the cell pellet in sterile ASW and incubate for 15 min at RT to enhance endocytic activity.

2. Add 50 mg/mL of FITC-dextran stock solution to the cells to achieve a final concentration of 5 mg/mL (e.g., mix 100 μL of the cell suspension with 10 μL of FITC-dextran stock).
3. Incubate the cells with FITC-dextran for a selected time point (e.g., 1, 2, 5, 15, or 30 min) at RT, depending on the experimental objective. From this point on, protect the samples from light.
4. After incubation, centrifugate the cells at 1500–2000 × *g* for 10 min, discard the supernatant to remove excess dextran, and then resuspend the pellet in PBS (*see* **Note 13**).
5. Fix the cells in 4% PFA as described in Subheading 3.3.1 *Formaldehyde and PFA fixation* and allow them to adhere to microscope slides following the procedure in Subheading 3.4.
6. Apply a drop of antifade mounting medium over the sample area and place a coverslip on top, avoiding air bubbles. Store the prepared slides at 4 °C in the dark until imaging.
7. Visualize the cells using a fluorescence microscope equipped with filters appropriate for FITC and DAPI detection.

3.9 Immunofluorescence Assay

Tube-Based Workflow:

Note that processing in tubes may lead to cell loss during centrifugation.

1. Harvest cells (> 2 × 10^6/sample) as described in Subheading 3.1.
2. Fix the cells with 4% PFA following **steps 1** and **2** in Subheading 3.3.1 *Formaldehyde and PFA fixation.*
3. Permeabilize the cells using a commercial buffer, as described in Subheading 3.5 for cell suspension.
4. Resuspend the permeabilized cells in 200–500 μL of 1× permeabilization buffer (*see* **Note 14**) and add the primary antibody at the recommended concentration. Incubate for 2 h or overnight at 4 °C.
5. After incubation, centrifugate the cells at 1500–2000 × *g* for 10 min at RT. Discard the supernatant, resuspend the pellet in distilled water, and repeat the centrifugation. Repeat the wash step twice.
6. Dilute the secondary antibody in 1× permeabilization buffer to the appropriate concentration and add it to the washed cell pellet. Incubate for 1 h at RT in the dark.

7. Wash the cells as described in **step 3** and resuspend in 200 µL of PBS or ultrapure water. Transfer 30 µL of the cell suspension to the microscopy slide and allow to adhere as instructed in Subheading 3.4, protecting from light (*see* **Note 15**).
8. Proceed to the mounting steps described below.

Slide-Based Workflow:

1. Harvest cells and fix them with 4% PFA according to Subheadings 3.1 and 3.3.1 *Formaldehyde and PFA fixation*, respectively. Allow cells to attach to slides as directed in Subheading 3.4, then permeabilize with methanol as described in Subheading 3.5.
2. Place the slide in a humid chamber and apply 250 µL of 5.5% (v/v) FBS in PBS-T or 5% (w/v) milk to block non-specific antibody binding. Cover the slide with parafilm and incubate at RT for 45 min.
3. Remove the parafilm and rinse the slide gently using PBS.
4. Apply 250 µL of primary antibody diluted to the appropriate concentration in 3% BSA in PBS-T or 5% milk. Cover with parafilm and incubate in a humid chamber overnight at 4 °C.
5. Remove the parafilm and wash the slide three times gently using PBS-T and twice with PBS to eliminate unbound antibody.
6. Apply 250 µL of secondary antibody diluted to the appropriate concentration in 3% BSA in PBS-T or 5% milk. Cover with parafilm and incubate at RT for 1 h in the dark.
7. Remove the parafilm and wash the slide three times using PBS-T and twice using PBS.
8. Tilt the slide and gently tap its edge onto a tissue paper to remove PBS.

Mounting:

9. Apply a drop of antifade mounting medium with or without nucleic acid stain, depending on experimental requirements, and place a coverslip on top, avoiding air bubbles.
10. Store the prepared slide at 4 °C in the dark until imaging.

3.10 Fluorescence In Situ Hybridization (FISH) with Bacterial Eub338 Probe for Endosymbiont Screening

In addition to the experimental slide, prepare both positive and negative controls. If an endosymbiont-bearing diplonemid is not available, any free-living bacterial species possessing a 16S rRNA region complementary to the Eub338 can be used as a positive control. The negative control is a slide of the diplonemid species in question, incubated under identical conditions but without the Eub338 probe.

1. Harvest diplonemid cultures as described in Subheading 3.1.
2. Fix cells with 4% PFA according to Subheading 3.3.1 *Formaldehyde and PFA fixation* and adhere them to microscopy slides following Subheading 3.4.
3. Permeabilize the affixed cells as described in Subheading 3.5.3.
4. Preheat the hybridization oven to 46 °C.
5. Thaw the Eub338 probe on ice, protecting it from light (*see* **Note 16**).
6. Prepare hybridization vessels by placing a folded, 3-ply tissue paper strip at the bottom of a 50 mL Falcon tube positioned horizontally. Spread 1.5–2 mL of hybridization buffer onto the tissue until it is fully soaked, then seal the tube until use.
7. For each slide, prepare 250 μL of hybridization buffer containing 250 nM Eub338 probe. Apply this solution onto the center of a dry slide (ensure no residual ethanol remains) with the fixed cells. Immediately cover the sample area with parafilm, avoiding air bubbles and ensuring complete coverage. From this point on, minimize the exposure of slides to light.
8. Place the slide above the soaked tissue paper inside the hybridization vessel and close securely.
9. Place the vessel in the hybridization oven and incubate at 46 °C for 1.5 to 3 h. During the incubation, ensure that the slides are completely protected from light.
10. For each slide, prepare a Falcon tube containing 50 mL of washing buffer, preheated to 48 °C in a water bath. After hybridization, remove the slide from the vessel, discard the parafilm, and quickly rinse with ~10 mL of warm washing buffer.
11. Immerse the slides inside the Falcon tubes with remaining washing buffer. Place the tubes in the oven on a horizontal shaker set to 50–60 rpm and incubate at 48 °C for 25 min in the dark.
12. Briefly rinse the slide by dipping it into deionized water, then allow to air-dry in the dark.
13. Mount the sample by applying two drops of antifade mounting medium and place a coverslip, avoiding air bubbles. Store the slide at 4 °C in the dark until imaging.

4 Notes

1. The temperature and boiling time for preparing agarose may vary depending on the specific type and brand of agarose used. To streamline the overall workflow, agarose preparation can be

started in parallel with cell incubation and washing (**steps 4–8** of Subheading 3.2). However, be careful not to overheat the agarose solution, as this can degrade its gelling properties.

2. Add 2 M NaOH dropwise to aid paraformaldehyde dissolution, as it requires an alkaline environment to fully dissolve.
3. A centrifugation speed of 3000 × *g* is sufficient to pellet cells of most cultured diplonemids without causing mechanical stress or compromising cell integrity. After centrifugation, you may inspect the supernatant for residual cells and the pellet for cell integrity under the microscope. If cells remain in the supernatant or the pellet appears loose or damaged, adjust the centrifugation speed and/or time. For larger cells, consider reducing the centrifugation speed to 1000–2000 × *g* to minimize cell damage. Due to the poor adhesion of *P. papillatum* to microscopy slides, we recommend to spin down at least 20–30 mL of exponentially growing cells for slide-based IFA method, while for tube-based workflow, about 5–10 mL of diplonemid culture is usually sufficient.
4. For species cultivated at a different temperature, set the centrifuge accordingly.
5. Ensure that the cell density is suitable for imaging. Typically, a concentration of 1–5 × 10^6 cells/mL is recommended for optimal visualization without overcrowding.
6. Fluorescence dyes are often supplied as ready-to-use stock solutions; however, some may require dilution in appropriate solvents. If needed, prepare a fresh stock solution according to the manufacturer's instructions. Ensure that the dye is fully dissolved before use, and protect it from light to prevent photobleaching.
7. When combining the agarose and stained cells, it is essential to add the cell suspension to the agarose, not the other way around. This minimizes the risk of premature gelling and ensures even distribution. Mix quickly and gently to avoid clumping.
8. The agarose-cell mixture will solidify within seconds to minutes after application, depending on the layer's thickness and temperature. If the gelling temperature of the agarose ranges between 8 °C and 17 °C, the sample can be cooled for a few minutes in a refrigerator or on ice for the solidification process.
9. Fixed cells are more fragile and require moderate centrifugation.
10. The fixative must be freshly prepared just before use. Long-term storage of 4% OsO_4 stock solution must follow institutional safety regulations, including refrigeration in sealed, labelled containers inside secondary containment.

11. Depending on the downstream application, slides may be rinsed briefly or rehydrated by applying PBS dropwise or dipping the slide in PBS for up to 10 min, respectively.
12. For longer incubation periods, cover the slide with Parafilm and place it in a humid chamber.
13. *P. papillatum* is the only known diplonemid species that tolerates PBS in a live state; other marine diplonemids burst due to osmotic shock.
14. No blocking step is required when using commercial permeabilization buffer.
15. The remaining resuspended cells can be stored in the dark at 4 °C, and additional slides can be prepared as needed.
16. Minimize probe exposure to light throughout the procedure.

Acknowledgments

This work was supported by Czech Science Foundation grant 23-06479X to J.L. and a Wellcome Discovery Award (227243/Z/23/Z to B.A.).

References

1. Tashyreva D, Prokopchuk G, Yabuki A et al (2018) Phylogeny and morphology of new diplonemids from Japan. Protist 169:158–179. https://doi.org/10.1016/j.protis.2018.02.001
2. Tashyreva D, Prokopchuk G, Votýpka J et al (2018) Life cycle, ultrastructure, and phylogeny of new diplonemids and their endosymbiotic bacteria. MBio 9. https://doi.org/10.1128/mbio.02447-17
3. Prokopchuk G, Tashyreva D, Yabuki A et al (2019) Morphological, ultrastructural, motility and evolutionary characterization of two new Hemistasiidae species. Protist 170:259–282. https://doi.org/10.1016/j.protis.2019.04.001
4. Marande W, Lukeš J, Burger G (2005) Unique mitochondrial genome structure in diplonemids, the sister group of kinetoplastids. Eukaryot Cell 4:1137–1146. https://doi.org/10.1128/EC.4.6.1137-1146.2005
5. Valach M, Moreira S, Petitjean C et al (2023) Recent expansion of metabolic versatility in *Diplonema papillatum*, the model species of a highly speciose group of marine eukaryotes. BMC Biol 21:99. https://doi.org/10.1186/s12915-023-01563-9
6. Tashyreva D, Týč J, Horák A et al (2023) Ultrastructure and 3D reconstruction of a diplonemid protist (Diplonemea) and its novel membranous organelle. MBio 14:e0192123. https://doi.org/10.1128/mbio.01921-23
7. Lukeš J, Wheeler R, Jirsová D et al (2018) Massive mitochondrial DNA content in diplonemid and kinetoplastid protists. IUBMB Life 70: 1267–1274. https://doi.org/10.1002/iub.1894
8. Prokopchuk G, Korytář T, Juricová V et al (2022) Trophic flexibility of marine diplonemids - switching from osmotrophy to bacterivory. ISME J 16:1409–1419. https://doi.org/10.1038/s41396-022-01192-0
9. Akiyoshi B, Faktorová D, Lukeš J (2025) Discovery of unique mitotic mechanisms in *Paradiplonema papillatum*. Open Biol 15:250096. https://doi.org/10.1098/rsob.250096

Chapter 4

Protocol for the Molecular Identification of Autotrophic Euglenids in Environmental Samples

Katarzyna Jankowska, Maja Łukomska-Kowalczyk, and Bożena Zakryś

Abstract

The majority of eukaryotic microbial biodiversity, particularly within certain groups, remains underexplored. Efforts to address this knowledge gap are often limited, as universal methods fail to capture the full scope of microbial diversity. Tailored metabarcoding approaches, in combination with modern techniques, such as the application of high-throughput sequencing, offer a more effective solution.

Autotrophic euglenids represent one such group that is frequently overlooked due to their habitat preferences and complex genetic features, causing the inadequacy of universal primers. To address these challenges, we present a customized workflow using group-specific primers targeting the V2 region of the 18S rDNA, optimized biomass harvesting, and refined DNA extraction protocols. This methodology was validated on 190 samples from 16 Polish water bodies, resulting in improved detection and a clearer picture of euglenid diversity and distribution. Our guide provides a comprehensive step-by-step protocol—from sampling to bioinformatics analysis, specifically designed to enhance the accuracy and resolution of euglenid biodiversity assessments.

Key words Autotrophic euglenids, Biodiversity, Biomass harvesting, DNA extraction methods, DNA metabarcoding, Next-generation sequencing, V2 18S rDNA

1 Introduction

The true extent of eukaryotic microbial biodiversity and community composition remains largely unknown [1]. Although large-scale sampling initiatives such as TARA Oceans or Ocean Sampling Day aim to explore microbial diversity, their broad focus generally relies on universal methods designed to capture a wide range of organisms [2–5]. Furthermore, these efforts are primarily focused on marine environments, which has led to a significant underrepresentation of studies targeting inland waterbodies [6]. It results in the underestimation of the abundance of certain organisms, particularly those that are difficult to detect using broad sampling techniques or are not typically found in aquatic habitats [7, 8]. Such organisms may also be overlooked due to their unique nucleic acid

Paul A. M. Michels et al. (eds.), *Euglenozoa: Methods and Protocols, Volume 1*, Methods in Molecular Biology, vol. 3013, https://doi.org/10.1007/978-1-0716-5142-1_4,

sequences or the need for specialized sample processing. Therefore, it is essential to evaluate specific methodologies for sampling, biomass collection, isolation, and purification in order to effectively process DNA from environmental samples (eDNA) [9].

The application of metabarcoding combined with methods tailored to the specific needs of overlooked groups of organisms offers a significant improvement in uncovering their true diversity [9, 10]. Its combination of high-throughput sequencing and cost-effective workflows makes it particularly suited to a variety of ecological studies [11]. One group that serves as a clear example of methodological limitations is autotrophic euglenids. These organisms are most commonly found in stagnant eutrophic and hypereutrophic ponds—habitats that may not only be difficult to sample (e.g., plankton net clogging from biogenic material, limited access to the waterbodies and their seasonal drying) but also prone to contamination and spatial variability [12]. Moreover, their 18S rDNA sequences are unusually long and highly variable, making them poorly suited to commonly used universal primers [13]. As a result, euglenids are often underrepresented in eDNA surveys, despite their ecological importance and high species richness—with even over 100 species reported from a single pond [14].

To address this issue, group-specific primers targeting various regions of 18S rDNA were analyzed, and it was proved that the use of V2-V3 and V4 fragments significantly improves the detection of euglenids [13]. Further refinement of this approach—including the design and testing of primers targeting the V2 region, along with the evaluation of optimal biomass harvesting and DNA extraction methods [14]—has established a reliable methodology for exploring euglenid diversity in freshwater ecosystems. Subsequently, this method was applied to a large dataset of 190 samples collected from 16 water bodies in Poland, resulting in a comprehensive analysis of the diversity and distribution [15].

In this guide, we present a detailed, step-by-step protocol for processing environmental samples to assess the biodiversity of autotrophic euglenids. This includes instructions for sample collection, laboratory processing, and bioinformatic analysis. Each step is specifically optimized for this group, resulting in a significantly higher detection rate of euglenid species compared to standard metabarcoding approach that relies on universal methods.

2 Materials

All of the solutions have to be prepared using nuclease-free water. Reagents should be stored and procedures should be conducted at room temperature unless stated otherwise.

2.1 Water Sampling

1. Glass containers (1 L).
2. Distilled water.
3. 50 mL Falcon tubes.
4. Nuclease-free water.
5. 1.5 mL Eppendorf tubes.
6. Centrifuge (suitable for Falcon and Eppendorf tubes).

2.2 DNA Isolation

1. Sterile laminar flow cabinet.
2. Cell disruptor (e.g., TissueLyser LT, Qiagen).
3. GeneMATRIX Soil DNA Purification Kit (EURx, Poland).
4. Thermoblock.

2.3 DNA Sequencing

1. Spectrophotometer for DNA quality and quantity assessment (NanoPhotometer NP80 (Implen) or similar).
2. Electrophoresis gel chamber.
3. 1× TAE buffer (diluted from 50× TAE buffer: 242.3 g Tris-base, 57.2 mL glacial acetic acid, 100 mL 0.5 M EDTA (pH 8.0), volume adjusted to 1 L with distilled H_2O).
4. 1% agarose gel prepared using 1× TAE buffer containing 0.2% Midori Green DNA stain (Nippon) or similar.
5. DNA ladder.
6. PCR tubes.
7. High fidelity DNA polymerase kit (e.g., Phusion DNA polymerase kit (Thermo Scientific)).
8. Primers with tags (*see* Table 1).
9. Nuclease-free water.
10. 1.5 mL Eppendorf tubes.
11. Thermocycler.
12. GeneMATRIX PCR/DNA Clean-Up Purification Kit (EURx).

2.4 Data Analysis

1. Computer with a Linux operating system. The process shown below was conducted using Ubuntu server 16.04.7 LTS with 2× Intel Xeon CPU E5–2620 and 192 GB RAM.
2. FastQC—a quality control tool for high throughput sequencing data, version v0.11.5, https://www.bioinformatics.babraham.ac.uk/projects/fastqc/ [16].
3. QIIME2—microbiome bioinformatics platform, https://qiime2.org/ [17]. The Cutadapt plugin should be integrated into the QIIME 2 environment prior to starting the bioinformatics analysis. We conducted our analysis on version 2022.2.0.
4. Statistic software R version 4.4.1 and R studio 2024.04.2, https://www.r-project.org [18, 19].

Table 1
Primers used for PCR amplification

Primer Name	Sequence 5' → 3' [Nextera adapter][tag] [rDNA complementary sequence]	Type
18S_V2i_F1	TCGTCGGCAGCGTCAGATGTGTATAAGAGACAG **CCTCTCTT**CTGTGAATGGCTCCTTACATCAG	Forward
18S_V2i_F2	TCGTCGGCAGCGTCAGATGTGTATAAGAGACAG **AACTCCGC**CTGTGAATGGCTCCTTACATCAG	Forward
18S_V2i_F3	TCGTCGGCAGCGTCAGATGTGTATAAGAGACAG **TTGTGTGG**CTGTGAATGGCTCCTTACATCAG	Forward
18S_V2i_F4	TCGTCGGCAGCGTCAGATGTGTATAAGAGACAG **TATGGCAC**CTGTGAATGGCTCCTTACATCAG	Forward
18S_V2i_F5	TCGTCGGCAGCGTCAGATGTGTATAAGAGACAG **TAAACGCG**CTGTGAATGGCTCCTTACATCAG	Forward
18S_V2i_F6	TCGTCGGCAGCGTCAGATGTGTATAAGAGACAG **GGACATTA**CTGTGAATGGCTCCTTACATCAG	Forward
18S_V2i_F7	TCGTCGGCAGCGTCAGATGTGTATAAGAGACAG **GTCCTTCG**CTGTGAATGGCTCCTTACATCAG	Forward
18S_V2i_F8	TCGTCGGCAGCGTCAGATGTGTATAAGAGACAG **CAAAGTGT**CTGTGAATGGCTCCTTACATCAG	Forward
18S_V2i_F9	TCGTCGGCAGCGTCAGATGTGTATAAGAGACAG **GTTACCGA**CTGTGAATGGCTCCTTACATCAG	Forward
18S_V2i_F10	TCGTCGGCAGCGTCAGATGTGTATAAGAGACAG **TTCCCATC**CTGTGAATGGCTCCTTACATCAG	Forward
18S_V2i_F11	TCGTCGGCAGCGTCAGATGTGTATAAGAGACAG **GTGTAAAC**CTGTGAATGGCTCCTTACATCAG	Forward
18S_V2i_F12	TCGTCGGCAGCGTCAGATGTGTATAAGAGACAG **TGCGTCTG**CTGTGAATGGCTCCTTACATCAG	Forward
18S_V2i_F13	TCGTCGGCAGCGTCAGATGTGTATAAGAGACAG **CTAGTAGC**CTGTGAATGGCTCCTTACATCAG	Forward
18S_V2i_F14	TCGTCGGCAGCGTCAGATGTGTATAAGAGACAG **AGACTGAC**CTGTGAATGGCTCCTTACATCAG	Forward
18S_V2i_F15	TCGTCGGCAGCGTCAGATGTGTATAAGAGACAG **ATTAGACC**CTGTGAATGGCTCCTTACATCAG	Forward
18S_V2i_F16	TCGTCGGCAGCGTCAGATGTGTATAAGAGACAG **TAGCTAGA**CTGTGAATGGCTCCTTACATCAG	Forward
18S_V2i_R1	GTCTCGTGGGCTCGGAGATGTGTATAAGAGACA **GCACTTGA**GCTsCCTCTCCGGAATCrAACC	Reverse
18S_V2i_R2	GTCTCGTGGGCTCGGAGATGTGTATAAGAGACA **GGCATCAT**GCTsCCTCTCCGGAATCrAACC	Reverse

(continued)

Table 1 (continued)

Primer Name	Sequence 5' → 3' [Nextera adapter][tag] [rDNA complementary sequence]	Type
18S_V2i_R3	GTCTCGTGGGCTCGGAGATGTGTATAAGAGACA **GAGGACAT**TCTsCCTCTCCGGAATCrAACC	Reverse
18S_V2i_R4	GTCTCGTGGGCTCGGAGATGTGTATAAGAGACA **GTCATTTC**CCTsCCTCTCCGGAATCrAACC	Reverse
18S_V2i_R5	GTCTCGTGGGCTCGGAGATGTGTATAAGAGACA **GGAGCCTA**TCTsCCTCTCCGGAATCrAACC	Reverse
18S_V2i_R6	GTCTCGTGGGCTCGGAGATGTGTATAAGAGACA **GGACTGAC**ACTsCCTCTCCGGAATCrAACC	Reverse
18S_V2i_R7	GTCTCGTGGGCTCGGAGATGTGTATAAGAGACA **GTGACGAC**TCTsCCTCTCCGGAATCrAACC	Reverse
18S_V2i_R8	GTCTCGTGGGCTCGGAGATGTGTATAAGAGACA **GCTCCGCA**ACTsCCTCTCCGGAATCrAACC	Reverse
18S_V2i_R9	GTCTCGTGGGCTCGGAGATGTGTATAAGAGACA **GCCATAGG**ACTsCCTCTCCGGAATCrAACC	Reverse
18S_V2i_R10	GTCTCGTGGGCTCGGAGATGTGTATAAGAGACA **GAGTTACC**GCTsCCTCTCCGGAATCrAACC	Reverse

3 Methods

3.1 Water Sampling

1. Equipment preparation: it is important to ensure that the equipment used in the process is properly cleaned to avoid contamination from microorganisms, DNA, or chemical residues. Therefore, follow a procedure of washing non-sterile or reusable equipment with a detergent for laboratory use, rinsing thoroughly with distilled water, and then sanitizing it using UV light (*see* **Note 1**).
2. Collect water from the sunlit surface using a sterile, 1 L glass container (*see* **Notes 2–5**).
3. Centrifuge 50–400 mL of the sample (*see* **Note 6**) in 50 mL aliquots using sterile Falcon tubes at 4000 × *g* for 20 min (*see* **Note 7**). After each centrifugation, discard the pellet and add the next 50 mL aliquot. Once the entire sample volume has been processed, retain the final pellet and transfer it to a sterile 1.5 mL microcentrifuge tube. Centrifuge the pellet again at 4000 × *g* for 4 min, then carefully discard the majority of the supernatant. For long-term storage, preserve the pellet at –80 °C.

3.2 DNA Isolation

1. Resuspend the pellet in nuclease-free water, using a volume of 200–1500 μL (*see* **Note 8**).
2. Use 200 μL from the previous step for DNA isolation with the GeneMATRIX Soil DNA Purification Kit (EURx). All steps should be performed in a sterile laminar flow cabinet. At the end of the isolation, elute the DNA in 50–60 μL of nuclease-free water (*see* **Note 9**).
3. To assess the DNA concentration and purity of the obtained DNA, use a spectrophotometer (e.g., NanoPhotometer NP80, Implen) (*see* **Note 10**). Perform standard agarose gel electrophoresis with a 1% agarose gel prepared using 1× TAE buffer, staining with 0.2% Midori Green DNA stain (Nippon). Load 5 μL of DNA mixed with dye into the gel chamber and run the gel at 80 V for 50 min. Compare the visible bands with the appropriate DNA ladder (*see* **Note 11**).

3.3 PCR Amplification and DNA Sequencing

1. Amplify the V2 region of 18S rDNA on specially designed primers universal for autotrophic euglenids ended with tags (*see* **Note 12**). The process should be conducted with high fidelity DNA polymerase, in our case Phusion polymerase (Thermo Scientific) (Tables 1, 2, and 3) (*see* **Note 13**).
2. Perform standard agarose gel electrophoresis with a 1% agarose gel prepared using 1× TAE buffer, staining with 0.2% Midori Green DNA stain (Nippon). Load 5 μL of DNA into the gel chamber and run the gel at 80 V for 30 min. Compare the visible bands with the appropriate DNA ladder (*see* **Notes 14** and **15**).
3. Purify PCR products (amplicons) using the GeneMATRIX PCR/DNA Clean-Up Purification Kit (EURx).
4. To assess the density and purity of the amplified product, use a spectrophotometer (*see* **Note 16**).

Table 2
Components of the PCR reaction mix using Phusion DNA polymerase kit

Reaction Component	Volume	
Water	× μL	Up to 25 μL
Buffer 1×	5 μL	
Primer R	0.5 μL	10 pmol
Primer F	0.5 μL	10 pmol
dNTPs	0.5 μL	0.2 mM
Phusion polymerase	0.25 μL	0.5 U
DNA	y μL	Up to 50 ng

Table 3
Thermocycler program settings

Temperature	Time	Cycles
98 °C	30″	
98 °C 55 °C/57 °C/59 °C 72 °C	20″ 20″ 30″	×35
72 °C	10′	

5. Prepare a mixture of the selected amplicons, ensuring that each contributes 100 ng of DNA. Subsequently, dilute the mixture to a final concentration of approximately 10–15 ng/μL using nuclease-free water (*see* **Notes 17** and **18**).
6. Depending on available resources, perform sequencing in-house or outsource it using the Illumina MiSeq platform with 2 × 300 bp paired-end reads (in this study, sequencing was outsourced to Genomed S.A.; for comparable applications, library preparation may be performed using kits such as the Nextera XT DNA Library Prep Kit).

3.4 Data Analysis

1. Beginning at this point, use the command line interface for further operations.
2. Activate QIIME2 (replace qiime2-2025.4 with your environment name if different).

```
conda activate qiime2-2025.4
```

3. Demultiplex the forward and reverse raw reads (ensure they are in.fastq.gz format) with cutadapt tool using command below (*see* **Note 19**).

```
cutadapt \
-e 0 --no-indels \
-g file:TagF.fas \
-G file:TagR.fas \
-o {name1}-{name2}.1.fastq.gz \
-p {name1}-{name2}.2.fastq.gz \
NGS-Euglena_S7_L001_R1_001.fastq.gz \
NGS-Euglena_S7_L001_R2_001.fastq.gz \
-m 293 -M 293
```

4. Substitute sample names in a folder containing demultiplexed sample files. The original and modified filenames should be separated by a tab and each pair should appear on a separate line (e.g., F1-R1.1.fastq.gz 17_IK2.1.fastq.gz).

```
mmv < file.txt
```

5. Prepare a manifest file that describes the used samples in the format: 'sample-id,absolute-filepath,direction' (e.g., 920_IK1, $PWD/raw/920_IK1.1.fastq.gz,forward) and save it as a.csv file. The first row should include the column names in this format. For more information, see: https://docs.qiime2.org/2024.10/tutorials/importing/ section "Fastq manifest" formats.
6. Combine all the samples into a single.qza file.

```
qiime tools import \
--type 'SampleData[PairedEndSequencesWithQuality]' \
--input-format PairedEndFastqManifestPhred33 \
--input-path manifest.csv \
--output-path reads_NGS1.qza
```

7. Filter sequences using the command below (*see* **Note 20**).

```
qiime dada2 denoise-paired \
--i-demultiplexed-seqs reads_NGS1.qza \
--p-trunc-len-f 0 \
--p-trunc-len-r 0 \
--p-trim-left-f 23 \
--p-trim-left-r 23 \
--p-chimera-method consensus \
--p-trunc-q 9 \
--o-denoising-stats den-stats.qza \
--o-representative-sequences rep-seqs.qza \
--o-table filtered-table.qza \
--p-n-threads 16 \
--verbose
```

8. After analyzing the number of reads for amplicon sequence variants (ASVs) in the samples, we observed "leaking" of reads from one sample to another within a single run. This issue is documented and has been discussed in previous research [9, 20]. To address this, we conducted a series of

control samples with a known set of species alongside the study samples. Based on that we set a threshold of 0.5% of the total reads for each ASV and subtracted this value from each sample in the run. To do this, follow the steps below.

(a) Extract the filtered table file in biom format:

```
mkdir extracted_halfper
qiime tools export \
--input-path filtered-table.qza \
--output-path extracted_halfper
```

(b) Convert to.tsv file

```
biom convert \
-i./extracted_halfper/feature-table.biom \
-o./extracted_halfper/filtered-table.txt \
--to-tsv
```

(c) Before performing any calculations, you need to manually remove the part "#Constructed from biom file #OTU ID" as it interferes with assigning column names to the table.

(d) Subtract 0.5% of the total reads for each ASV in R using a custom script halfper.r available at https://github.com/Mayakk/Euglenid_identification/blob/main/halfper.r (no additional packages are required—base R is sufficient for these calculations).

(e) Convert to.biom file:

```
biom convert \
-i filtered-table-halfper.txt \
-o filtered-table-halfper.biom \
--to-hdf5
```

(f) Import to.qza format:

```
qiime tools import \
--input-path filtered-table-halfper.biom \
--output-path filtered-table-halfper.qza \
--type "FeatureTable[Frequency]"
```

9. If multiple NGS runs were conducted, merge the tables with the 0.5% threshold subtracted into a single table, along with the corresponding representative sequences files.

```
qiime feature-table merge \
--i-tables \
filtered-table-halfper-NGS1.qza \
filtered-table-halfper-NGS2.qza \
filtered-table-halfper-NGS3.qza \
filtered-table-halfper-NGS4.qza \
--o-merged-table \
filtered-table-halfper-NGS1234.qza \
--p-overlap-method sum
qiime feature-table merge-seqs \
--i-data \
rep-seqs-NGS1.qza \
rep-seqs-NGS2.qza \
rep-seqs-NGS3.qza \
rep-seqs-NGS4.qza \
--o-merged-data \
rep-seqs-NGS1234.qza
```

10. Prepare a reference sequence database required for the classification of the obtained sequence reads.
    (a) Download the latest SILVA reference database specifically formatted for QIIME2 from the following link:
        https://www.arb-silva.de/no_cache/download/archive/qiime/Silva_132_release.zip
    (b) Unzip the downloaded archive. Two essential files are required for downstream analysis (*see* **Note 21**):
        Reference sequences file:

```
Silva_132_release\SILVA_132_QIIME_release\rep_set\rep_set_all
\97\silva132_97.fna
```

Taxonomy file:

```
Silva_132_release\SILVA_132_QIIME_release\taxonomy\taxono-
my_all\97\majority_taxonomy_7_levels.txt
```

    (c) Remove autotrophic euglenid sequences from the database. Use the grep command to exclude taxonomy entries associated with autotrophic euglenids (Phacaceae and Euglenaceae) (*see* **Note 22**):

```
grep -vE "(Phacaceae|Euglenaceae|)" majority_taxonomy_7_le-
vels.txt > no_Eugl_taxonomy.txt
```

    (d) Download the curated V2 Hypervariable Region of 18S rDNA database, which contains 714 sequences from 196 euglenid species with verified taxonomy.

File name: V2_Euglenid_barcodes_09_2023.fas (*see* **Note 23**):

Download link: https://repod.icm.edu.pl/dataset.xhtml?persistentId=doi:10.18150/9MKYCN

(e) Use the Python script Tax_and_Seq.py available at https://github.com/Mayakk/Euglenid_identification/blob/main/Tax_and_Seq.py to process the downloaded V2_Euglenid_barcodes_09_2023.fas file and convert it into the format required by QIIME2. The script should generate two separate output files:

Reference sequences file (FASTA file containing only the nucleotide sequences):

```
SE_V2_Euglenid_barcodes_09_2023.fas
```

Taxonomy file (text file with sequence IDs and their corresponding taxonomy):

```
TX_V2_Euglenid_barcodes_09_2023.fas
```

(f) Combine taxonomy files: Append the taxonomy (TX_V2_Euglenid_barcodes_09_2023.fas) to the filtered SILVA taxonomy (no_Eugl_taxonomy.txt) to create the final taxonomy file:

```
cat no_Eugl_taxonomy.txt TX_V2_Euglenid_barcodes_09_2023.fas
> reference_taxonomy.txt
```

(g) Combine sequence files: Append the EuglRef sequences to the SILVA reference sequences to generate the full sequence reference file

```
cat silva132_97.fna SE_V2_Euglenid_barcodes_09_2023.fas > re-
ference_sequences.fas
```

(h) Import the combined taxonomy and sequence files into QIIME2:

```
qiime tools import --type 'FeatureData[Taxonomy]' --input-
format HeaderlessTSVTaxonomyFormat --input-path reference_-
taxonomy.txt --output-path reference_taxonomy.qza
qiime tools import --type 'FeatureData[Sequence]' --input-path refer-
ence_sequences.fas --output-path reference_sequences.qza
```

(i) Extract the V2 region from reference sequences using the appropriate primers sequences:

```
qiime feature-classifier extract-reads --i-sequences referen-
ce_sequences.qza --p-f-primer TGTGAATGGCTCCTTACATCAG --p-r-
primer GCTSCCTCTCCGGAATCRAACC --p-trunc-len 800 --o-reads V2_
reference_sequences.qza
```

(j) Train the Naive Bayes classifier on the trimmed reference data:

```
qiime feature-classifier fit-classifier-naive-bayes --i-re-
ference-reads V2_ reference_sequences.qza --i-reference-tax-
onomy reference_taxonomy.qza --o-classifier classifier.qza
```

11. Classify sequences using prepared database (*see* **Note 24**):

```
qiime feature-classifier classify-sklearn \
--i-classifier classifier.qza \
--i-reads rep-seqs-NGS1234.qza \
--o-classification taxonomy.qza \
--p-n-jobs 12
```

12. Filter the table to retain the target group of organisms for analysis.

```
qiime taxa filter-table \
--i-table filtered-table-halfper-NGS1234.qza \
--i-taxonomy taxonomy.qza \
--p-include "Euglenida" \
--o-filtered-table filtered-table-halfper-NGS1234-eugl.qza
```

13. Generate a barplot to visualise the data (*see* **Note 25**).

```
qiime taxa barplot \
--i-table filtered-table-NGS1234-halfper-eugl.qza \
--i-taxonomy taxonomy.qza \
--m-metadata-file metadata.tsv \
--o-visualization barplot-eugl.qzv
```

4 Notes

1. Ideally, the detergent should be easily washable and low in phosphorus as well as other chemicals particularly when analyzing physicochemical parameters and their potential impact on the euglenid biodiversity.

2. To address the challenge of low plankton density in water bodies, you can use a plankton net with a mesh size of 10 μm. Depending on the microorganism density in the water, filter 10–100 L into a 1 L glass container.
3. It can be highly beneficial to use a larger, sterile filter (mesh size 200 μm) before transferring the water to the final container to prevent small animals, like those that prey on the target group, from being included. However, their bodies are often used by *Colacium* as a substrate for colonization.
4. If immediate processing of the samples is not possible, adding ethanol (10–20% of the final volume, ideally use the absolute ethanol for its purity) can effectively preserve them. DNA extracted from such samples maintains comparable quality and yield. To simplify long-distance transportation, including air travel, it is acceptable to use plastic containers (e.g., 500 mL containers filled to half capacity).
5. Euglenids are most commonly found in shallow, small water bodies, typically eutrophic or hypereutrophic, such as farm ponds, park ponds, field ponds, and fish ponds).
6. A satisfactory pellet is typically achieved after three to five centrifugation cycles depending on the density of the sample.
7. If the centrifuge is equipped with a high capacity rotor holding up to 250 mL, it can significantly improve the efficiency of the centrifugation process.
8. This helps maintain an appropriate sample density, as overly concentrated samples may affect amplification efficiency.
9. To improve DNA yield, we recommend adding an additional step of freezing the samples, as suggested in the manual.
10. A260/230 and A260/280 ratios should oscillate around 2.0–2.2 and 1.8, respectively. However, we observed that at this stage, the A260/230 ratio tends to be significantly lower than predicted, which did not influence the outcome.
11. At this point, the DNA on the gel image may appear as a smear due to the presence of many fragments of varying sizes. If the spectrophotometer or gel indicates no trace of DNA continue with the process, as this does not necessarily mean the outcome will be negative.
12. We tagged the 5′ end of our primers with 8 bp sequences (*see* Table 1) to enable multiplexing of samples within a single sequencing run (tags and adapters used in the sequencing process are typically provided by the manufacturer of the sequencing kit). This approach is more cost-effective than sequencing each sample individually. In total, we used 16 tagged forward primers and 10 tagged reverse primers, resulting in 160 unique forward-reverse primer combinations.

13. To ensure obtaining the whole range of genetic variability we amplified each sample at three temperatures (55, 57, 59 °C) and then pulled them together in the first steps of purification. If the DNA concentration measured in the previous point is higher than 50 ng per microliter, we recommend adding 0.5 μL instead of 1 to the PCR mixture.
14. The expected size of gel bands for euglenid V2 18S rDNA typically ranges from 300 to 400 bp. However, certain species, such as *Euglena velata*, may produce sequences up to 550 bp.
15. It is best to control the purity of the products by conducting negative controls.
16. A260/230 and A260/280 ratios should oscillate around 2.0–2.2 and 1.8, respectively.
17. To obtain an appropriate number of reads in one run, we propose sequencing no more than 150 samples.
18. This step can be omitted by preparing separate libraries for each sample. While this approach is more straightforward, it is associated with increased costs.
19. We chose not to allow insertions and deletions in the starters and added X at the beginning of the starter in each Tag fasta file to ensure that the program searches for the starter only at the beginning of the sequence. We trimmed sequences to 293 nucleotides and discarded shorter ones after careful consideration of sequences quality in fastqc. While using a higher number of jobs or threads is not required, it can be beneficial, as the process is computationally intensive.
20. We did not cut the sequences, as the "q" parameter already handles that. Instead, we trimmed the primers and tags. By setting the "q" parameter high enough, we ensure that the expected errors [ee] do not exceed 2 (the default), while also keeping the sequences long enough for proper pairing. The quality parameter was set separately for each NGS run, varying between 8 and 10 in our study. This was determined based on the denoising statistics file. To evaluate this file, it needs to be converted into a.qzv file.

```
qiime metadata tabulate \
--m-input-file den-stats.qza \
--o-visualization den-stats.qzv
```

21. You may choose alternative options such as different sequence identity thresholds (e.g., 99%) or taxonomy assignment strategies (e.g., consensus taxonomy). For more details, refer to:

```
Silva_132_release\SILVA_132_QIIME_release\Silva_132_notes.txt
```

22. For other datasets, consider also removing additional autotrophic euglenid groups such as **Eutreptiales** and **Rapaza** by extending the pattern:

```
grep -vE "(Phacaceae|Euglenaceae|Eutreptiales|Rapaza)" majority_taxonomy_7_levels.txt > no_Eugl_taxonomy.txt
```

23. You should always use the latest available version (currently version 2).
24. While using a higher number of jobs or threads is not required to obtain a classifier, it can be beneficial, as the process is computationally intensive.
25. If you intend to continue the analysis in R, possible approaches include exporting the data from the bar plot available at https://view.qiime2.org/ in.csv format, or importing the data directly into R using the qiime2R package. When using qiime2R, proceed with the filtered feature table file.

Acknowledgments

This work was supported by the OPUS 2016/23/B/NZ8/00919 grant from the National Science Centre, Poland.

References

1. Singer D, Seppey CVW et al (2021) Protist taxonomic and functional diversity in soil, freshwater and marine ecosystems. Environ Int 146:106262. https://doi.org/10.1016/j.envint.2020.106262
2. Pawlowski J, Audic S, Adl S et al (2012) Barcoding eukaryotic richness beyond the animal, plant, and fungal kingdoms. PLoS Biol 10: e1001419. https://doi.org/10.1371/journal.pbio.1001419
3. Pesant S, Not F, Picheral M et al (2015) Open science resources for the discovery and analysis of Tara Oceans data. Sci Data 2:150023. https://doi.org/10.1038/sdata.2015.23
4. De Vargas C, Audic S, Henry N et al (2015) Eukaryotic plankton diversity in the sunlit ocean. Science 348:1–12. https://doi.org/10.1126/science.1261605
5. Kopf A, Bicak M, Kottmann R et al (2015) The ocean sampling day consortium. Gigascience 4: 1. https://doi.org/10.1186/s13742-015-0066-5
6. Burki F, Roger AJ, Brown MW et al (2021) Diversity and ecology of protists revealed by metabarcoding. Curr Biol 31:R1267–R1280. https://doi.org/10.1016/j.cub.2021.08.015
7. Simon M, Jardillier L, Deschamps P et al (2015) Complex communities of small protists and unexpected occurrence of typical marine lineages in shallow freshwater systems. Environ Microbiol 17:3610–3627. https://doi.org/10.1111/1462-2920.12591
8. De Luca D, Piredda R, Sarno D et al (2021) Resolving cryptic species complexes in marine protists: phylogenetic haplotype networks meet global DNA metabarcoding datasets. ISME J 15:1931–1942. https://doi.org/10.1038/s41396-021-00895-0
9. Deiner K, Bik HM, Mächler E et al (2017) Environmental DNA metabarcoding: transforming how we survey animal and plant communities. Mol Ecol 26:5872–5895. https://doi.org/10.1111/mec.14350
10. Taberlet P, Coissac E, Pompanon F et al (2012) Towards next-generation biodiversity assessment using DNA metabarcoding. Mol Ecol 21(8):2045–2050. https://doi.org/10.1111/j.1365-294X.2012.05470.x

11. Shokralla S, Spall JL, Gibson JF, Hajibabaei M (2012) Next-generation sequencing technologies for environmental DNA research. Mol Ecol 21:1794–1805. https://doi.org/10.1111/j.1365-294X.2012.05538.x
12. Zakryś B, Milanowski R, Karnkowska A (2017) Evolutionary origin of euglena. Adv Exp Med Biol 979:3–17. https://doi.org/10.1007/978-3-319-54910-1_1
13. Łukomska-Kowalczyk M, Karnkowska A et al (2016) DNA barcoding in autotrophic euglenids: evaluation of COI and 18s rDNA. J Phycol 52:951–960. https://doi.org/10.1111/jpy.124394
14. Gumińska N, Łukomska-Kowalczyk M, Chaber K et al (2021) Evaluation of V2 18S rDNA barcode marker and assessment of sample collection and DNA extraction methods for metabarcoding of autotrophic euglenids. Environ Microbiol 23:2992–3008. https://doi.org/10.1111/1462-2920.15495
15. Jankowska K, Łukomska-Kowalczyk M, Milanowski R et al (2024) Biodiversity of autotrophic euglenids based on the group specific DNA metabarcoding approach. Protist 175:126024. https://doi.org/10.1016/j.protis.2024.126024
16. Andrews S (2010) FastQC: a quality control tool for high throughput sequence data. Available at: http://www.bioinformatics.babraham.ac.uk/projects/fastqc. Accessed 30 Apr 2025
17. Bolyen E, Rideout JR, Dillon MR et al (2019) Reproducible, interactive, scalable and extensible microbiome data science using QIIME 2. Nat Biotechnol 37:852–857. https://doi.org/10.1038/s41587-019-0209-9. Accessed 30 Apr 2025
18. R Core Team (2023) R: a language and environment for statistical computing. R Foundation for Statistical Computing, Vienna, Austria. https://www.R-project.org/
19. RStudio Team (2024) RStudio: integrated development for R. RStudio, PBC. https://www.rstudio.com/. Accessed 30 Apr 2025
20. Zizka VMA, Elbrecht V, Macher J, Leese F (2019) Assessing the influence of sample tagging and library preparation on DNA metabarcoding. Mol Ecol Resour 19:893–899. https://doi.org/10.1111/1755-0998.13018

Chapter 5

Culture and Genetic Modification of the Monoxenous Trypanosomatids *Crithidia fasciculata* and *Crithidia bombi*

Blyssalyn V. Bieber, Faith A. St. Clair, Jesse Quatse, Shane Denecke, Michael Povelones, and Megan L. Povelones

Abstract

Monoxenous trypanosomatids exclusively parasitize insects and are powerful systems for exploring trypanosomatid biology, the evolution of parasitism, and host–parasite interactions. In addition, they are tractable early-diverging eukaryotes that can serve as model organisms for a variety of conserved processes. We have elaborated on existing methods for culturing the mosquito parasite *Crithidia fasciculata* and the bumblebee parasite *Crithidia bombi*, including quantitative assessment of surface attachment and genetic modification. In addition, as cultured strains retain their ability to infect their insect hosts, we describe use of a luciferase-based reporter line to quantitate infection intensities in laboratory mosquitoes.

Key words *Crithidia*, Adhesion, Attachment, Nucleofection, Luciferase

1 Introduction

For many years, *Crithidia fasciculata* and other non-human-infective trypanosomatids have served as important models for shared processes including kDNA replication [1–3], RNA editing [4–6], and metabolism [7–12]. These species are appealing because they have rapid doubling times and reach high densities in serum-free medium. In addition, as Biosafety Level 1 organisms, they are suitable to use in a variety of settings, including undergraduate teaching labs [13].

Monoxenous (single-host) trypanosomatids are widespread and diverse parasites [14, 15]. Species such as *Lotmaria passim*, *Crithidia bombi*, and *Crithidia mellificae* parasitize bees and may be a contributing factor in pollinator decline [16, 17]. As such, these species have long been of interest to the ecological community, which has extensively studied the prevalence and intensity of infections, transmission dynamics, host specificity, and the impacts of parasitism on insect host physiology [18].

Paul A. M. Michels et al. (eds.), *Euglenozoa: Methods and Protocols, Volume 1*, Methods in Molecular Biology, vol. 3013, https://doi.org/10.1007/978-1-0716-5142-1_5,

Now, with complete genome sequences for *C. fasciculata* [19] and *C. bombi* [20, 21], as well as the expansion of accessible and readily adaptable genetic tools for manipulation of trypanosomatids [22, 23], a wealth of opportunities exist for detailed cell and molecular investigations into the biology of these important species.

By combining previously established conditions for axenic culture [1, 24] and transfection of related parasites [25] with newly developed assays, we have generated a standard set of protocols for genetic modification as well as quantitative techniques to measure growth, surface attachment, and colonization of tissues in the insect host.

2 Materials

All solutions are prepared using ultrapure water (e.g., MilliQ >18 MΩ) and tissue-culture or molecular biology grade reagents. Unused parasites in culture medium should be killed with bleach or another disinfectant and all contaminated plastics subjected to autoclaving or incineration. All solutions for cell culture are sterilized by filtering or autoclaving and live cells are handled in a biosafety cabinet to prevent contamination.

2.1 Culturing C. fasciculata *(Adapted from Ref. [1])*

1. BHI incomplete medium: Dissolve 18.5 g brain heart infusion (BHI) powder in 500 mL water and sterilize (*see* **Note 1**).
2. Hemin stock (2 mg/mL): dissolve 1 g hemin in 500 mL 50 mM NaOH (*see* **Note 2**). Sterile-filter and store 4 mL aliquots at −20 °C.
3. BHI complete medium: Add 4 mL 2 mg/mL hemin stock to 500 mL incomplete BHI medium and swirl to mix. Optional, complete medium can be filtered using 0.2 μm bottle-top filters.
4. Non-vented 25 cm^2 tissue culture flasks (non-vented T25 *see* **Note 3**).
5. Temperature-controlled orbital shaker or orbital shaker placed in an incubator.
6. 3% formalin: dilute formalin in water.
7. Gentian Violet Stain: dissolve 4 g NaCl, 0.1 g EDTA, 0.75 mL Gentian Violet in water to 500 mL.
8. Neubauer hemacytometer.

2.2 Culturing C. bombi (Adapted from Ref. [24])

1. FP stock: In a beaker, add 800 mL water and dissolve 2.8 g NaCl, 0.4 g KCl, 10 g $Na_2HPO_4.2\ H_2O$, 10 g Tryptose broth, 2 g Liver Infusion Broth. Once fully dissolved, adjust pH to 5.8. Add water to 1 L and transfer to autoclave safe bottles. Autoclave and store at 4 °C.
2. 10× FB stock: In a beaker, add 150 mL FP stock and dissolve 3.6 g D-(−)fructose, 0.578 g L-proline, 0.0134 g thiamine hydrochloride, 0.0088 g folic Acid. Once fully dissolved, add FP stock up to 200 mL total volume (*see* **Note 4**). Store 5 mL aliquots at −20 °C.
3. Hemin stock (2 mg/mL) prepared as above. Sterile filter and aliquot in 1.5 mL tubes. Store at −20 °C.
4. FP-FB medium: Combine 36 mL FP stock, 4 mL 10× FB stock, 4.5 mL heat-inactivated fetal bovine serum, 44.5 μL hemin stock. Sterile filter into a 50 mL tube and store at 4 °C for up to 3 weeks.
5. Vented tissue culture flasks (vented T25 *see* **Note 3**).
6. Temperature-controlled orbital shaker with CO_2 or orbital shaker placed in a CO_2 incubator.
7. 3% formalin, Gentian Violet Stain, and hemacytometer (as above).

2.3 Growth Curves for C. bombi and C. fasciculata

1. Non-vented (*C. fasciculata*) or vented (*C. bombi*) T25 tissue culture flasks.
2. Hemacytometer and counting solutions (above).

2.4 C. fasciculata Attachment Assays

1. 60 mm plastic dishes or tissue culture 12- or 24-well plates.
2. BHI complete medium.
3. 1× PBS: in 800 mL water, add 137 mM NaCl, 2.7 mM KCl, 10 mM Na_2HPO_4, 1.8 mM KH_2PO_4. Adjust pH to 7.4 with HCl and fill to 1 L using water and autoclave or sterile filter.

2.5 C. bombi Attachment Assays

1. 60 mm plastic dishes.
2. FP-FB medium.
3. 1× PBS.

2.6 Nanoluciferase Assay for Cell Density Measurements of Cultured Crithidia sp.

1. Nanoluciferase (NLuc) assay buffer and substrate (Promega Nano-Glo cat. no. N1110).
2. White 96-well polystyrene plates, flat-bottom (chimney well).
3. BHI complete or FP-FB medium.

2.7 Transfection of Crithidia sp.

1. Tb-BSF transfection buffer [26]: 67 mM Na_2HPO_4, 23 mM NaH_2PO_4, 5 mM KCl, 0.15 mM $CaCl_2$, 50 mM HEPES, pH 7.3 (*see* **Note 5**).
2. Lonza Nucleofector 2b device with Lonza cuvettes and transfer pipettes.

2.8 NLuc to Quantify Infection of Laboratory Mosquitoes with C. fasciculata

1. Paper food cups (8 or 16 oz) and fine mesh netting for housing mosquitoes.
2. Cotton balls.
3. 10% sucrose in water.
4. 1× PBS.
5. BHI complete medium for diluting standard curve.
6. NLuc buffer and substrate.
7. White 96-well polystyrene plates, flat-bottom (chimney well).

2.9 Isolating C. bombi *from Bees Using Limiting Dilution*

1. Ringer's solution: made from tablets at ¼ strength in water.
2. Pestle for microcentrifuge tubes.
3. Modified Mäser mix: Combine 44 mL of FP-FB medium, 450 μL penicillin-streptomycin, and 450 μL of 250 μg/mL amphotericin B.
4. Freezing medium: 45 mL FP-FB medium (no drug), 5 mL 100% glycerol.
5. 96- and 24-well non-treated tissue culture plates.

3 Methods

3.1 Culturing C. fasciculata

1. For standard culturing, cells are maintained in 5 mL of BHI complete medium in non-vented T25 flasks on an orbital shaker at 27 °C (*see* **Note 6**). A typical dilution scheme would be to dilute 200× in 5 mL (final) on Monday and Wednesday by adding 25 μL of cells from the starting flask to 5 mL of BHI complete medium in a T25 flask. On Friday, cells are split 500× by taking 10 μL of cells from the starting flask and adding them to 5 mL of BHI complete medium.
2. Count cells by combining in order 25 μL 3% formalin, 25 μL culture, and 200 μL Gentian Violet Stain to a 1.5 mL tube. Flick gently to mix.
3. Load 10 μL of fixed, stained cells on a Neubauer hemacytometer and count at least 100 cells. If cells are sparse, count the full 25 squares in the central grid. In this case, use this equation to determine cell density:

$$\text{Number of cells counted} \times 10^4 \times 10 \text{ (dilution factor)} = \text{cells/mL}$$

4. If cell samples are dense, count five squares in the grid (four corners and center). Use the following equation to determine cell density (A–E are the # of cells counted in each of the squares):

$$(A + B + C + D + E) \times 5 \times 10^4 \times 10 \text{ (dilution factor)} = \text{cells/mL}$$

5. By counting a culture at two time points 8–24 h apart, you can use the following equation to determine doubling time (T_d, *see* **Note 7**). T_2–T_1 is the elapsed time (in hours) between counts, C_1 is the initial density of cells, C_2 is the density of cells at the second time point (both in cells/mL).

$$T_d = (T_2 - T_1) \times \frac{\ln(2)}{\ln\left(c_2/c_1\right)}$$

6. Once doubling time is known, you can dilute cells to a starting density so that they grow to reach a desired density for an experiment (target density, usually in log phase) by using the following equation (*see* **Note 8**):

$$\text{Starting density} = \frac{\text{Target density}}{2^{(\text{predicted\#doublings})}}$$

3.2 Culturing C. bombi

1. Count cells using the method described in Subheading 3.1.
2. For standard culturing, cells are grown in 5 mL of FP-FB medium in vented T25 flasks on a shelf at 27 °C in 3% CO_2 (*see* **Note 9**).
3. They are maintained on Monday and Wednesday by diluting them 10× in 5 mL medium (final) i.e., 500 μL of cells from the starting flask are added to 4.5 mL of FP-FB medium in a fresh vented T25 flask. On Friday, cells are diluted 25× by taking 200 μL of cells from the starting flask and adding them to ~5 mL of FP-FB medium.

3.3 Growth Curves for C. bombi and C. fasciculata

1. Count cells and dilute them in the appropriate medium to 5×10^5 cells/mL in 5 mL total volume in vented (*C. bombi*) or non-vented (*C. fasciculata*) T25 flasks. Determine density in each flask after dilution to acquire an initial time point (0 h).
2. Count cells every 12–24 h for 3–4 days until cells reach stationary phase (terminal growth curve). Alternatively, cumulative growth curves can be used to provide more information on growth rate of cells in log phase. To do this, cells are diluted back to 5×10^5 cells/mL every day. Subsequent counts are then multiplied by the dilution factor to calculate cumulative cells/mL over multiple days of primarily log-phase growth.

3.4 Standard C. fasciculata Attachment Assay

1. Add 2 mL of cells from a log-phase culture at a density of 1×10^7 cells/mL in complete BHI medium to a 60 mm (untreated) plastic tissue culture dish.
2. Place plate on shelf in incubator at 27 °C for 2 h.
3. Take plate out of incubator and remove medium containing unattached cells using a vacuum or pipette (*see* **Note 10**). Perform three washes with 2 mL sterile 1× PBS.

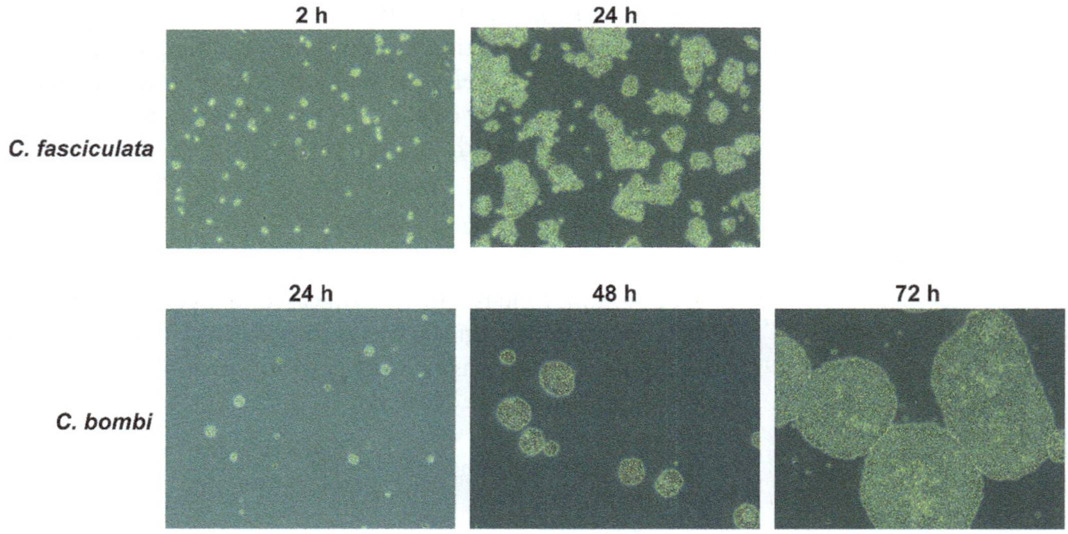

Fig. 1 Example images from attachment assays with *C. fasciculata* and *C. bombi*. *C. fasciculata* are allowed to adhere for 2 h, while *C. bombi* is allowed to adhere for 24 h before washing. Following washing and imaging, attached cells are incubated in medium for a further 24–48 h. Washes are repeated prior to imaging each time point to remove unattached cells

4. Wash plate once with 2 mL of BHI complete medium, then add 2 mL fresh BHI complete medium.
5. Image 25 representative fields at 20× magnification to quantify attachment (Fig. 1 and *see* **Note 11**).
6. Seal plate by wrapping its edge with Parafilm or electrical tape and return it to the incubator shelf.
7. After 24 h, repeat washes and imaging as described above.

3.5 Standard C. bombi *Attachment Assay*

1. Dilute cells from a log-phase culture to 5×10^6 cells/mL in FP-FB medium in a final volume of 2 mL. Add diluted cells to a 60 mm plastic dish.
2. Place plate on shelf in incubator at 27 °C, 3% CO_2 for 24 h.
3. Remove medium and swimming cells with vacuum or pipette. Perform three washes using 2 mL sterile 1× PBS per wash (*see* **Note 10**).
4. Wash plate once with 2 mL of FP-FB, then add 2 mL fresh FP-FB.
5. Image 25 representative fields at 20× magnification to quantify attachment (*see* **Note 11**).
6. Put back on shelf in incubator for another 24 h.
7. Repeat the washes and imaging as described above at 48 h. Repeat process for as many time points as needed (*see* **Note 12**).

3.6 Nanoluciferase Assay for Cell Density Measurements of Cultured Crithidia sp.

1. Carefully count cells that contain a NLuc expressing plasmid (*see* **Note 13**). Spin down 1×10^7 cells/mL for each sample in a 1.5 mL tube at 800 rcf for 5 min.
2. Remove supernatant and resuspend pellet in 1 mL of BHI complete (*C. fasciculata.*) or FP-FB (*C. bombi*) medium. This will serve as the starting point for serial dilutions.
3. Perform 10× serial dilutions from 1×10^2 to 1×10^7 cells/mL. There will be an additional 2× dilution when adding the NLuc reagent, final cell densities will therefore be 1×10^1 to 1×10^6 cells/well.
4. Transfer 100 μL of each of the serial dilutions and 100 μL of each unknown sample to separate wells in a white 96-well plate (*see* **Note 14**).
5. Prepare the NLuc reagent fresh each time by combining the NLuc buffer and substrate according to the manufacturer's instruction (*see* **Note 15**). Make enough reagent to have 100 μL for each sample (*see* **Note 16**).
6. Add 100 μL of the NLuc reagent each of the wells. Read samples in plate reader (emission at 460 nm) as quickly as possible, ideally within 10–15 min.

3.7 Transfection of Crithidia sp.

1. Prepare 4–10 μg of purified DNA to be transfected (between 0.5 and 2 μg/μL). Circular plasmids are concentrated by standard ethanol precipitation. DNA linearized by restriction endonucleases is first subjected to phenol:chloroform extraction prior to ethanol precipitation. DNA is resuspended in sterile water.
2. Count cells using hemacytometer. 3×10^7 cells are needed per *C. bombi* transfection. 5×10^7 cells are needed per *C. fasciculata* transfection. For each selecting drug, use the same number of cells for a mock transfection.
3. For *C. bombi*, spin down 3×10^7 cells/sample (transfections and mock). For *C. fasciculata*, spin down 5×10^7 cells. If less than 1.5 mL volume, use a microcentrifuge and spin at 800 rcf, 5 min. If more than 1.5 mL, place in 15 mL tube and centrifuge at 800 rcf, 10 min.
4. For 1.5 mL tubes, remove supernatant leaving behind ~20 μL. If you have a larger volume (in a 15 mL tube) it is best to transfer the resuspended pellet (in ~500 μL residual liquid) to a microcentrifuge tube and spin again to remove all but ~20 μL of the supernatant.
5. Resuspend pellets for transfection in 100 μL Tb-BSF buffer (mock samples can be resuspended in Tb-BSF buffer or medium).

6. Add 4–10 μg DNA to transfection samples and flick gently to mix.
7. Transfer transfection samples to Lonza cuvettes. Tap cuvette so liquid is touching both sides.
8. Place cuvette in Lonza nucleofector 2b and run program X-001. Make sure carousel spins and that you see the "OK" message.
9. Transfer the cuvette into the culture hood and use the Lonza pipettes to transfer the nucleofected cells to 5 mL media without selecting drug in a T25 flask. Transfer mock samples (in Tb-BSF buffer or medium) directly to 5 mL media without selecting drug.
10. Incubate for 18–24 h. For *C. fasciculata*, place flasks on an orbital shaker set to ~70 rpm at 27 °C.
11. Transfer transfected and mock cultures to 15 mL tubes and spin at 800 rcf, 10 min.
12. Discard supernatant and resuspend in 10 mL media including selecting drugs (*see* **Note 17**).
13. Transfer back to flask and put back in incubator (*C. fasciculata* on orbital shaker, *C. bombi* on shelf). For *C. fasciculata*, it is beneficial to perform another 2× split in 10 mL final volume after 24 h of selection. Cells will recover in 10–14 days. For *C. bombi*, cells can recover in as little as ~5 days.

3.8 NLuc to Quantify Infection of Laboratory Mosquitoes with Crithidia fasciculata

1. We have infected laboratory strains of *Aedes aegypti* and *Anopheles gambiae* male and female mosquitoes. Typically, we infect non-bloodfed female mosquitoes. Put 50–70 mosquitoes 1–3 days post-eclosion into paper cups covered with fine mesh and lined at the bottom with filter paper. Maintain the mosquitoes on a diet of 10% sucrose via a saturated cotton ball. The day before infection, remove the sugar solution to encourage feeding on sucrose containing *C. fasciculata* the next day (*see* **Note 18**).
2. On the day of the feed, prepare *C. fasciculata* containing sucrose by centrifuging 1×10^8 log-phase NLuc-expressing *C. fasciculata* at 800 rcf 5 min.
3. Remove the medium and wash the pellet 1× with 10 mL 1× PBS.
4. Mix 4 mL 1× PBS with 8 mL 10% sucrose (in water).
5. Remove supernatant and resuspend pellet in 12 mL sucrose/PBS solution. Mix well.
6. Divide this solution over four cotton balls and place two cotton balls on top of each cup of mosquitoes. Cotton balls should be fully saturated but not dripping.

7. Allow mosquitoes to feed for 24 h before replacing *C. fasciculata* containing cotton balls with cotton balls containing standard 10% sucrose.
8. At the desired time points post-infection, the hindgut (including the rectum), from 20–50 mosquitoes are dissected and placed individually into labeled 0.2 mL tubes containing 50 μL of PBS. Samples can be frozen at −80 °C prior to assaying. We assay NLuc in each tissue individually and perform at least three biological replicates with independent mosquito generations and *C. fasciculata* cultures.
9. Once all samples are collected and ready to assay, thaw all tubes on ice.
10. Prepare NLuc reagent as described in Subheading 3.6 making enough to add 50 μL to each tube.
11. Add 50 μL of NLuc reagent to each tube. Pipette up-and-down to mix and then transfer the entire contents to a well of a white 96-well plate. Each well will contain 100 μL total: 50 μL of tissue in PBS mixed with 50 μL of NLuc reagent.
12. Create a standard curve of twofold serial dilutions of cultured NLuc-expressing *C. fasciculata* as described in Subheading 3.6 and add them to the same plate in a final volume of 50 μL/well. Prepare duplicate wells for each dilution.
13. Prepare NLuc reagent as described in Subheading 3.6 and add 50 μL of reagent per well.
14. Read samples in plate reader (emission 460 nm) as quickly as possible.

3.9 Isolating C. bombi from Bees Using Limiting Dilution

1. Dissect intestinal tracts from infected bees and place in 1.5 mL tube with 300 μL Ringer's solution. Macerate with pestle and let sit at room temperature for 3–4 h.
2. Take 150 μL of liquid from the top of the tube and leave the settled tissue behind. Transfer to 15 mL tube with 5 mL FP-FB medium.
3. Store samples at 4 °C until ready to isolate; use within 5 days.
4. Spin down cells in 15 mL tubes at 800 rcf for 5 min. If there is no pellet, repeat spin for 5 more minutes at 800 rcf.
5. Remove supernatant and resuspend each sample in 1 mL modified Mäser mix. Transfer to one well of a 24-well plate and place in incubator for 24 h.
6. Passage the most viable samples (those with minimal contamination of bacteria/fungi and plenty of *Crithidia* sp. cells) by diluting samples 1:10 in 1 mL final volume in modified Mäser mix in a new well in a 24-well plate. Place in incubator for 24 h.

7. Carefully count cell density by the counting method described in Subheading 3.1.
8. Perform serial dilutions in modified Mäser mix to decrease cell density to 0.5 cells/well and 0.1 cells/well in a 96-well plate with 200 μL in each well (*see* **Note 19**).
9. Place the 96-well plate unwrapped in incubator and begin checking for cloudy wells after 1 week (*see* **Note 20**).
10. Scale up cloudy wells to 1 mL final volume, keeping cells in modified Mäser mix in a 24-well plate.
11. After 4–5 days, scale up to 5 mL FP-FB in vented T25 flasks.
12. Once cells are fully recovered, make frozen stocks (also called stabilates) by placing cells growing at a density between 2.5×10^7 and 3×10^7 cells/mL in cryovials in 10% glycerol (final) in FP-FB medium without drug and freezing at −80 °C. Vials can be transferred to liquid nitrogen for long-term storage (*see* **Note 21**).

4 Notes

1. Combine BHI powder with water in an autoclave safe bottle, shake to mix, and autoclave on a liquids cycle. The mixture will dissolve during the sterilization process. Sterile filtration through a 0.22 μm filter can also be used.
2. Make 50 mM NaOH first to dissolve the hemin powder. Transfer some of the 50 mM NaOH directly to the supplied hemin bottle and pipet back into the beaker containing the remainder of the hemin stock. When it is sterile filtered, there will be some residual hemin that gets stuck in the filter. This recipe makes a large stock solution. For convenience we freeze ~450 mL in 50 mL aliquots and the remainder in 4 mL or 1 mL aliquots. When depleted, a 50 mL aliquot is thawed to generate more 4 mL or 1 mL aliquots.
3. 25 cm^2 and 75 cm^2 flasks are used depending on volume required for experiments. Both tissue-culture-treated and non-treated culture vessels can be used, although we have found that attachment rates differ on these two surfaces [27], so materials should be kept consistent between experiments.
4. FP stock must be kept sterile. It is helpful to measure out 200 mL in a graduated cylinder inside the hood and then use that outside the hood to create 10× FB stock. The final FP-FB medium is filter-sterilized.
5. Make stock solutions of each: 0.5 M Na_2HPO_4, 0.5 M NaH_2PO_4, 1 M KCl, 1 M $CaCl_2$, 1 M HEPES (free acid), pH 7.7. In a beaker with ~50 mL H_2O, add 13.4 mL 0.5 M

Na_2HPO_4, 4.6 mL 0.5 M NaH_2PO_4, 500 μL 1 M KCl, 15 μL 1 M $CaCl_2$, 5 mL 1 M HEPES (pH 7.7). Once all is combined, fill to 100 mL with water and pH to 7.3. Sterile filter and aliquot. Store at −20 °C. Aliquots should only be thawed once and unused solution discarded.

6. *C. fasciculata* readily attaches to tissue culture plastic to form rosettes. Keeping cells on an orbital shaker set to ~60–70 rpm prevents most attachment and keeps cells in their swimming form. Settings might need adjusting depending on the specific brand of shaker used. *C. fasciculata* should be maintained in culture between 2×10^5 and 1×10^8 cells/mL. For cells that have been genetically modified and are grown in drug-containing media, cells can be split 100–200× on Monday and Wednesday and 200–500× on Friday depending on growth rate.
7. *C. fasciculata* doubling time is around 6 h. *C. bombi* doubling time ranges from 12 to 18 h. Cells should be in log (exponential) growth for doubling time measurements.
8. It is best to use cells when they are mid-log (exponential growth) phase for experiments. For *C. fasciculata* this is 2×10^7 to 8×10^7 cells/mL. For *C. bombi*, this is 9×10^6 to 1.5×10^7 cells/mL. For example, if you are planning to collect a *C. bombi* sample in 24 h, you first need to know how many doublings your cells will undergo in that time (predicted number of doublings), e.g., 24 h/18 h doubling time = 1.3 doublings. If your target density in 24 h is 1×10^7, you can now calculate your starting density i.e., the density the cells need to be diluted to today to divide and reach 1×10^7 cells/mL in 24 h, e.g., ${}^{1\times10^7}/_{2^{1.3}} \cong 4 \times 10^6$cells/mL. If you count your cells and they are currently at 3×10^7 cells/mL, you can use C1V1 = C2V2 (C1 = 3×10^7 cells/mL; C2 = 4×10^6 cells/mL; V2 = 5 mL). Solving for V1 will give you 0.68 mL of cells, which you would combine with 4.3 mL of medium.
9. *C. bombi* will also attach to plastic, so flasks for experiments requiring only swimming cells should be maintained on an orbital shaker at ~60–70 rpm. For standard (maintenance) culture, flasks can be kept on an incubator shelf. *C. bombi* are typically maintained between 5×10^5 and 5×10^7 cells/mL. For cells that have been genetically modified and are grown in drug-containing medium, cells can be split 5–10× on Monday and Wednesday and 15–25× on Friday.
10. Although attached cells are firmly stuck to the plastic, add and remove medium/washes from the edge of the plate/flask by tilting the dish on its edge and placing the pipette tip/vacuum on the edge of the bottom of the dish. For each wash step, add the sterile 1× PBS and rock the dish a few times in both

directions before aspirating off the liquid. Instructions are provided for a single plate, however up to five plates can be processed at the same time. If processing larger numbers of samples (>5), or reducing culture medium volume is a priority (for example, in drug study), assays can be performed in a multi-well plate and a vacuum aspirator used to remove medium/PBS. In this case, the volume of the assay is adjusted to 0.5 mL of log-phase cells at a density of 1×10^7 cells/mL for attachment and 1 mL per wash. In this format, each well contains 5×10^6 *C. fasciculata* cells.

11. Quantify attachment by categorizing attachment events as singlets (one attached cell), doublets (two cells), triplets (three cells), and rosettes (four or more cells). At later time points, merged rosettes may appear where rosettes are large enough to contact each other. This category can be counted separately.
12. As *C. bombi* typically has a slower doubling time than *C. fasciculata*, and forms larger rosettes (Fig. 1), we have found that 24 h and 48 h are good time points. 72 h can also be informative.
13. For the culture used to make the standard curve, it is important that the initial count is accurate. Make 2 separate dilutions of cells in formalin/stain, count both, and use the average.
14. The NLuc assay can be used to perform growth curves by measuring cell density after different durations of growth. Since the entire plate must be read at once, along with the standard curve created by the serial dilutions, such cultures must be staggered and processed on the same day. The standard curve is used to relate cells/mL with relative light units, creating a line of best fit with an R^2 value and an equation allowing for calculation of cells/mL in unknown samples. We create technical duplicates of each sample (standard curve and unknown) and graph the average.
15. We use the NanoGlo® Luciferase Assay System (Promega), adding 200 μL of substrate for every 10 mL of buffer to make 10 mL total of reagent. Make enough NLuc reagent to account for pipetting error. Other luciferase assays may vary, requiring adaptation of this protocol according to the manufacturer's instructions.
16. It is possible to use a smaller volume of cells and NLuc reagent. NLuc assays have been successfully performed on *C. fasciculata* using 25 μL of cells and 25 μL of NLuc reagent per sample.
17. For episomal (extrachromosomal plasmids), we find that 80 μg/mL hygromycin or neomycin works well for selection, with neomycin generally correlated with higher expression

levels of transgenes [28]. For hygromycin, increasing to 200 μg/mL can select for higher plasmid copy number and increased expression [29, 30]. For constructs that integrate into the genome, lower drug concentrations are used (40 μg/mL for hygromycin and 20 μg/mL for neomycin).

18. We typically use *A. aegypti* Liverpool strain (LVP-IB12) female mosquitoes but both male and female mosquitoes are susceptible to infection. We have also successfully infected *A. gambiae* (G3) and *A. stephensi* (STE2). All strains were obtained from BEI Resources. Other investigators have used similar protocols to infect *Culex pipiens molestus* [31].

19. Make 100× serial dilutions ("A" and "B") and then use C1V1 = C2V2 so that the third dilution ("C") is 0.5 cells/well and the fourth dilution ("D") is 0.1 cells/well. To conserve medium, make "A" and "B" dilutions in 5 mL total. "C" and "D" will require 30 mL each. "C" and "D" can be plated at 200 μL/well in separate 96-well plates using a multichannel micropipette.

20. Cells will not likely come up until 2 weeks later, but it is best to check early before they overgrow. Examine plates for cloudy wells and then check those with an inverted tissue culture microscope to ensure wells contain *Crithidia* sp. and are free from contamination.

21. If cells are ~3×10^7 cells/mL in culture, you can combine 800 μL cells with 200 μL 50% glycerol in a cryovial. If cells are at a lower density, spin down enough cells to get 2.5×10^7 to 3×10^7 per mL at 800 rcf for 5 min and remove supernatant. Resuspend pellets in 1 mL freezing medium/vial to create frozen stocks. Freezing medium is made by adding 5 mL 100% glycerol to 45 mL FP-FB medium in a 50 mL conical tube. This solution is mixed thoroughly and sterile-filtered using a 0.2 μm filter. Freezing medium can be stored at −20 °C for future use (multiple freeze thaw cycles do not affect its performance). Transfer 1 mL of cells resuspended in freezing medium to each cryovial. Place cryovials in an insulated container that will prevent the cells from freezing too quickly (e.g., Styrofoam box, Mr. Frosty) and place at −80 °C for at least 1 day. Although they are likely stable at −80 °C for months-years, we typically transfer frozen stocks to liquid nitrogen for long-term storage.

References

1. Englund PT (1978) The replication of kinetoplast DNA networks in *Crithidia fasciculata*. Cell 14:157–168. https://doi.org/10.1016/0092-8674(78)90310-0

2. Lukes J, Guilbride DL, Votypka J, Zikova A, Benne R, Englund PT (2002) Kinetoplast DNA network: evolution of an improbable

structure. Eukaryot Cell 1:495–502. https://doi.org/10.1128/ec.1.4.495-502.2002

3. Liu B, Liu Y, Motyka SA, Agbo EEC, Englund PT (2005) Fellowship of the rings: the replication of kinetoplast DNA. Trends Parasitol 21: 363–369. https://doi.org/10.1016/j.pt.2005.06.008
4. Van der Spek H, Speijer D, Arts GJ, Van den Burg J, Van Steeg H, Sloof P, Benne R (1990) RNA editing in transcripts of the mitochondrial genes of the insect trypanosome *Crithidia fasciculata*. EMBO J 9:257–262. https://doi.org/10.1002/j.1460-2075.1990.tb08103.x
5. Landweber LF, Gilbert W (1993) RNA editing as a source of genetic variation. Nature 363: 179–182. https://doi.org/10.1038/363179a0
6. Hajduk S, Ochsenreiter T (2010) RNA editing in kinetoplastids. RNA Biol 7:229–236. https://doi.org/10.4161/rna.7.2.11393
7. Parkin DW, Horenstein BA, Abdulah DR, Estupiñán B, Schramm VL (1991) Nucleoside hydrolase from *Crithidia fasciculata*. Metabolic role, purification, specificity, and kinetic mechanism. J Biol Chem 266:20658–20665. https://doi.org/10.1016/S0021-9258(18)54759-1
8. Parodi AJ, Quesada Allue LA, Cazzulo JJ (1981) Pathway of protein glycosylation in the trypanosomatid *Crithidia fasciculata*. Proc Natl Acad Sci 78:6201–6205. https://doi.org/10.1073/pnas.78.10.6201
9. Nogoceke E, Gommel DU, Kieß M, Kalisz HM, Flohé L (1997) A unique cascade of oxidoreductases catalyses trypanothione-mediated peroxide metabolism in *Crithidia fasciculata*. Biol Chem 378:827–836. https://doi.org/10.1515/bchm.1997.378.8.827
10. Le Trant N, Meshnick SR, Kitchener K, Eaton JW, Cerami A (1983) Iron-containing superoxide dismutase from *Crithidia fasciculata*. Purification, characterization, and similarity to Leishmanial and trypanosomal enzymes. J Biol Chem 258:125–130. https://doi.org/10.1016/S0021-9258(18)33229-0
11. Tiwari K, Dubey VK (2018) Fresh insights into the pyrimidine metabolism in the trypanosomatids. Parasites Vectors 11:87. https://doi.org/10.1186/s13071-018-2660-8
12. Shames SL, Fairlamb AH, Cerami A, Walsh CT (1986) Purification and characterization of trypanothione reductase from *Crithidia fasciculata*, a new member of the family of disulfide-containing flavoprotein reductases. Biochemistry 25:3519–3526. https://doi.org/10.1021/bi00360a007
13. Andanje R, Greene AS (2022) A biochemistry course undergraduate research experience (CURE) on *Crithidia fasciculata* parasites. FASEB J 36. https://doi.org/10.1096/fasebj.2022.36.S1.0R444
14. Frolov AO, Kostygov AY, Yurchenko V (2021) Development of Monoxenous Trypanosomatids and Phytomonads in insects. Trends Parasitol 37:538–551. https://doi.org/10.1016/j.pt.2021.02.004
15. Lukeš J, Butenko A, Hashimi H, Maslov DA, Votýpka J, Yurchenko V (2018) Trypanosomatids are much more than just trypanosomes: clues from the expanded family tree. Trends Parasitol 34:466–480. https://doi.org/10.1016/j.pt.2018.03.002
16. Goulson D, Nicholls E, Botías C, Rotheray EL (2015) Bee declines driven by combined stress from parasites, pesticides, and lack of flowers. Science 347:1255957. https://doi.org/10.1126/science.1255957
17. Brown MJF (2022) Complex networks of parasites and pollinators: moving towards a healthy balance. Philos Trans R Soc B Biol Sci 377: 20210161. https://doi.org/10.1098/rstb.2021.0161
18. Tiritelli R, Cilia G, Gómez-Moracho T (2025) The trypanosomatid (Kinetoplastida: Trypanosomatidae) parasites in bees: a review on their environmental circulation, impacts and implications. Curr Res Insect Sci 7:100106. https://doi.org/10.1016/j.cris.2025.100106
19. Warren WC, Akopyants NS, Dobson DE, Hertz-Fowler C, Lye L-F, Myler PJ, Ramasamy G, Shanmugasundram A, Silva-Franco F, Steinbiss S, Tomlinson C, Wilson RK, Beverley SM (2021) Genome assemblies across the diverse evolutionary spectrum of *Leishmania* protozoan parasites. Microbiol Resour Announc 10. https://doi.org/10.1128/mra.00545-21
20. Schmid-Hempel P, Aebi M, Barribeau S, Kitajima T, du Plessis L, Schmid-Hempel R, Zoller S (2018) The genomes of *Crithidia bombi* and *C. expoeki*, common parasites of bumblebees. PLoS One 13:e0189738. https://doi.org/10.1371/journal.pone.0189738
21. Schmid-Hempel P, Reber Funk C (2004) The distribution of genotypes of the trypanosome parasite, *Crithidia bombi*, in populations of its host, *Bombus terrestris*. Parasitology 129:147–158. https://doi.org/10.1017/s0031182004005542
22. Yagoubat A, Corrales RM, Bastien P, Lévêque MF, Sterkers Y (2020) Gene editing in

Trypanosomatids: tips and tricks in the CRISPR-Cas9 era. Trends Parasitol 36:745–760. https://doi.org/10.1016/j.pt.2020.06.005

23. Paterou A, Sáez Conde J, Týč J, Sunter JD, Vaughan S, Gull K, Dean S (2025) A comprehensive toolkit for protein localization and functional analysis in trypanosomatids. Open Biol 15:240361. https://doi.org/10.1098/rsob.240361
24. Salathé R, Tognazzo M, Schmid-Hempel R, Schmid-Hempel P (2012) Probing mixed-genotype infections I: extraction and cloning of infections from hosts of the Trypanosomatid *Crithidia bombi*. PLoS One 7:e49046. https://doi.org/10.1371/journal.pone.0049046
25. Burkard G, Fragoso CM, Roditi I (2007) Highly efficient stable transformation of bloodstream forms of Trypanosoma brucei. Mol Biochem Parasitol 153:220–223. https://doi.org/10.1016/j.molbiopara.2007.02.008
26. Burkard GS, Jutzi P, Roditi I (2011) Genome-wide RNAi screens in bloodstream form trypanosomes identify drug transporters. Mol Biochem Parasitol 175:91–94. https://doi.org/10.1016/j.molbiopara.2010.09.002
27. Denecke S, Malfara MF, Hodges KR, Holmes NA, Williams AR, Gallagher-Teske JH, Pascarella JM, Daniels AM, Sterk GJ, Leurs R, Ruthel G, Hoang R, Povelones ML, Povelones M (2024) Adhesion of *Crithidia fasciculata* promotes a rapid change in developmental fate driven by cAMP signaling. mSphere 9:e00617–24. https://doi.org/10.1128/msphere.00617-24
28. Bieber BV, Lockett SG, Glasser SK, St. Clair FA, Portillo NO, Adler LS, Povelones ML (2024) Genetic modification of the bee parasite *Crithidia bombi* for improved visualization and protein localization. Exp Parasitol 262:108789. https://doi.org/10.1016/j.exppara.2024.108789
29. Tetaud E, Lecuix I, Sheldrake T, Baltz T, Fairlamb AH (2002) A new expression vector for *Crithidia fasciculata* and *Leishmania*. Mol Biochem Parasitol 120:195–204. https://doi.org/10.1016/S0166-6851(02)00002-6
30. DiMaio J, Ruthel G, Cannon JJ, Malfara MF, Povelones ML (2018) The single mitochondrion of the kinetoplastid parasite *Crithidia fasciculata* is a dynamic network. PLoS One 13:e0202711. https://doi.org/10.1371/journal.pone.0202711
31. Votýpka J, Ray DS, Lukes J (2001) *Crithidia fasciculata*: a test for genetic exchange. Exp Parasitol 99:104–107. https://doi.org/10.1006/expr.2001.4648

Part III

Omics

Chapter 6

Poly(A) Profiling for Euglenozoans Using Oxford Nanopore Direct RNA Sequencing, Illustrated by the Example of *Euglena gracilis*

Natalia Gumińska, Agnieszka Czarnocka-Cieciura, Bożena Zakryś, and Rafał Milanowski

Abstract

There are numerous methods to study polyadenylation. These techniques offer insights into mRNA abundance and the dynamic regulation of poly(A) tail length and nucleotide content. The Oxford Nanopore Technologies (ONT) platform is particularly noteworthy among them. It allows analysis of the whole transcriptome at the single-molecule level without the amplification biases introduced by PCR-based methods.

Most current studies in this area have focused on well-established model organisms such as humans, yeast, *Arabidopsis*, and *Caenorhabditis elegans*. In contrast, the poly(A) dynamics remains largely unexplored in non-model organisms like euglenids. These organisms possess a range of unique evolutionary traits, and investigating their poly(A) metabolism may elucidate the evolution of the enzymatic machinery that controls mRNA processing, stability, and turnover.

Here, we present a step-by-step protocol for measuring poly(A) tail lengths and characterizing the nucleotide makeup of poly(A) tails in *Euglena gracilis* using ONT Direct RNA Sequencing.

Key words Nanopore, Direct RNA sequencing, Poly(A), Non-adenosine, *Euglena gracilis*, Euglenids

1 Introduction

In eukaryotes, RNA molecules typically end with a polyadenylated 3′ tail, which plays a critical role in regulating their fate: stability, intracellular transport, and translational efficiency [1–4]. Investigating post-transcriptional gene regulation in non-model organisms such as microeukaryotes presents a considerable challenge [5–7]. This group displays a remarkable diversity in genome size and structure, which complicates the application of standardized methodologies. Furthermore, many microeukaryotic genomes are rich in repetitive sequences, making transcript assembly and annotation particularly difficult [7–9]. In addition, for a substantial number of

Paul A. M. Michels et al. (eds.), *Euglenozoa: Methods and Protocols, Volume 1*, Methods in Molecular Biology, vol. 3013, https://doi.org/10.1007/978-1-0716-5142-1_6,

species, high-quality reference genomes are lacking, which significantly impedes accurate transcriptome reconstruction [10].

In this chapter, we outline the use of Oxford Nanopore Technologies (ONT) Direct RNA Sequencing (DRS) to investigate poly (A) tails without the need for reverse transcription or PCR [11–13]. This method avoids biases introduced by amplification, although it comes with certain limitations, such as reduced throughput and higher cost [3, 14]. Although Oxford Nanopore sequencing was once associated with low read accuracy, recent developments have significantly improved its performance. Advances in basecalling—particularly the shift from Albacore and Guppy to the transformer-based Dorado (launched in 2024)—have raised accuracy to Q26 (99.75%). Improvements in flow cell design, such as the R10 pore type, and the availability of post hoc error correction tools have further enhanced data quality.

Nanopore-based approaches have already been applied successfully to assess poly(A) tail landscape in several model organisms, including humans, yeast, plants, and mammals [2, 6, 12, 13, 15–24]. In contrast, polyadenylation in non-model species remains largely unexplored [5, 7]. Among them, euglenids such as *Euglena gracilis* represent a phylogenetically distinct group, providing a valuable system to examine mRNA regulation in an early-diverging lineage [25–27] with unique evolutionary features [28–31]. This offers a rare opportunity to gain insight into the evolution of poly (A) metabolism and the molecular machinery governing mRNA processing.

Standard sequencing library preparation protocols, designed primarily for model organisms such as mammalian cells, are often unsuitable for microeukaryotes [32]. Many of these species possess robust cell walls or accumulate compounds that are difficult to eliminate during nucleic acid extraction [33–35]. Such substances can affect nanopore performance by altering buffer conditions or directly interacting with nucleic acids, potentially blocking pores and reducing sequencing yield. Therefore, it is crucial to tailor sample preparation and library construction methods to the specific biological characteristics of non-model organisms [10].

In this protocol, we apply DRS to explore transcriptome-wide poly(A) tail length and composition in *E. gracilis*. We provide RNA isolation and DRS library preparation protocols specifically optimized for euglenid species. Additionally, we describe a step-by-step bioinformatic analysis of the sequencing data, including poly(A) tail length profiling using Nanopolish [13] and poly(A) tail nucleotide makeup analysis with Ninetails [12], a neural network-based tool capable of detecting non-adenosine residues within poly(A) tails based on characteristic signal deviations in raw nanopore signals. While methods like TAIL-seq [36], FLAM-seq [37], and PAIso-seq [38] have advanced poly(A) research, their reliance on reverse transcription and amplification can introduce artifacts. In contrast, the direct signal-based analysis used by Ninetails enables a more accurate view of poly(A) composition [12, 39].

We believe that the proposed framework for investigating the complexity of poly(A) tails in non-model systems will contribute to a better understanding of post-transcriptional regulation in euglenids.

2 Materials

2.1 Cultivation

1. *Euglena* cultures, e.g., *E. gracilis* strain "Z" (SAG 1224-5/25) from the Culture Collection of Algae at the University of Göttingen, Germany.
2. Cramer-Myers liquid medium, supplemented with 0.8% (v/v) ethanol and 1% soil extract.
3. Fitotron chamber.

2.2 RNA Extraction and Preparation for Library Synthesis

1. TRIzol reagent.
2. Chloroform.
3. Isopropanol.
4. 70% ethanol.
5. TURBO DNAse (Invitrogen).
6. Nuclease-free water.
7. DNA LoBind Tubes.
8. Micro-volume spectrophotometer (e.g., Nanodrop).

2.3 Nanopore Direct RNA Sequencing

1. Oligo-d$T_{(25)}$ beads.
2. KAPA Pure Beads (Roche).
3. DNA LoBind Tubes.
4. Nuclease-free water.
5. Magnetic separation rack.
6. Direct RNA Sequencing Kit (SQK-RNA002, SQK-RNA004; ONT) (*see* **Note 1**).
7. SuperScript III Reverse Transcriptase (Invitrogen) (*see* **Note 2**).
8. T4 DNA ligase (highly concentrated; 2000 U/μL).
9. Fluorometer (e.g., Qubit or equivalent).
10. Nanopore sequencing flow cell (ONT).
11. Nanopore sequencer: either MinION, GridION, or PromethION (ONT) (*see* **Note 3**).

2.4 Computational Software

All analysis steps are expected to be performed in a Linux environment. Installation instructions and dependencies are available in the documentation (user manual) of each of the following software.

1. Guppy basecaller from ONT (https://nanoporetech.com/software/other/guppy) (*see* **Note 4**).
2. PycoQC (https://a-slide.github.io/pycoQC/).
3. Minimap2 (https://github.com/lh3/minimap2).
4. Samtools (https://www.htslib.org/).
5. Nanopolish (https://github.com/jts/nanopolish) or Tailfindr (https://github.com/adnaniazi/tailfindr).
6. Rstudio (Posit; https://posit.co/download/rstudio-desktop/).
7. Ninetails (https://github.com/LRB-IIMCB/ninetails).
8. Nanotail (https://github.com/LRB-IIMCB/nanotail).

3 Methods

3.1 Cell Culture and Sample Collection

Axenic *E. gracilis* cultures are crucial for molecular biology applications, especially high-throughput sequencing, where even minimal bacterial contamination can affect the results. For the axenization instructions, please refer to the "Protocol for Efficient Ribodepletion of *Euglena gracilis* RNA" in Chap. 2 in Vol. II of this book. Here, we focus on nanopore sequencing and data analysis; thus, we assume that RNA samples are derived from axenic cultures. Below, general guidance is provided for culturing and preparing samples for RNA isolation. These recommendations serve as a useful starting point for experimental planning, though specific conditions may need to be adjusted depending on the context. For instance, in studies involving 4-thiouridine (4sU) labeling, stress exposure, or other transcriptomic changes, media composition and light conditions should be tailored accordingly.

1. Grow *E. gracilis* statically in liquid Cramer-Myers medium [40].
2. Maintain cultures at 18 °C with a 16/8 h light/dark cycle and a light intensity of 27 μmol photons m^{-2} s^{-1}.
3. When cultures reach an optical density of OD_{600nm} ~ 0.6, centrifuge twice (5 min; 5000× *g*).
4. Wash cell pellets with nuclease-free water and either store at −80 °C, or proceed with RNA isolation.

3.2 RNA Extraction

1. Resuspend cell pellets in TRIzol (1 mL per 100 g of wet weight cell pellet).
2. Vortex briefly and incubate for 15 min at room temperature.
3. Extract RNA by adding chloroform (1/5 volume of TRIzol used).
4. Vortex for 2 min and incubate for 5 min at room temperature.
5. Centrifuge (12,000× *g*, 4 °C; 15 min).
6. Transfer the aqueous phase and precipitate RNA with an equal volume of isopropanol.
7. Incubate for 10 min at room temperature and centrifuge under the same conditions.
8. Wash the RNA pellet with 1 mL of 70% ethanol and centrifuge (7500× *g*; 4 °C).
9. Remove the supernatant, dry the pellet, and resuspend in nuclease-free water.
10. Treat RNA with TURBO DNase according to the manufacturer's instructions.
11. Measure the RNA concentration using spectrophotometer. Assess the RNA quality and purity based on the 260/280 and 260/230 absorbance ratios.

3.3 Library Preparation and Direct RNA Sequencing

A DRS library can be prepared from total RNA (*see* **Note 5**). However, enrichment for polyadenylated RNA using oligo-$dT_{(25)}$ magnetic beads and a commercial kit or ribodepletion according to the instruction provided in the "Protocol for Efficient Ribodepletion of *Euglena gracilis* RNA" (Chap. 2 in Vol. II) is generally recommended.

We usually use up to 5 μg of total RNA or 200 ng of poly(A)-enriched RNA, along with 50 ng of spike-ins [11, 14, 22]. We strongly recommend including spike-in controls (e.g., RNA Calibration Strand (RCS) from ONT or in-house in vitro-transcribed standards) to assess sequencing performance. Altogether, this gives a higher input than the amount recommended by ONT for library preparation, which is 500 ng of total RNA or 50 ng of poly (A) + RNA. In our experience, these increased ratios result in higher output.

It is also important to note that key reagents, such as T4 DNA ligase and SuperScript III Reverse Transcriptase, are not included in the sequencing kit and must be purchased separately. Library preparation should follow ONT official protocols, though minor modifications can improve the yield. We suggest extending all ligation steps to 30 min and performing the final bead-based purification (using KAPA Pure Beads) twice to efficiently remove short RNA fragments and adapter sequences [11, 14].

As nanopore sequencing continues to evolve, with frequent updates to both protocols and the MinKNOW software, a fixed procedure for loading the library and starting the run is not provided here. For up-to-date instructions, consult the ONT's Community Website (https://nanoporetech.com/community).

1. Bring the suggested amount of RNA to the final volume of 11 μL concentrating with isopropanol or using magnetic beads.
2. Measure RNA concentration using 1 μL of the sample with a fluorometer (e.g., Qubit or equivalent).
3. Prepare the sequencing library using 9 μL of RNA sample following ONT's "Direct RNA Sequencing" protocol.
4. Load the DRS library onto the flow cell and start the run using MinKNOW.
5. Once the desired number of reads is reached or most pores become inactive, stop the sequencing run and proceed with computational analysis.

We also recommend several best practices to improve sequencing outcomes:

1. Avoid introducing air bubbles during library loading, as they can damage flow cell pores and reduce sequencing yield.
2. Check the flow cell beforehand to ensure a sufficient number of active pores. ONT guarantees a minimum of 800, but in our experience, starting with more than 1400 active pores yields better results. Pore activity gradually declines over time, and a typical 72-h run produces one to three million reads.
3. Monitor performance using MinKNOW throughout the run. A healthy run shows a gradual decline in active pores. A sudden drop in "sequencing" or "strand" state pores during the first 2 h may indicate a problem with library preparation [14].
4. Conducting small-scale pilot experiments using a cost-effective Flongle flow cell can help to estimate the number of reads required for given experiment based on the output produced for transcript(s) of interest.

3.4 Basecalling

Currently, ONT recommends using the Dorado basecaller, which generates output in the POD5 format, while the legacy basecaller, Guppy, is being gradually phased out. However, Guppy produces metadata required for downstream analyses, and thus is recommended for evaluating poly(A) tail length and composition. Empirical testing has confirmed that Guppy and the other tools in this workflow remain compatible with the latest ONT's chemistry (SQK-RNA004) and R10 pore type. Finally, since basecalling is the most computationally intensive step in the pipeline it should be performed on systems equipped with graphics processing units (GPUs).

1. Basecall the DRS raw reads using Guppy version 6.0.0 is as follows:

```
guppy_basecaller -i input_directory -s output_directory -re-
cursive
-records_per_fastq 0 --fast5_out --flowcell FLO-MIN106 --kit
SQK-RNA002 -x "cuda:all"
```

An overview of the program parameters is provided below:

- `-i input_directory`: directory containing the raw (uncalled) fast5 files. Using the `-recursive` option allows processing of files located within all subdirectories as well.
- `-s output_directory`: directory where the basecalled fastq and fast5 files will be saved.
- `--fast5_out`: ensures that, in addition to fastq files, the fast5 files storing basecalled data are also produced, which is necessary for downstream analyses.
- `-x "cuda:all"`: enables basecalling on all available GPU devices. To run the process on a specific device, replace with -x "cuda:n" where *n* is the GPU index (e.g., 0, 1, 2,...).
- `--flowcell FLO-MIN106` and `--kit SQK-RNA002`: define the flow cell and sequencing kit types used for the experiment. If SQK-RNA004 is used, provide SQK-RNA002 instead. The results should still be reliable, even though ONT does not officially support backward compatibility.

Upon completion, the output_directory will include several key files and folders. The sequencing_summary.txt file provides basic information for each read, including its file association within the FAST5 dataset. The "pass" folder contains FASTQ files for reads that passed quality filtering, while the "fail" folder includes reads that did not meet the quality threshold. These failed reads are typically very short and generally not suitable for downstream analyses. However, they can still be used with caution, keeping in mind their reduced quality. The workspace folder holds the basecalled FAST5 files with additional data blocks within the `Basecall_1D_000` group structure, such as move table, introduced by Guppy basecaller. This table is essential for non-adenosine profiling within poly(A) tails [12].

2. Perform QC of the basecall with PycoQC:

```
pycoQC -f sequencing_summary.txt -o report.html
```

The parameters used are defined as follows:

`-f`: (path to) the sequencing_summary.txt file generated by the Guppy basecaller.

`-o`: name of the output HTML file containing the interactive quality control report.

We suggest running quality control after basecalling, even though it is not mandatory. It can help evaluate various aspects of the sequencing run, such as the distribution of read lengths, changes in performance during the run, and general read quality. We recommend using PycoQC [41], as it is user-friendly and generates interactive QC reports.

3.5 Mapping

Assignment of reads to their corresponding transcripts requires alignment to a suitable reference. Selecting an appropriate reference is critical for reliable results. Mapping to a reference transcriptome, rather than the genome, ensures better identification (*see* **Note 6**). We recommend using the *E. gracilis* reference transcriptome available in the European Nucleotide Archive (ENA) under study accession PRJEB38787 [8]. Minimap2 is well suited for mapping direct RNA sequencing (DRS) reads [42]. Following basecalling, multiple FASTQ files are typically generated in both "pass" and "fail" directories. These should be merged into a single file before downstream analysis, which can be done using standard Linux commands (e.g., cat). Aligned reads can be converted into BAM format, and indexed using Samtools [43]. Here is an example of the command syntax:

```
minimap2 -ax map-ont -k 14 secondary=no -t 30 index reads.
fastq | samtools view -b -F 2320 -@30 | samtools sort -@ 30 -o
mapping.bam
```

An overview of the used parameters is provided below:

- `-ax`: specifies SAM output format (default is PAF) and enables setting custom alignment presets.
- `map-ont`: preset for aligning noisy long reads (~10% error rate), typical for ONT data.
- `-t`: defines the number of threads for Minimap2.
- `index`: (path to) the reference sequences in FASTA format or a prebuilt Minimap2 index.
- `reads.fastq`: (path to) the input FASTQ file containing merged basecalled reads.
- `--secondary=no`: reports only the best alignment for each read, suppressing secondary hits.

- `-k 14`: sets the k-mer size to 14, improving alignment accuracy for error-prone DRS reads.
- `-F 2320`: filters out reads mapped to the reverse strand from the output. Since DRS is inherently strand-specific, reads aligning to the reverse strand can be excluded as most likely artifacts, unless antisense transcripts are also of interest.
- `-b`: outputs alignment in BAM format.
- `-@`: specifies the number of threads used by Samtools.

This command generates a BAM file (mapping.bam) with alignments sorted by reference position. The resulting file should be indexed before being used in downstream analyses.

3.6 Determination of Poly(A) Lengths

Currently, three main tools are available for measuring poly(A) tail lengths from nanopore sequencing data: Nanopolish polya [13], Tailfindr [17], and the estimate-poly-a module of Dorado basecaller (https://github.com/nanoporetech/dorado?tab=readme-ov-file#polya-tail-estimation). Each uses a different approach to detect poly(A) signals in the raw nanopore signal.

Although Dorado is the tool recommended by ONT, its poly (A) estimation module does not provide tail coordinates by default, which limits further analysis of nucleotide content. The other two tools—Nanopolish and Tailfindr—are more suitable for such downstream tasks and differ mainly in computational speed and resource requirements (Nanopolish is more computationally efficient). In our experience, all three methods show a good correlation in poly(A) length estimation, though Dorado may slightly underestimate shorter tails. For detailed instructions on using Dorado or Tailfindr, see the respective protocols [44, 45]. In the following steps, we will use Nanopolish polya.

1. Link reads with the raw signal data using `nanopolish index`:

```
nanopolish index -d workspace -s sequencing_summary.txt
reads.fastq
```

An overview of the software parameters is provided below:

- `-d workspace`: (path to) the directory containing the FAST5 files. This folder is typically located within the directory produced by Guppy during basecalling.
- `-s sequencing_summary.txt`: (path to) the sequencing summary file, also found in the Guppy basecalling output directory.
- `reads.fastq`: (path to) the merged FASTQ file used previously for mapping with Minimap2.

2. Estimate poly(A) tail lengths using `nanopolish polya`:

```
nanopolish polya -b mapping.bam -r reads.fastq -t 20 -g
reference.fasta > nanopolish_output.tsv
```

An explanation of the used parameters is provided below:

- `mapping.bam`: (path to) the alignment file produced with Minimap2 and sorted with Samtools.
- `reads.fastq`: (path to) the merged FASTQ file used previously for mapping with Minimap2.
- `reference.fasta`: (path to) the reference transcriptome FASTA file used for mapping with Minimap2.

The output is a tab-separated (TSV) file that provides detailed information for each sequencing read. Among the most relevant fields are the unique read identifier (read_id), the reference transcript assigned during alignment (contig), and the estimated poly (A) tail length (polya_length).

Additionally, the qc_tag column indicates the confidence level of the poly(A) length estimation, where values such as "PASS" and "SUFFCLIP" are considered the most trustworthy. For a full explanation of the file structure and tags, refer to the official documentation of the Nanopolish polya module (https://nanopolish.readthedocs.io/en/latest/quickstart_polya.html).

3.7 Determination of Poly(A) Nucleotide Content

Ninetails is the only available tool that enables analysis of poly (A) tail composition from DRS data [12]. The program requires two main types of input: output files from the Guppy basecaller, including basecalled reads and `sequencing_summary.txt` file, and an output generated by the Nanopolish polya. Ninetails is an R package and utilizes computer vision to recognize signal perturbations introduced by non-adenosine residues. Installation instructions can be found in the official documentation (https://github.com/LRB-IIMCB/ninetails).

Ninetails relies on additional components, as its neural network module is built on the TensorFlow backend. To ensure compatibility, it is recommended to run the software within a dedicated virtual environment that includes all necessary dependencies. The preferred method is to create a Conda environment using the provided configuration file (r-ninetails.yml), which is available on GitHub (https://github.com/LRB-IIMCB/ninetails_processing).
Detailed setup and usage instructions are available in the official tutorial (https://github.com/LRB-IIMCB/ninetails/wiki). The main analysis steps are outlined below:

1. Open RStudio and load the required libraries (ensure they are installed).

```
library(rhdf5)
library(reticulate)
library(ninetails)
```

2. Activate Python virtual environment from R:

```
reticulate::use_condaenv("r-ninetails")
```

3. Run Ninetails analysis function `check_tails()` to obtain the non-adenosine predictions:

```
results <- ninetails::check_tails(nanopolish = "/path/to/
nanopolish_output.tsv", sequencing_summary = "/path/to/se-
quencing_summary.txt", workspace = "/path/to/folder/with/
basecalled/fast5", num_cores = 2, basecall_group = "Base-
call_1D_000", pass_only = FALSE, save_dir ="/path/to/output/
directory")
```

An explanation of the used parameters is provided below:

- `nanopolish`: (path to) the output file with poly (A) measurements produced by Nanopolish (or Tailfindr).
- `sequencing_summary`: (path to) the sequencing summary file, also found in the Guppy basecalling output directory.
- `workspace`: (path to) the directory containing the FAST5 files. This folder is typically located within the directory produced by Guppy during basecalling.
- `basecall_group`: the specific data block in the FAST5 file that contains the basecalling data used in this analysis. Since the FAST5 format can hold several basecall groups in one file, it is important to select the correct one. By default, this is set to Basecall_1D_000, but if there are multiple basecalls, the number will increase (e.g., Basecall_1D_000, Basecall_1D_001).
- `pass_only`: a logical (TRUE or FALSE) option. If set to TRUE, the analysis will only include reads marked as "PASS" by Nanopolish. If set to FALSE, it will also consider reads labeled as "SUFFCLIP".
- `save_dir`: output folder where the results of non-adenosine predictions will be saved.

This function produces two tab-separated (TSV) files, stored in directory specified by the user. The file with the *read_classes suffix contains general read classification. Key columns include unique read identifier (readname), classification result (class) indicating whether the read contains non-adenosine residue(s) within its poly(A) tail, and a detailed comment (comments) explaining the basis of the classification. Additional columns, such as polya_-length, contig, and qc_tag, are inherited from the Nanopolish output.

The file with *nonadenosine_residues suffix provides details on each non-adenosine detected. Important fields include the unique read identifier (readname), the predicted type of non-adenosine residue (prediction; C, G, or U), and its estimated position within the poly(A) tail, reported from the 5′ end (est_nonA_pos). This file also includes selected metadata from Nanopolish, such as polya_-length and qc_tag. For a complete description of the outputs, consult the Ninetails documentation (https://github.com/LRB-IIMCB/ninetails/wiki).

3.8 Poly(A) Data Analysis

With poly(A) tail lengths assigned to all reads and non-adenosine residues identified, we can proceed to both exploratory and statistical analyses. From this point onward, several analytical paths are available, depending on the specifics of the experimental design. In this section, we focus on the basic exploratory data visualizations. All downstream processing is carried out in the R environment, utilizing the Ninetails for postprocessing and visualization of non-adenosine data, and NanoTail, which we developed specifically for handling poly(A) lengths data [12, 14, 39]. For more comprehensive guidance, we refer the reader to the respective user manuals.

1. Open RStudio and load the required libraries (ensure they are installed).

```
library(reticulate)
library(ninetails)
```

2. Load the poly(A) data into R environment:

   The data can be imported into R using standard I/O functions such as `read.table()` from the utils package or faster alternatives like vroom (https://www.tidyverse.org/blog/2019/05/vroom-1-0-0/). Alternatively, wrapper functions from the NanoTail can be used for loading and preprocessing of poly(A) data. For this purpose, two main functions are provided:

   - `read_polya_single()`—for loading data from a single sequencing run,
   - `read_polya_multiple()`—for loading results from multiple runs simultaneously.

   The latter requires a metadata table describing each sample (e.g., conditions, replicate, species), which should be passed as an argument. Detailed usage instructions are available in previous publications [14] and in the NanoTail package documentation (https://github.com/LRB-IIMCB/nanotail).

To import a single Nanopolish poly(A) output file with NanoTail, use:

```
polya_data_table <- nanotail::read_polya_single("/path/to/
nanopolish/output.tsv")
```

3. Generate a basic QC plot for Nanopolish tags:

   First get the summary of the respective qc_tags:

```
nanopolish_qc <- nanotail::get_nanopolish_processing_info
(polya_data = polya_data_table, grouping_factor = NA)
```

   Then plot the aggregated data:

```
nanotail::plot_nanopolish_qc(nanopolish_processing_info =
nanopolish_qc, frequency = FALSE)
```

   This QC plot displays the frequency of Nanopolish poly (A) qc_tags assigned to each read, providing an overview of the general quality of tail length estimation (Fig. 1a).

4. Visualize the global distribution of poly(A) tail lengths:

   A density plot is used to illustrate the overall variability in poly(A) lengths across all transcripts (Fig. 1b). It can be done with `plot_polya_distribution()` function:

```
nanotail::plot_polya_distribution(polya_data = polya_data_-
table, grouping_factor = NA, show_center_value = "median")
```

5. Load the Ninetails output files into R environment:

   Ninetails provides dedicated functions for reading its output files into R. For single DRS run, the following functions are used:

   - read_class_single()—loads a read_classes data frame,
   - read_residue_single()—loads a nonadenosine_residues data frame.

   Example usage:

```
class_data <- ninetails::read_class_single("/directory/with/
ninetails/read_classes.tsv")
residue_data <- ninetails::read_residue_single("/directory/
with/ninetails/nonadenosine_residues.tsv")
```

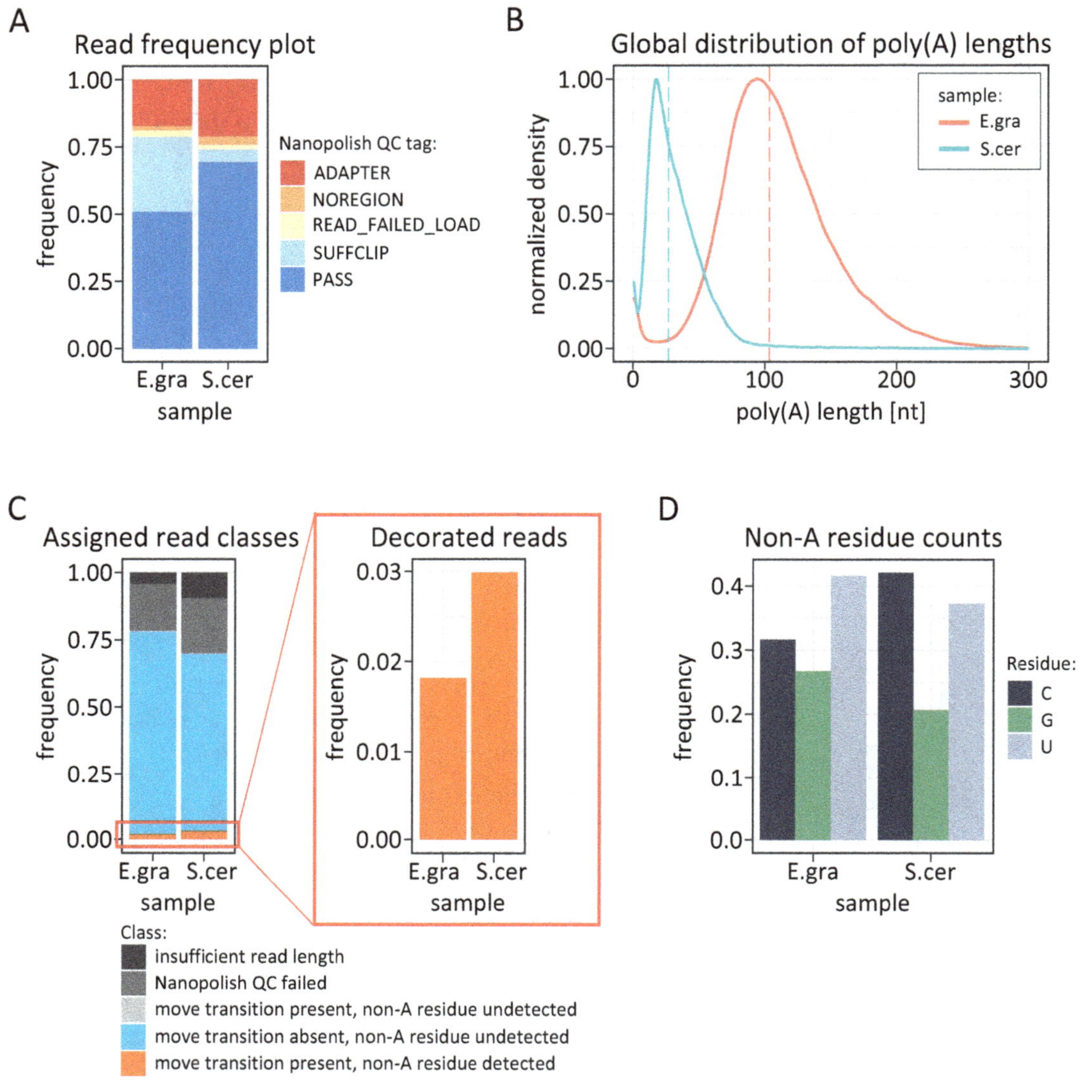

Fig. 1 Exploratory data analysis of DRS run from *Euglena gracilis* "Z" compared to *Saccharomyces cerevisiae* S288. (**a**) Nanopolish QC tag plot. (**b**) Global distribution of poly(A) tails. Dashed lines represent medians. (**c**) Read classification results—non-adenosine content with respect to the total run (left) and decorated reads only (right). (**d**) Frequency of the respective non-adenosine residues as reported by Ninetails. Sample labels: E. gra—*E. gracilis*, S.cer—*S. cerevisiae*. Plots prepared in R 4.1.2 with Ninetails and NanoTail packages

For processing multiple output files simultaneously, Ninetails provides dedicated functions `read_class_multiple()` and `read_residue_multiple()`, which also support user-defined metadata for sample annotation. For detailed usage and examples, please refer to the Ninetails user manual (https://github.com/LRB-IIMCB/ninetails/wiki/5.-Data-postprocessing).

6. Display general classification of reads by Ninetails.

```
ninetails::plot_class_counts(class_data = class_data, grouping_factor = NA, frequency = TRUE, type = "R")
```

Reads are grouped into classes assigned by the Ninetails package based on whether they are decorated with non-adenosines, blank or unclassified due to various reasons (Fig. 1c).

7. Plot the global frequency of non-adenosine residues (C, G, U).

```
ninetails::plot_residue_counts(residue_data=residue_data, grouping_factor=NA, frequency=TRUE, by_read=FALSE)
```

This plot quantifies how often each type of non-adenosine base is found within the poly(A) region across all reads, providing a general overview of mixed tail composition (Fig. 1d).

4 Notes

1. Nanopore chemistry and pore types evolve rapidly. Currently, ONT is transitioning from SQK-RNA002 to SQK-RNA004 chemistry and from R9 to R10 pore types, with the former being phased out. This protocol is compatible with both chemistries and pore types. However, future generations may not be reverse compatible. Always verify the chemistry and pore type available before starting the experiment.
2. Although optional, we highly recommend performing reverse transcription after adapter ligation. While only RNA is sequenced, generating a cDNA strand can improve yield, likely by reducing RNA secondary structure that may block RNA translocation through the pore. In our experience, SuperScript III gives higher overall yield than SuperScript IV and ONT's recently recommended Induro (NEB). However, Induro produces longer reads with better 5′ coverage. The choice of reverse transcriptase should depend on experimental goals—balancing yield and read length. For 3′-end-focused studies, SuperScript III is usually sufficient.
3. The sequencing is controlled by MinKNOW software installed on the host computer (for MinION) or the instrument internal system (for GridION). MinKNOW saves the data in either FAST5 (older) or POD5 (newer) format to a user-specified directory. It can also perform real-time basecalling, if enabled. However, data generated by live basecalling often lack certain details important for downstream analysis. For this reason, we

recommend performing basecalling separately using Guppy, ideally version 6.0.0. If MinKNOW cannot produce FAST5 files, which may happen in newer versions, POD5 data can be easily converted to FAST5 using the *pod5* Python library provided by ONT. Detailed instructions are available in the pod5 documentation (https://github.com/nanoporetech/pod5-file-format/blob/master/python/pod5/README.md#pod5-convert-to_fast5).

4. Guppy is a legacy basecaller developed by ONT, which will soon be retired in favor of the next-generation software, Dorado. However, Dorado is still under active development and currently lacks certain features required for non-adenosine poly(A) profiling, as used by the existing Ninetails pipeline. A new version of Ninetails is in progress and is expected to support Dorado in future releases. Until then, we recommend using Guppy rather than Dorado to ensure full compatibility with the current pipeline.
5. Preparing a library without poly(A) enrichment using oligo $(dT)_{(25)}$ often results in lower input yield from a single library. This is caused by the presence of non-polyadenylated RNAs, which reduces the efficiency of adapter ligation and may lead to pore clogging due to secondary RNA structures.
6. If noncoding RNAs are to be included in the analysis, they should be incorporated into the existing reference and annotation. It should also be verified whether the selected transcriptome reference includes annotated 3′ UTRs, as these may be missing in some versions.

Acknowledgments

We thank Professor Andrzej Dziembowski for providing the infrastructure to perform the DRS and the computing server to perform the analyses.

Authors Contribution NG and ACC contributed to the conception and design of the protocol. BZ axenized, provided, and maintained the cell cultures. NG and BZ performed the experiments. NG, RM, ACC, and BZ drafted the manuscript. All authors read and approved the submitted version.

Conflict of Interest Statement The authors declare no conflict of interests.

References

1. Nicholson AL, Pasquinelli AE (2019) Tales of detailed poly(A) tails. Trends Cell Biol 29:191–200. https://doi.org/10.1016/j.tcb.2018.11.002
2. Shin J, Paek KY, Chikhaoui L et al (2022) Oppositional poly(A) tail length regulation by FMRP and CPEB1. RNA 28:756–765. https://doi.org/10.1261/rna.079050.121
3. Brouze A, Krawczyk PS, Dziembowski A, Mroczek S (2023) Measuring the tail: methods for poly(A) tail profiling. Wiley Interdiscip Rev RNA 14:e1737. https://doi.org/10.1002/WRNA.1737
4. Czarnocka-Cieciura A, Poznański J, Turtola M et al (2024) Modeling of mRNA deadenylation rates reveal a complex relationship between mRNA deadenylation and decay. EMBO J 43: 6525–6554. https://doi.org/10.1038/s44318-024-00258-3
5. Kruse E, Göringer HU (2023) Nanopore-based direct RNA sequencing of the Trypanosoma brucei transcriptome identifies novel lncRNAs. Genes 14:610. https://doi.org/10.3390/genes14030610
6. Guegan F, Rajan KS, Bento F et al (2022) A long noncoding RNA promotes parasite differentiation in African trypanosomes. Sci Adv 8: 2706. https://doi.org/10.1126/SCIADV.ABN2706
7. Ye F, Chen X, Li Y et al (2025) Comprehensive genome annotation of the model ciliate Tetrahymena thermophila by in-depth epigenetic and transcriptomic profiling. Nucleic Acids Res 53:gkae1177. https://doi.org/10.1093/nar/gkae1177
8. Cordoba J, Perez E, Van Vlierberghe M et al (2021) De novo transcriptome meta-assembly of the mixotrophic freshwater microalga Euglena gracilis. Genes 12:842. https://doi.org/10.3390/genes12060842
9. Ebenezer TE, Zoltner M, Burrell A et al (2019) Transcriptome, proteome and draft genome of Euglena gracilis. BMC Biol 17:11. https://doi.org/10.1186/s12915-019-0626-8
10. Ungaro A, Pech N, Martin J-F et al (2017) Challenges and advances for transcriptome assembly in non-model species. PLoS One 12: e0185020. https://doi.org/10.1371/journal.pone.0185020
11. Bilska A, Kusio-Kobiałka M, Krawczyk PS et al (2020) Immunoglobulin expression and the humoral immune response is regulated by the non-canonical poly(A) polymerase TENT5C. Nat Commun 11. https://doi.org/10.1038/S41467-020-15835-3
12. Gumińska N, Matylla-Kulińska K, Krawczyk PS et al (2025) Direct profiling of non-adenosines in poly(A) tails of endogenous and therapeutic mRNAs with Ninetails. Nat Commun 16: 2664. https://doi.org/10.1038/s41467-025-57787-6
13. Workman RE, Tang AD, Tang PS et al (2019) Nanopore native RNA sequencing of a human poly(A) transcriptome. Nat Methods 16:1297–1305. https://doi.org/10.1038/S41592-019-0617-2
14. Krawczyk PS, Tudek A, Mroczek S, Dziembowski A (2024) Transcriptome-wide analysis of mRNA adenylation status in yeast using nanopore sequencing. In: Methods Mol Biol Clifton NJ, vol 2723, pp 193–214. https://doi.org/10.1007/978-1-0716-3481-3_12
15. Liu Y, Zhao H, Shao F et al (2023) Remodeling of maternal mRNA through poly(A) tail orchestrates human oocyte-to-embryo transition. Nat Struct Mol Biol 30:200–215. https://doi.org/10.1038/s41594-022-00908-2
16. Krawczyk PS, Mazur M, Orzeł W et al (2025) Re-adenylation by TENT5A enhances efficacy of SARS-CoV-2 mRNA vaccines. Nature:1–9. https://doi.org/10.1038/s41586-025-08842-1
17. Krause M, Niazi AM, Labun K et al (2019) Tailfindr: alignment-free poly(A) length measurement for Oxford Nanopore RNA and DNA sequencing. RNA 25:1229–1241. https://doi.org/10.1261/rna.071332.119
18. Soneson C, Yao Y, Bratus-Neuenschwander A et al (2019) A comprehensive examination of Nanopore native RNA sequencing for characterization of complex transcriptomes. Nat Commun 10:3359. https://doi.org/10.1038/s41467-019-11272-z
19. Parker MT, Knop K, Sherwood AV et al (2020) Nanopore direct RNA sequencing maps the complexity of Arabidopsis mRNA processing and m6A modification. elife 9. https://doi.org/10.7554/ELIFE.49658
20. Brouze M, Czarnocka-Cieciura A, Gewartowska O et al (2024) TENT5-mediated polyadenylation of mRNAs encoding secreted proteins is essential for gametogenesis in mice. Nat Commun 15:5331. https://doi.org/10.1038/s41467-024-49479-4
21. Liudkovska V, Krawczyk PS, Brouze A et al (2022) TENT5 cytoplasmic noncanonical poly (A) polymerases regulate the innate immune response in animals. Sci Adv 8. https://doi.org/10.1126/SCIADV.ADD9468
22. Gewartowska O, Aranaz-Novaliches G, Krawczyk PS et al (2021) Cytoplasmic polyadenylation by TENT5A is required for proper bone formation. Cell Rep 35. https://doi.org/10.1016/J.CELREP.2021.109015

23. Mulroney L, Wulf MG, Schildkraut I et al (2022) Identification of high-confidence human poly(A) RNA isoform scaffolds using nanopore sequencing. RNA 28:162–176. https://doi.org/10.1261/rna.078703.121
24. Cozzuto L, Liu H, Pryszcz LP et al (2020) MasterOfPores: a workflow for the analysis of Oxford Nanopore direct RNA sequencing datasets. Front Genet 11:211. https://doi.org/10.3389/fgene.2020.00211
25. Adl SM, Bass D, Lane CE et al (2019) Revisions to the classification, nomenclature, and diversity of eukaryotes. J Eukaryot Microbiol 66:4–119. https://doi.org/10.1111/jeu.12691
26. Cavalier-Smith T (2009) Kingdoms protozoa and Chromista and the eozoan root of the eukaryotic tree. Biol Lett 6:342–345. https://doi.org/10.1098/rsbl.2009.0948
27. Burki F (2014) The eukaryotic tree of life from a global phylogenomic perspective. Cold Spring Harb Perspect Biol 6:a016147. https://doi.org/10.1101/cshperspect.a016147
28. Novák Vanclová AMG, Zoltner M, Kelly S et al (2020) Metabolic quirks and the colourful history of the Euglena gracilis secondary plastid. New Phytol 225:1578–1592. https://doi.org/10.1111/nph.16237
29. Gumińska N, Hałakuc P, Zakryś B, Milanowski R (2024) Circular extrachromosomal DNA in Euglena gracilis under normal and stress conditions. Protist 175:126033. https://doi.org/10.1016/j.protis.2024.126033
30. Gumińska N, Zakryś B, Milanowski R (2021) A new type of circular RNA derived from nonconventional introns in nuclear genes of Euglenids. J Mol Biol 433:166758. https://doi.org/10.1016/j.jmb.2020.166758
31. Gumińska N, Płecha M, Zakryś B, Milanowski R (2018) Order of removal of conventional and nonconventional introns from nuclear transcripts of Euglena gracilis. PLoS Genet 14:e1007761. https://doi.org/10.1371/journal.pgen.1007761
32. Chiari Y, Howard L, Moreno N et al (2023) Influence of RNA-Seq library construction, sampling methods, and tissue harvesting time on gene expression estimation. Mol Ecol Resour 23:803–817. https://doi.org/10.1111/1755-0998.13757
33. Tiwari S, Vis ML (2024) Fine-tuning RNA extraction protocols for polysaccharide-rich freshwater red algae (Rhodophyta). MicroPubl Biol. https://doi.org/10.17912/micropub.biology.001180
34. Mundt F, Heinrich S, Hanelt D (2019) RNA isolation from taxonomically diverse photosynthetic protists. Limnol Oceanogr Methods 17:190–199. https://doi.org/10.1002/lom3.10299
35. Jensen T, Saleh L, Bents D et al (2023) Optimised protocols for RNA extraction from a broad taxonomic range of algae. J Appl Phycol 35:1743–1753. https://doi.org/10.1007/s10811-023-02980-7
36. Chang H, Lim J, Ha M, Kim VN (2014) TAIL-seq: genome-wide determination of poly (A) tail length and 3′ end modifications. Mol Cell 53:1044–1052. https://doi.org/10.1016/J.MOLCEL.2014.02.007
37. Legnini I, Alles J, Karaiskos N et al (2019) FLAM-seq: full-length mRNA sequencing reveals principles of poly(A) tail length control. Nat Methods 169(16):879–886. https://doi.org/10.1038/s41592-019-0503-y
38. Liu Y, Nie H, Liu H, Lu F (2019) Poly (A) inclusive RNA isoform sequencing (PAIso-seq) reveals wide-spread non-adenosine residues within RNA poly(A) tails. Nat Commun 10. https://doi.org/10.1038/S41467-019-13228-9
39. Czarnocka-Cieciura A, Brouze M, Gumińska N et al (2025) Comprehensive analysis of poly (A) tails in mouse testes and ovaries using nanopore direct RNA sequencing. Sci Data 12:43. https://doi.org/10.1038/s41597-024-04226-8
40. Cramer M, Myers J (1952) Growth and photosynthetic characteristics of euglena gracilis. Arch Für Mikrobiol 17:384–402. https://doi.org/10.1007/BF00410835
41. Leger A, Leonardi T (2019) pycoQC, interactive quality control for Oxford nanopore sequencing. J Open Source Softw 4:1236. https://doi.org/10.21105/joss.01236
42. Li H (2018) Minimap2: pairwise alignment for nucleotide sequences. Bioinformatics 34: 3094–3100. https://doi.org/10.1093/bioinformatics/bty191
43. Danecek P, Bonfield JK, Liddle J et al (2021) Twelve years of SAMtools and BCFtools. Gigascience 10:giab008. https://doi.org/10.1093/gigascience/giab008
44. Oxford Nanopore Technologies (2025) Dorado Basecaller – Oxford Nanopore Technologies. https://github.com/nanoporetech/dorado. Accessed 29 Jun 2025
45. Niazi AM, Krause M, Valen E (2021) Transcript isoform-specific estimation of poly (A) tail length by nanopore sequencing of native RNA. Methods Mol Biol 2284:543–567. https://doi.org/10.1007/978-1-0716-1307-8_30

Chapter 7

Euglenid Extrachromosomal DNA: Assembly and Annotation

Paweł Hałakuc, Kacper Maciszewski, and Anna Karnkowska

Abstract

Euglenids (Euglenozoa) contain several forms of extrachromosomal DNA (ecDNA) in their cells, including the ribosomal DNA operon (rDNA), the mitochondrial genome (mtDNA), and, in photosynthetic species, the plastid genome (ptDNA). These ecDNA elements can be easily and accurately assembled and annotated even from limited sequencing data, such as single-cell genomic or metagenomic datasets. They are an important source of information for phylogenomic analyses, metabarcoding and evolutionary studies. In this chapter, we present a robust and adaptable bioinformatics pipeline for the identification, assembly, and annotation of extrachromosomal DNA from whole-genome datasets. The pipeline was developed with euglenids in mind and takes into account their unique genomic features, but can also be adapted for other Euglenozoa (and protists). This approach enables the recovery of organellar and rDNA sequences with high confidence and supports both targeted studies and large-scale environmental analyses.

Key words Euglenida, Euglenozoa, Ribosomal operon, Plastid genome, Mitochondrial genome, Next-generation sequencing

1 Introduction

Euglenids are a diverse lineage within the Euglenozoa, comprising both heterotrophic and phototrophic species. They have been known since the earliest microscopic observations, and their morphological and ecological diversity is now well established [1]. Many species, especially phototrophic ones, are maintained in culture and provide valuable material for laboratory studies. However, genomic research on euglenids is still in its infancy. This is mainly due to the considerable genome size, variable ploidy, and the presence of non-canonical introns [2], all of which complicate assembly and annotation. In contrast, extrachromosomal DNA (ecDNA), including the rDNA operon, mitochondrial DNA, and plastid DNA, has been sequenced more frequently, either through targeted approaches or as part of total genomic DNA assemblies. ecDNA can often be reconstructed not only from genomic datasets but also from metagenomic and single-cell sequencing data

Paul A. M. Michels et al. (eds.), *Euglenozoa: Methods and Protocols, Volume 1*, Methods in Molecular Biology, vol. 3013,
https://doi.org/10.1007/978-1-0716-5142-1_7,

[3, 4]. Since nuclear genomes are difficult to obtain, ecDNA offers a more accessible and reliable source for evolutionary, ecological, and taxonomic studies.

Each type of extrachromosomal DNA (ecDNA) has its own peculiarities and challenges. As early as the 1980s, it was discovered that in euglenids, the ribosomal DNA (rDNA) operon is encoded on the ecDNA molecules, which are present in multiple copies in the cell [5]. This high copy number made the rDNA relatively easy to amplify, and thus the operon, especially the 18S rDNA, became a valuable molecular marker [6] that is frequently used in phylogenetic analyses and is of central importance for euglenid taxonomy (e.g., [7–9]) and diversity assessments (e.g., [10, 11]). Recent advances have made it possible to assemble rDNA operons from genomic datasets, revealing a broader spectrum of sequence diversity and enabling more detailed reconstructions of their evolutionary history [12].

The structure of mitochondrial DNA (mtDNA) in euglenids remains poorly understood, primarily due to the difficulty in isolating mitochondrial DNA fractions. The only mitochondrial genome that has been sequenced directly from an isolated mitochondrial fraction is that of *Euglena gracilis* and found to be composed of linear fragments of 5–8 kbp, including only seven protein-coding genes, fragmented across these pieces of DNA [13]. However, there is evidence that the mtDNA in some lineages, including taxa branching next to the main trunk of euglenids, may have a typical circular architecture [3]. This structural diversity adds complexity to genome assembly and increases the likelihood of assembly errors.

In contrast to mtDNA, the ptDNA in euglenids has not undergone particularly divergent or drastically reductive evolution, and its organization bears a close resemblance to that of most eukaryotic lineages carrying secondary plastids. Complete euglenid plastid genome sequences, thus far obtained for over 30 species representing all known phototrophic genera (e.g., [14–16]), are invariably recovered as single circular molecules, ranging between approximately 60 and 140 kbp in size, and carrying a slightly reduced subset of the coding contents of green algal plastid genomes [17]. Euglenids display very little variability of the genetic complement in their ptDNA, with the only prominent exceptions being the ribosomal operon (pt-rDNA), which can be present in single, double, or multiple identical or nearly identical copies [16], and the number of group II intron maturases [15]. Additionally, plastid genomes of euglenids are distinctly rich in introns, whose quantities and length account for a major part of the observed interspecific variability in ptDNA size. Furthermore, these genetic elements are group II introns, which, in contrast with nuclear introns, are not excised by the spliceosome and do not have strictly conserved boundaries—hence, their accurate annotation can be challenging.

2 Materials

2.1 Software

UNIX-based command-line environment (e.g., Linux, macOS, or Windows Subsystem for Linux), fastp [18], SPAdes [19], metaSPAdes [20], BLAST+ [21], BlobTools [22], Tiara [23], Bandage (installed in a graphical environment) [24], Barrnap [25], RNAmmer [26], RNAfold [27], MFAnnot [28], NOVOPlasty [29], DOGMA [30], cpGAVAS [31], GeneiousPrime (www.geneious.com), InterProScan [32]. In general, it is recommended to use the latest available version of the software and, for inexperienced users, to stick to the default settings.

2.2 Genomic DNA

To generate sequencing reads, genomic DNA must first be sequenced. Extrachromosomal DNA (ecDNA) can be identified not only through targeted genome sequencing but also from single-cell data and metagenomic samples. Therefore, we do not specify the input material here (*see* **Note 1**). However, the pipeline is optimised for Illumina short-read data, which provides the best results at a lower cost (*see* **Note 2**).

3 Methods

3.1 Raw Reads Quality Check and Filtering

Check the quality of the raw reads with fastp using the following command:

```
fastp -i [reads_1.fastq.gz] -I [reads_2.fastq.gz] -o [reads_trimmed_1.fastq.gz] -O [reads_trimmed_2.fastq.gz] -q 20 --detect_adapter_for_pe
```

In addition to generating trimmed FASTQ files, fastp provides an HTML report summarizing read quality before and after trimming. The most critical parameter to assess is the filtering results: as long as the percentage of reads passing the filters remains above 70%, the data are considered suitable for downstream assembly.

3.2 Genome Assembly and Quality Assessment

De novo assembly of trimmed reads into contigs is performed using SPAdes. While alternative assemblers may also be used, SPAdes has consistently demonstrated high-quality results (*see* **Note 3**).

```
spades -t [threads] -k [21,33,55,77,99,121] -1 [reads_trimmed_1.fastq.gz] -2 [reads_trimmed_2.fastq.gz] --meta -o [output_directory]
```

The basic parameters are defined in the command line above, with the metaSPAdes mode recommended due to the typically uneven coverage between nuclear and extrachromosomal DNA. Additional parameters can be adjusted as needed, depending on the characteristics of the input data. The resulting assembly is saved in the contigs.fasta file, which includes sequences from both the nuclear genome and extrachromosomal DNA.

3.3 Identification of Extrachromosomal DNA

In euglenids, rDNA is typically encoded on extrachromosomal plasmid-like molecules present in relatively high copy numbers. As a result, it is the easiest of the three ecDNA types discussed in this chapter to assemble and identify. Based on our experience, rDNA recovery is most straightforward in phototrophic euglenids; however, even in heterotrophic species, rRNA genes can often be identified from lower-quality data, which might be insufficient to detect mtDNA.

In general, three approaches can be used to identify extrachromosomal DNA (ecDNA): homology-based methods (e.g., BLAST), coverage-based methods (relying on the typically higher copy number of ecDNA compared to the nuclear genome), or a combination of both. The effectiveness of each method may vary depending on the assembly quality and the species under study. Notably, contigs encoding mitochondrial proteins are sometimes misclassified as bacterial by popular binning tools such as BlobTools or Tiara and may be inadvertently discarded.

3.4 rDNA Operon Assembly and Annotation

rDNA typically assembles into either one (full rDNA) or two contigs (separate SSU and LSU). The simplest way to identify it is by using any euglenid rDNA sequence as a query in a BLASTn search (e.g., *Euglena gracilis* rDNA—accession X53361.2) (*see* **Note 4**). We recommend using Bandage to verify the circularity and completeness of the assembled rDNA (*see* **Note 5**).

The user should load the assembly graph file (e.g., assembly_-graph.fastg) into Bandage. Next, create a BLASTn database, load query sequences (18S, 5S, and 28S rRNA) from a FASTA file, run the BLAST search, and visualize regions surrounding the hits (Fig. 1). Ideally, this process yields a single circular contig 12–20 kbp in length. If so, the next stage is to confirm that it represents euglenid rDNA and proceed to annotation. Other scenarios we have encountered include: (1) two circles connected by a repetitive sequence; (2) a disordered single circle; (3) two disordered circles; (4) a completely fragmented or messy graph (*see* **Note 6**).

To confirm that the identified sequence is indeed rDNA, it is best to perform an alignment followed by reconstruction of phylogeny (*see* **Note 7**).

```
mafft --genafpair --maxiterate 1000 –add [all_sequences.fasta]
> [einsi_aligned_sequences.fasta]
trimal -automated1 -in [einsi_aligned_sequences.fasta] -out
[trimmed_einsi_aligned_sequences.fasta]
```

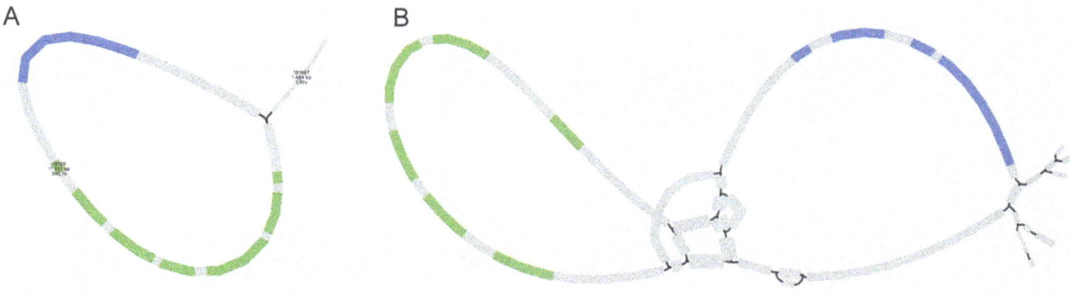

Fig. 1 Example Bandage graph illustrating rDNA assembly scenarios in euglenids. The assembly graph is visualized in Bandage after BLASTn search using 18S, 5S, and 28S rRNA queries. (**a**) almost circular high-quality assembly of *Euglena agilis*; (**b**) a highly fragmented graph of *Trachelomonas volvocina* assembly

```
iqtree -T 8 -m TEST -B 1000 -s [trimmed_einsi_aligned_sequences.fasta]
```

We recommend using the reference alignment for rDNA from Hałakuc et al. [12] and MAFFT with the *einsi* algorithm (--genafpair --maxiterate 1000), as it handles long expansion segments and additional ITS regions in euglenid rDNA particularly well.

```
mafft --genafpair --maxiterate 1000 –add [new_sequences.fasta] [previously_aligned_sequences.aln.fasta] > [einsi_aligned_sequences.fasta]
```

The simplest approach for the annotation of rDNA is to use annotations provided by rDNA identification software such as barrnap or RNAmmer; however, these are often unreliable for the elongated rDNA typical of euglenids. For this reason, we recommend manually identifying the ends of the SSU and LSU based on sequence alignment, and ideally verifying them by detecting conserved secondary structures, for example, using RNAfold.

3.5 mtDNA Assembly and Annotation

In general, euglenid mitochondrial DNA (mtDNA) does not assemble into a single contig, but rather into several contigs, each typically encoding one or two proteins. The gene repertoire is highly reduced—in all analyzed cases, we identified no more than eight protein-coding genes (*atp6, cob, cox1, cox2, cox3, nad1, nad4,* and *nad5*) and two rRNA genes (*rnl* and *rns*), which are often split into two separate fragments each.

Proper identification of mitochondrial genes in the assembly is challenging due to their high divergence. They show low similarity to mitochondrial genes from other euglenozoans, discobans, or eukaryotes in general. For this reason, we recommend identifying mitochondrial genes by running tBLASTn search against the assembly graph using known mitochondrial proteins as queries and visualizing the results in Bandage (Fig. 2) (*see* **Note 8**).

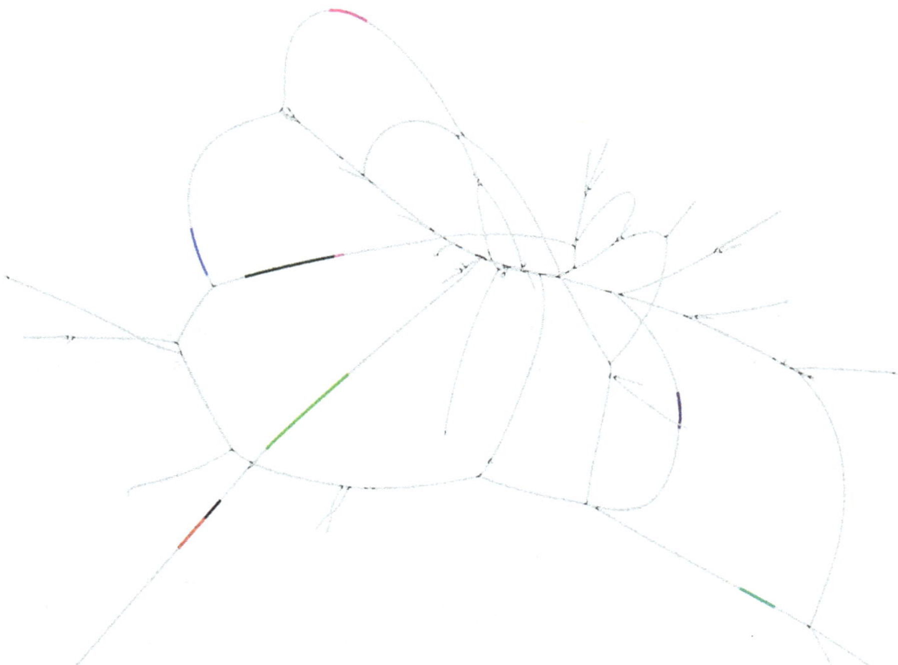

Fig. 2 Example Bandage graph illustrating typical mtDNA assembly in euglenids. The assembly graph of *Trachelomonas volvocina* is visualized in Bandage after tBLASTn search using eight protein-coding genes (*atp6, cob, cox1, cox2, cox3, nad1, nad4,* and *nad5*)

Annotation of protein-coding genes is generally easier than that of rRNA genes, thanks to the presence of readily identifiable start and stop codons. One potential challenge is the presence of group I introns interrupting the gene sequences. Annotation can begin with MFannot, which should be provided with all identified mitochondrial contigs.

3.6 ptDNA Assembly and Annotation

Due to the large number of ptDNA copies in euglenid cells, the assembly of its complete sequence with satisfactory coverage is possible even from a library of only several million paired-end Illumina HiSeq or MiSeq reads. This can be achieved using a two-step assembly method, starting with de novo assembly using SPAdes (*see* Subheading 3.2), followed by seeded assembly using NOVOPlasty.

After the first step is completed, plastid genome fragments can be identified among the assembled contigs (contigs.fasta) via BLASTN search, using the publicly available ptDNA-derived sequences of closest relatives as queries, and the contigs.fasta file as the database. True matches to the plastid genome-derived query are likely to stand out in the BLAST output in default format due to their relatively high (usually >100) coverage in the assembly, and their identity can be confirmed via BLAST search against the NCBI non-redundant protein database. Next, the longest identified plastid fragment can be used as the seed sequence for the second step of the assembly using NOVOPlasty. All settings, including paths to the input files, are read from the config.txt file, whose template is included in the software package. We recommend starting with the default settings and, if the resulting assembly is not circularized, adjusting the config.txt file (*see* **Note 9**).

```
perl NOVOPlasty4.3.5.pl -c config.txt
```

Once the ptDNA contig has been successfully circularized, its annotation can be performed. While automatic annotation of euglenid plastid genome sequences using browser-based tools (DOGMA, cpGAVAS) has yielded moderate success, the most convenient method for obtaining draft annotation is, in our experience, the "Annotate From…" function, implemented in the Geneious Prime software suite. This method enables quick construction of a custom database of reference plastid genomes from data available in various formats (e.g., Genbank, GFF, BED, EMBL, and many others), manual setting of sequence identity cutoff value, instantaneous annotation accuracy checks by comparing the boundaries of transferred annotations and predicted ORFs, and seamless progression from draft to manually curated annotation within one software suite (*see* **Note 10**). Naturally, any automatic annotation will require manual curation, e.g., using homology-based search methods to determine the precise location of exon/intron boundaries; however, as a tool for obtaining accurate and informative draft annotations, Geneious Prime remains unmatched in its efficiency.

What is crucial to mention in this section is that the euglenid ptDNA has one component that defies the usual rules for performing genome annotation—the aforementioned group II intron maturases, which constitute a rare example of intron-encoded proteins (IEPs). Even with the awareness that a protein-coding gene may be located inside an intron within another, unrelated gene, group II intron maturases are notoriously challenging to identify and annotate—with the exception of *ycf13/mat1*, they are not found in conserved loci (i.e., their location in the genome varies between taxa). What is more, these genes are extremely fast-evolving and often cannot be identified using homology-based search methods. Instead, an IEP candidate can be more accurately identified as an intron maturase through a protein domain-targeted, predictive model-based search method, such as InterProScan. In addition to its general usefulness in manual curation of genomic annotations, the PFAM database, used by the InterProScan web service, contains three protein domains found in group II intron maturases: a reverse transcriptase domain, a group II intron maturase domain, and a H-N-H endonuclease domain, and even though many maturases found in euglenid ptDNA have differentially lost some of their domains, each of them is sufficient for positive identification.

4 Notes

1. Based on our experience, both axenic and non-axenic cultures, which are not overgrown with bacteria, are well-suited for obtaining extrachromosomal DNA (ecDNA). Another recommended source of DNA is single cells, where DNA has been amplified using whole genome amplification techniques. For Illumina sequencing, we recommend starting with a few million 250 bp reads or longer. In most cases, deeper sequencing yields limited benefits: if the biological sample is of good quality, reliable results can be achieved with shallow sequencing; conversely, poor-quality samples are unlikely to yield interpretable data, regardless of sequencing depth.
2. We recommend using Illumina sequencing over long-read technologies. Given the size and complexity of euglenid nuclear genomes, obtaining sufficient high molecular weight DNA and achieving the necessary depth of Nanopore or PacBio sequencing is often cost- and labor-prohibitive. Furthermore, in our experience, the improvements in completeness or resolution of ecDNA assemblies are typically minimal and do not justify the additional investment. However, if long-read data are available, they can also produce high-quality results and can be beneficial for nuclear genome assembly.
3. Most genome assemblers for Illumina data use de Bruijn graphs. We recommend SPAdes due to its versatility and built-in modes, particularly metaSPAdes and scSPAdes. Standard genome assembly assumes even sequencing coverage, which often does not hold true for extrachromosomal DNA. In such cases, metaSPAdes or scSPAdes typically yield better assemblies of rDNA, mtDNA, and ptDNA. Based on our experience, metaSPAdes generally performs best.

   We advise against adjusting parameters unless the user is experienced. One critical setting is the choice of *k-mers*: we recommend using the default set (21,33,55,77,99,121), ensuring the largest *k-mer* is smaller than half the read length. If not explicitly set, SPAdes may choose suboptimal values, reducing assembly quality. Users are encouraged to test multiple modes and settings to determine the best outcome.

   Assembly quality can be evaluated with metrics such as total length, average coverage, and N50 (e.g., using QUAST). However, these technical metrics may not reflect suitability for specific goals. If the objective is to recover and analyze ecDNA, the most practical assessment is whether the assembly enables the successful identification and analysis of those elements.

4. Alternatively, it is possible to use rDNA identification software such as Barrnap or RNAmmer. However, in our experience, these tools often overlook fragments of euglenid rDNA, particularly the LSU. We believe this is due to the presence of very long expansion segments. Therefore, we recommend using those tools only as a last resort to verify that no rDNA sequences were missed within the assembly.
5. If the assembled rDNA is of high quality, its circularity can often be inferred directly from the sequence: the first *k* nucleotides (where *k* is the highest *k-mer* size used in SPAdes) will exactly match the last *k* nucleotides.
6. We do not recommend investing effort in recovering rDNA assemblies that appear highly fragmented in the assembly graph. Instead, we suggest obtaining higher-quality sequencing material.
7. The rDNA assembly may be fragmented, as assemblers sometimes struggle with the large expansion segments present in euglenid rDNA. In such cases, the rDNA can either be reassembled or, for the purposes of alignment and phylogenetic analysis, the assembled fragments can be manually joined in a text editor. We recommend marking any gaps with a few *Ns* to ensure they are easily identifiable later. It is important not to deposit such edited sequences in any public repository—this is only a temporary workaround to facilitate alignment and tree construction.
8. Similar sequences often occur in non-coding regions of mitochondrial contigs, causing them to appear entangled in the assembly graph visualized with Bandage. This pattern confirms that the identified sequences are indeed mitochondrial rather than a few contaminants with coincidental similarity to euglenid mitochondrial proteins. Another reason to use Bandage is that highly divergent sequences may not be assembled by SPAdes into continuous contigs but instead spread across multiple short fragments, making identification within the assembly difficult or even impossible. In such cases, it is possible to manually select a path through the edges with the highest coverage, enabling reconstruction of the complete gene sequence (Output - > copy/save selected path sequence to clipboard/FASTA).
9. To address issues with circularization, the following changes can be made in the config.txt file:
   - (a) Trying different *k*-mer lengths—up to half the length of reads used in the assembly;
   - (b) Trying different seed sequences—while the longest SPAdes contig usually yields best results, it is recommended not to use contigs that end with an rRNA gene,

as these may be present in multiple copies and therefore cannot be unambiguously extended, or a ribosomal protein-coding gene, as their homologs are likely present in the nuclear genome and may lead to erroneous inclusion of a nuclear sequence in the assembly;

(c) Trying iterative NOVOPlasty runs, i.e., using the output contig of the first software run as the seed for the next run.

In the unlikely event that NOVOPlasty does not yield a circularized assembly under any settings, the incomplete plastid genome sequence can still be manually closed using the Bandage software tool, in a similar manner as a mitochondrial genome, as outlined in Subheading 3.5. This might occur if the analyzed ptDNA contains tandemly repeated copies of the ribosomal operon, and the assembly is based only on short reads (e.g., from Illumina). In such a case, the correct number of plastidial rDNA copies can be estimated from their coverage (which should be approximately equal to a multiple of the mean coverage of the remainder of the ptDNA) and assembled manually. Assuming the genetic complement of such manual assembly is annotated and identified as complete, the assembled structure can eventually be confirmed via PCR.

10. The true advantage of using Geneious Prime for annotating euglenid ptDNA sequences comes from its "perception" of interrupted coding sequences, such as genes riddled with group II introns. In many genes encoded within euglenid plastid genomes, the intron length may exceed the total length of the protein-coding sequence, which leads many automatic annotation tools to detect the exons as separate gene copies, or not to detect one or more exons at all. In contrast, the transferred annotations created in Geneious Prime will appear as CDS with multiple intervals, even if the sequence being annotated is the only one carrying introns, while the reference sequences contain none.

Acknowledgments

This work was supported by the National Science Centre, Poland (Preludium Grant 2021/41/N/NZ8/03760 to PH).

References

1. Kostygov AY, Karnkowska A, Votýpka J et al (2021) Euglenozoa: taxonomy, diversity and ecology, symbioses and viruses. Open Biol 11(3):200407. https://doi.org/10.1098/rsob.200407
2. Fields O, Hammond MJ, Xu X et al (2025) Advances in euglenoid genomics: unravelling the fascinating biology of a complex clade. Trends Genet 4:251–260. https://doi.org/10.1016/j.tig.2024.07.007

3. Záhonová K, Lax G, Sinha SD et al (2021) Single-cell genomics unveils a canonical origin of the diverse mitochondrial genomes of euglenozoans. BMC Biol 19:103. https://doi.org/10.1186/s12915-021-01035-y
4. Hollender M, Sałek M, Karlicki M et al (2024) Single-cell genomics revealed Candidatus Grellia alia sp. nov. as an endosymbiont of *Eutreptiella* sp. (Euglenophyceae). Protist 175(2): 126018. https://doi.org/10.1016/j.protis.2024.126018
5. Ravel-Chapuis P, Nicolas P, Nigon V et al (1985) Extrachromosomal circular nuclear rDNA in *Euglena gracilis*. Nucleic Acids Res 13:7529–7537. https://doi.org/10.1093/nar/13.20.7529
6. Kolisko M, Flegontova O, Karnkowska A et al (2020) EukRef-excavates: seven curated SSU ribosomal RNA gene databases. Database J Biol Databases Curation 2020:baaa080. https://doi.org/10.1093/database/baaa080
7. Lax G, Simpson AGB (2020) The molecular diversity of phagotrophic euglenids examined using single-cell methods. Protist 171:125757. https://doi.org/10.1016/j.protis.2020.125757
8. Rackevei AS, Karnkowska A, Wolf M (2023) 18S rDNA sequence–structure phylogeny of the *Euglenophyceae* (Euglenozoa, Euglenida). J Eukaryot Microbiol 70:e12959. https://doi.org/10.1111/jeu.12959
9. Fells A, Jiang X, Jankowska K et al (2023) Molecular and morphological delimitation of species in *Strombomonas* (Euglenaceae) including a protocol for DNA isolation utilising a chelating resin. Taxon 72:733–750. https://doi.org/10.1002/tax.12937
10. Łukomska-Kowalczyk M, Karnkowska A, Krupska M et al (2016) DNA barcoding in autotrophic euglenids: evaluation of COI and 18s rDNA. J Phycol 52:951–960. https://doi.org/10.1111/jpy.12439
11. Jankowska K, Łukomska-Kowalczyk M, Milanowski R (2024) Biodiversity of autotrophic euglenids based on the group specific DNA metabarcoding approach. Protist 175: 126024. https://doi.org/10.1016/j.protis.2024.126024
12. Hałakuc P, Karnkowska A, Milanowski R (2022) Typical structure of rRNA coding genes in diplonemids points to two independent origins of the bizarre rDNA structures of euglenozoans. BMC Ecol Evol 22:59. https://doi.org/10.1186/s12862-022-02014-9
13. Dobáková E, Flegontov P, Skalický T et al (2015) Unexpectedly streamlined mitochondrial genome of the euglenozoan. Genome Biol Evol 7:3358–3367. https://doi.org/10.1093/gbe/evv229
14. Hallick RB, Hong L, Drager RG et al (1993) Complete sequence of *Euglena gracilis* chloroplast DNA. Nucleic Acids Res 21:3537–3544. https://doi.org/10.1093/nar/21.15.3537
15. Maciszewski K, Dabbagh N, Preisfeld A et al (2022) Maturyoshka: a maturase inside a maturase, and other peculiarities of the novel chloroplast genomes of marine euglenophytes. Mol Phylogenet Evol 170:107441. https://doi.org/10.1016/j.ympev.2022.107441
16. Maciszewski K, Fells A, Karnkowska A (2022) Challenging the importance of plastid genome structure conservation: new insights from euglenophytes. Mol Biol Evol 39:msac255. https://doi.org/10.1093/molbev/msac255
17. Novák Vanclová AM, Nef C, Füssy Z et al (2024) New plastids, old proteins: repeated endosymbiotic acquisitions in kareniacean dinoflagellates. EMBO Rep 25:1859–1885. https://doi.org/10.1038/s44319-024-00103-y
18. Chen S, Zhou Y, Chen Y et al (2018) Fastp: an ultra-fast all-in-one FASTQ preprocessor. Bioinformatics 34:i884–i890. https://doi.org/10.1093/bioinformatics/bty560
19. Prjibelski A, Antipov D, Meleshko D et al (2020) Using SPAdes De Novo Assembler. Curr Protoc Bioinformatics 70:e102. https://doi.org/10.1002/cpbi.102
20. Nurk S, Meleshko D, Korobeynikov A et al (2017) metaSPAdes: a new versatile metagenomic assembler. Genome Res 27:824–834. https://doi.org/10.1101/gr.213959.116
21. Camacho C, Coulouris G, Avagyan V et al (2009) BLAST+: architecture and applications. BMC Bioinformatics 10:421. https://doi.org/10.1186/1471-2105-10-421
22. Laetsch DR, Blaxter ML (2017) BlobTools: interrogation of genome assemblies. F1000Res 6:1287. https://doi.org/10.12688/f1000research.12232.1
23. Karlicki M, Antonowicz S, Karnkowska A (2022) Tiara: deep learning-based classification system for eukaryotic sequences. Bioinformatics 38:344–350. https://doi.org/10.1093/bioinformatics/btab672
24. Wick RR, Schultz MB, Zobel J et al (2015) Bandage: interactive visualization of *de novo* genome assemblies. Bioinformatics 31:3350–3352. https://doi.org/10.1093/bioinformatics/btv383
25. Seemann T (2013) barrnap: rapid ribosomal RNA prediction. https://github.com/tseemann/barrnap. Accessed 30 June 2025

26. Lagesen K, Hallin P, Rødland EA et al (2007) RNAmmer: consistent and rapid annotation of ribosomal RNA genes. Nucleic Acids Res 35: 3100–3108. https://doi.org/10.1093/nar/gkm160
27. Gruber AR, Lorenz R, Bernhart SH et al (2008) The Vienna RNA websuite. Nucleic Acids Res 36:W70–W74. https://doi.org/10.1093/nar/gkn188
28. Lang BF, Beck N, Prince S et al (2023) Mitochondrial genome annotation with MFannot: a critical analysis of gene identification and gene model prediction. Front Plant Sci 14:1222186. https://doi.org/10.3389/fpls.2023.1222186
29. Dierckxsens N, Mardulyn P, Smits G (2016) NOVOPlasty: *de novo* assembly of organelle genomes from whole genome data. Nucleic Acids Res:gkw955. https://doi.org/10.1093/nar/gkw955
30. Wyman SK, Jansen RK, Boore JL (2004) Automatic annotation of organellar genomes with DOGMA. Bioinformatics 20:3252–3255. https://doi.org/10.1093/bioinformatics/bth352
31. Liu C, Shi L, Zhu Y et al (2012) CpGAVAS, an integrated web server for the annotation, visualization, analysis, and GenBank submission of completely sequenced chloroplast genome sequences. BMC Genomics 13:715. https://doi.org/10.1186/1471-2164-13-715
32. Blum M, Andreeva A, Florentino LC et al (2025) InterPro: the protein sequence classification resource in 2025. Nucleic Acids Res 53: D444–D456. https://doi.org/10.1093/nar/gkae1082

Chapter 8

Nuclear Genome Assembly and Annotation of Kinetoplastids

Kristína Záhonová, Rajendra Mandage, and Vyacheslav Yurchenko

Abstract

In this chapter, we describe a pipeline for the nuclear genome analysis of kinetoplastids. Our approach relies on a combination of short- and long-sequencing reads and can be universally applied to studies of other kinetoplastids' genomes.

Key words Kinetoplastea, Euglenozoa, Trypanosomatids, Next-generation sequencing

1 Introduction

Kinetoplastea (Euglenozoa) is a group of unicellular eukaryotes with widely variable lifestyles, incorporating free-living species found in marine and freshwater environments, intracellular symbionts of other protists, and medically important parasites of vertebrates [1]. The evolutionary success of kinetoplastids across a wide range of ecological niches makes their genomes a compelling subject of study offering insights into the genomic changes underlying lifestyle transitions within the group [2, 3]. Presently, hundreds of kinetoplastid genomes are available in public databases, with an overwhelming majority of them belonging to members of the obligately parasitic family Trypanosomatidae [4, 5].

Kinetoplastid genomes exhibit considerable variation in size and complexity overlaid over a unique organization mode [6]. The smallest known kinetoplastid genome (~9.5 Mb) belongs to *Perkinsela* sp., an intracellular symbiont of an amoebozoan [7], while the largest (~40 Mb) to the free-living eubodonid *Bodo saltans* [3]. Protein-coding genes in the relatively compact kinetoplastid genomes are arranged in strand-specific clusters, transcribed as polycistronic units [8], and typically lack *cis*-spliceosomal introns, with only three known exceptions [9]. Trypanosomatids are characterized by high levels of synteny, i.e., conservation of gene order [10] that greatly facilitates and simplifies the genome annotation

Paul A. M. Michels et al. (eds.), *Euglenozoa: Methods and Protocols, Volume 1*, Methods in Molecular Biology, vol. 3013, https://doi.org/10.1007/978-1-0716-5142-1_8,

[11]. Nevertheless, other features of kinetoplastid genomes may undesirably influence this process and must be carefully considered when evaluating the assembly quality. These include: (1) a high proportion of repetitive regions including those associated with the expansion of species- and/or lineage-specific protein families, exemplified by variant surface glycoproteins in *Trypanosoma brucei*, *trans*-sialidases in *Trypanosoma cruzi*, or amastins in *Leishmania* spp. [12–16]; (2) the prevalent presence of pseudogenes, particularly among genes encoding members of large protein families [17–19]; (3) widespread aneuploidy highlighting the need of having fully resolved phased haplotypes for proper analysis [20–22]; (4) presence of the short intergenic regions within polycistronic clusters [23, 24], as well as (5) presence of the clusters of tandemly duplicated genes [25, 26]. Genome assembly and annotation in some kinetoplastid lineages may be particularly challenging. A good example of this are members of the genus *Blastocrithidia*, in which all three universal stop codons have been reassigned to encode amino acids with UAA serving also as a genuine translation terminator [27–29]. All the concerns and restrictions mentioned above justify a standard practice of combining short- and long-sequencing reads for kinetoplastid genome assemblies that we describe in this chapter. We think that it will benefit the community of trypanosomatid researchers and facilitate further genome-guided investigations into the biology of these interesting and important parasites. Some parts of these protocols have been published before [14, 30, 31], but not in the pipeline form presented here.

2 Materials

2.1 Software

FastQC [32], BBDuk [33], Karect [34], SPAdes [35], Platanus [36], GapCloser from SOAPdenovo2 [37], Filtlong [38], proovread [39], Flye [40], BLAST+ [41], DIAMOND [42], QUAST [43], BUSCO [44], Bowtie2 [45], SAMtools [46], Hisat2 [47], Cufflinks [48], Gffread [49], tRNAscan-SE [50], ARAGORN [51], IGV [52], Artemis [53], Companion [54] (*see* **Note 1**).

3 Methods

3.1 Short Read Filtering

1. Check the quality of the raw short, i.e., Illumina paired-end, reads by FastQC using the following command:

```
fastqc -t 5 -o <output_directory> <short_reads_file.fastq.gz>
```

   where **-t** is the number of threads.

2. Trim the reads with BBDuk to remove adaptors and low-quality reads. It is not necessary to unpack the read files beforehand. The typical command for trimming of paired-end short reads is:

```
bbduk in1=<forward_1.fastq.gz> in2=<reverse_2.fastq.gz>
out1=<trimmed_1.fastq.gz> out2=<trimmed_2.fastq.gz> ref=a-
dapters.fa usejni=t qtrim=rl trimq=20 ktrim=r k=22 mink=11
hdist=2 tpe tbo t=10 2> <bbduk_report.log>
```

In case of interleaved reads, the command uses one input file (**in=<reads.fastq.gz>**) and produces one output file (**out=<trimmed.fastq.gz>**). The exact description of parameters can be found in the BBDuk manual (*see* **Note 2**).

3. Check the quality of trimmed reads by FastQC using the command from **step 1**.

4. Perform error-correction of reads by Karect (*see* **Note 3**):

```
karect -correct -threads=10 -matchtype=hamming
-celltype=diploid -inputfile=<trimmed_1.fastq.gz> -inputfi-
le=<trimmed_2.fastq.gz> 2> <karect_report.log>
```

3.2 Long Reads Filtering

1. Filter the long reads, i.e., PacBio or Nanopore, with Filtlong using the short reads with high coverage and high depth as a reference data set:

```
filtlong -1 <trimmed_1.fastq.gz> -2 <trimmed_2.fastq.gz> --
min_length 1000 --keep_percent 90 --verbose <long_reads.fastq.
gz> > <filtered_long_reads.fastq>
```

Adding --**verbose** provides information about the read length and quality. The description of other parameters can be found on the Filtlong GitHub page.

2. Perform error-corrections on long reads with proovread using a high coverage short reads dataset.

```
proovread -i <filtered_long_reads.fastq> -s <trimmed_1.fastq.
gz> -s <trimmed_2.fastq.gz> -o <corrected_long_reads.fastq>
```

where -**o** is a prefix of the output file.

3.3 Genome Assembly of Short Reads

1. Assemble trimmed reads de novo into contigs/scaffolds using SPAdes:

```
spades.py --pe1-1 <trimmed_1.fastq.gz> --pe1-2 <trimmed_2.
fastq.gz> -t 40 -o <spades_output_directory>
```

The description of other parameters can be found in the SPAdes manual. In the output directory, the resulting **contigs.fasta** and **scaffolds.fasta** files can be found. The **contigs.fasta** file should be scaffolded (*see* Subheading 3.3, **step 2** below), alternatively, the **scaffolds.fasta** can be used.

2. Scaffold contigs by Platanus:

```
platanus scaffold -o <platanus_rnd1> -c <contigs.fasta> -IP1
<trimmed_1.fastq> <trimmed_2.fastq> -t 40 2> <platanus_rnd1.
log>
```

where **-o** is a prefix of the output file. Note that the files of trimmed reads must be unpacked for this tool.

3. Use GapCloser module of SOAPdenovo2 for gap filling:

```
GapCloser -b <config_file> -a <platanus_rnd1_scaffold.fasta>
-o <platanus_rnd1_scaffold.gapcloser.fasta> -t 20 2> <plata-
nus_rnd1.gapcloser.log>
```

Note that this step requires a "**config_file**", an example of which can be found in the SOAPdenovo manual and should be modified according to your data.

4. Repeat **steps 2** and **3** once again using the produced scaffolds from the previous round (**platanus_rnd1_scaffold.gapcloser.fasta**).

3.4 Hybrid Assembly Using Both Short and Long Reads

1. Assemble filtered short and long reads into contigs using SPAdes:

```
spades.py --pe1-1 <trimmed_1.fastq.gz> --pe1-2
<trimmed_2.fastq.gz> --pacbio <corrected_long_reads.fastq> -t
40 -o <spades_output_directory>
```

2. Scaffold the contigs and polish the assembly by Flye:

```
Flye --pacbio-hifi <corrected_long_reads.fastq> <SPAdes_con-
tigs.fasta> --out-dir <flye_output_directory> --threads 40 --
iterations 3 --scaffold
```

where **--iterations** specifies assembly polishing number, **--scaffold** performs scaffolding of the assembly (*see* **Notes 4** and **5**).

3.5 Assembly Decontamination and Quality Assessment

1. Discard scaffolds <500 nt.
2. Screen scaffolds in BLASTn searches against the NCBI nucleotide database:

```
blastn -query <assembly.fasta> -db <NCBI_nt_database.fasta>
-out <blast_output.tsv> -outfmt 6 -evalue 1e-05 -num_threads
10
```

Discard scaffolds showing nucleotide identity >95% and query coverage >85% to non-euglenozoan sequences.

3. Screen scaffolds with non-euglenozoan hits below the specified thresholds in the previous step by DIAMOND in sensitive mode against the NCBI non-redundant database:

```
diamond blastx -q <updated_assembly.fasta> -d <NCBI_nr_data-
base.fasta> -o <diamond_output.tsv> -f 6 -evalue 1e-10 --
taxonmap prot.accession2taxid.gz --sensitive -p 10
```

where --**taxonmap** is a mapping file that maps NCBI protein accession numbers to taxon IDs (can be downloaded from NCBI: ftp://ftp.ncbi.nlm.nih.gov/pub/taxonomy/accession2taxid/prot.accession2taxid.gz). Discard scaffolds with non-euglenozoan sequences retrieved as best hits.

4. Evaluate the completeness of the assembly by BUSCO:

```
busco -i <final_assembly.fasta> -l <lineage> -o <base_name> -m
genome -c 5
```

where -**l** specifies the name of the BUSCO lineage dataset (euglenozoa_odb10 or eukaryota_odb10), and -**m** specifies the assessment mode (note that BUSCO can be run in genome, transcriptome, and proteins mode) (*see* **Note 6**).

5. Map the trimmed reads back to the genome assembly using Bowtie2 and SAMtools. First, index the genome assembly:

```
bowtie2-build <final_assembly.fasta> <base_name>
```

Perform the read mapping:

```
bowtie2 --very-sensitive -p 50 -x <base_name> -1 <trimmed_1.
fastq.gz> -2 <trimmed_2.fastq.gz> --un-gz <unmapped_unpaired.
fastq.gz> --un-conc-gz <unmapped_paired.fastq.gz> --al-conc-
gz <mapped.fastq.gz> -S <mapped_reads.sam> 2> <bowtie2.log>
```

where -**x** specifies the same name from the indexing step, -**1** and -**2** are trimmed reads, --**very-sensitive** is a preset of alignment parameters, -**S** specifies the produced SAM file with read mapping.

Convert the SAM file into BAM file with sorted and indexed reads:

```
samtools view -bS -@ 50 <mapped_reads.sam> > <mapped_reads.bam>
samtools sort -o <mapped_sorted_reads.bam> -@ 50 <mapped_reads.bam>
samtools index -b <mapped_sorted_reads.bam>
```

where **<mapped_reads.sam>** is the SAM file, **<mapped_reads.bam>** is the BAM file, **<mapped_sorted_reads.bam>** contains the sorted mapped reads. The last command produces indexed BAM file (**<mapped_sorted_reads.bam.bai>**).

6. Inspect read mapping rate in the **<bowtie2.log>** file and visually in a genome viewer program (e.g., IGV or Artemis) after importing the genome assembly and mapped sorted reads.

3.6 Transcriptome Assembly

1. Check quality and filter the RNA-Seq reads as in **steps 1** and **2** of the Subheading 3.1 above (*see* **Note 2**).
2. Map trimmed reads to the genome assembly using Hisat2 and SAMtools. First, index the genome assembly:

```
hisat2-build -p 50 <final_assembly.fasta> <base_name>
```

Perform the read mapping:

```
hisat2 --very-sensitive --dta --secondary -p 50 -x <base_name> -1 <RNA_trimmed_1.fastq.gz> -2 <RNA_trimmed_2.fastq.gz> --un-gz <RNA_unmapped_unpaired.fastq.gz> --un-conc-gz <RNA_unmapped_paired.fastq.gz> -S <RNA_mapped_reads.sam> 2> <hisat2.log>
```

where --**dta** reports alignments tailored for transcript assemblers, and --**secondary** reports secondary alignments.

Convert the SAM file into BAM file with sorted and indexed reads:

```
samtools view -bS -@ 50 <RNA_mapped_reads.sam> > <RNA_mapped_reads.bam>
samtools sort -o <RNA_mapped_sorted_reads.bam> -@ 50 <RNA_mapped_reads.bam>
samtools index -b <RNA_mapped_sorted_reads.bam>
```

3. Assemble reads into transcripts using Cufflinks:

```
cufflinks -p 40 -o <RNA_mapped_sorted_reads.bam>
```

This command produces **transcripts.gtf** file, which can be converted into gff file using Gffread:

```
gffread transcripts.gtf -o <organism_cufflinks.gff>
```

Using a perl script from Augustus [55] and awk command produces file with transcripts in fasta format:

```
perl getAnnoFasta.pl --seqfile=<final_assembly.fasta> --protein=off --codingseq=on <organism_cufflinks.gff>
awk 'BEGIN{FS=" "}{if(!/>/){print toupper($0)}else{print $1}}' <organism_cufflinks.mrna> > <organism_cufflinks.fasta>
```

3.7 Genome Annotation

1. Annotate protein-coding genes using Companion. In "Step 1: Basic job properties," specify properties of the run. "Step 2: Target sequence" asks for the genome assembly (**<final_assembly.fasta>**). In "Step 3: Transcript evidence," supply gtf file produced by Cufflinks in a previous step (**transcripts.gtf**; *see* Subheading 3.6 above). In "Step 4: Reference organism," choose a reference organism that is phylogenetically closest to yours (all available kinetoplastids can be found here: https://companion.gla.ac.uk/references). In "Step 5: Pseudochromosome contiguation," keep default settings. In "Step 6: Advanced settings," use the following options:
   - Yes, align reference proteins to target sequence.
   - Yes, perform pseudogene detection.
   - Yes, use RATT with the Species transfer type to transfer reference gene models.
2. Identify rRNA-coding genes by BLASTn searches using reference kinetoplastid sequences:

```
blastn -query <reference_rRNAs.fasta> -db <final_assembly.fasta> -out <blast_output.tsv> -outfmt 6 -num_threads 10
```

3. Identify tRNA-coding genes using tRNAscan-SE:

```
tRNAscan-SE -E -o <organism.tRNAscan_table.tsv> -a <organism.tRNAscan_seqs.fasta> --thread 10 <final_assembly.fasta>
```

where **-E** specifies eukaryotic genome, **-o** is tRNAscan-SE output in a tabular format and **-a** collects the predicted tRNA gene sequences. Alternatively, ARAGORN can be used:

```
aragorn -fo -o <organism.aragorn_seqs.fasta>
<final_assembly.fasta>
```

where **-fo** prints out sequences in fasta format only (without secondary structures).

3.8 Specific Cases

1. The genome annotation for *Blastocrithidia* spp. is more challenging as all three stop codons were reassigned as sense ones [27] and specific pipelines have been developed [29].
2. The genomes of diplonemids, a sister lineage of kinetoplastids [1], possess long repetitive regions, numerous transposable elements, or very long introns, i.e., features that require special caution and tinkering when assembling and annotating. The only high-quality genome of a diplonemid is that of *Paradiplonema papillatum* [56], renamed from *Diplonema papillatum* [57]. This assembly was produced by merging separate assemblies of short Illumina and long PacBio reads rather than using hybrid assembly as we describe in this chapter.

4 Notes

1. Please use latest versions of the software available.
2. When running BBDuk, filtering short reads with length <75 nt or <50 nt for DNA-Seq and RNA-Seq reads using **minlen=75** or **minlen=50** parameter, respectively, is recommended to improve the assembly.
3. Alternative error-correction tools for short reads (e.g., RECKONER [58], CARE [59]) are available.
4. Alternative hybrid assembly tools (e.g., MaSuRCA [60], WENGAN [61]) are also available.
5. Caution should be taken when performing rounds of the assembly scaffolding and polishing as over-polishing may introduce errors.
6. BUSCO can be run in the genome mode since kinetoplastids contain very few introns [9].

Acknowledgments

This work was supported by the European Union Operational Program "Just Transition" (LERCO CZ.10.03.01/00/22_003/

0000003) and the Czech Ministry of Education, Youth and Sports (MEYS CZ) (INTER-EXCELLENCE-LUASK22033) to V.Y. Computational resources were provided by the e-INFRA CZ project (ID:90254), supported by the MEYS CZ.

References

1. Kostygov AY, Karnkowska A, Votýpka J et al (2021) Euglenozoa: taxonomy, diversity and ecology, symbioses and viruses. Open Biol 11: 200407
2. Butenko A, Hammond M, Field MC et al (2021) Reductionist pathways for parasitism in euglenozoans? Expanded datasets provide new insights. Trends Parasitol 37(2):100–116
3. Jackson AP (2015) Genome evolution in trypanosomatid parasites. Parasitology 142(Suppl 1):S40–S56
4. Yurchenko V, Butenko A, Kostygov AY (2021) Genomics of Trypanosomatidae: where we stand and what needs to be done? Pathogens 10(9):1124
5. Briggs EM, Marques CA, Reis-Cunha J et al (2020) Next-generation analysis of trypanosomatid genome stability and instability. In: Methods Mol Biol, vol 2116, pp 225–262
6. Maslov DA, Opperdoes FR, Kostygov AY et al (2019) Recent advances in trypanosomatid research: genome organization, expression, metabolism, taxonomy and evolution. Parasitology 146(1):1–27
7. Tanifuji G, Cenci U, Moog D et al (2017) Genome sequencing reveals metabolic and cellular interdependence in an amoeba-kinetoplastid symbiosis. Sci Rep 7(1):11688
8. Clayton C (2019) Regulation of gene expression in trypanosomatids: living with polycistronic transcription. Open Biol 9(6):190072
9. Kostygov AY, Skýpalová K, Kraeva N et al (2024) Comprehensive analysis of the Kinetoplastea intron landscape reveals a novel intron-containing gene and the first exclusively *trans*-splicing eukaryote. BMC Biol 22(1):281
10. Ghedin E, Bringaud F, Peterson J et al (2004) Gene synteny and evolution of genome architecture in trypanosomatids. Mol Biochem Parasitol 134(2):183–191
11. Wu F, Mai Y, Chen C et al (2024) SynGAP: a synteny-based toolkit for gene structure annotation polishing. Genome Biol 25(1):218
12. Silva Pereira S, Jackson AP, Figueiredo LM (2022) Evolution of the variant surface glycoprotein family in African trypanosomes. Trends Parasitol 38(1):23–36
13. Pita S, Díaz-Viraqué F, Iraola G et al (2019) The tritryps comparative repeatome: insights on repetitive element evolution in trypanosomatid pathogens. Genome Biol Evol 11(2): 546–551
14. Albanaz ATS, Gerasimov ES, Shaw JJ et al (2021) Genome analysis of *Endotrypanum* and *Porcisia* spp., closest phylogenetic relatives of *Leishmania*, highlights the role of amastins in shaping pathogenicity. Genes 12(3):444
15. Jackson AP (2010) The evolution of amastin surface glycoproteins in trypanosomatid parasites. Mol Biol Evol 27(1):33–45
16. Freitas LM, dos Santos SL, Rodrigues-Luiz GF et al (2011) Genomic analyses, gene expression and antigenic profile of the *trans*-sialidase superfamily of *Trypanosoma cruzi* reveal an undetected level of complexity. PLoS One 6(10):e25914
17. Abrahim M, Machado E, Alvarez-Valin F et al (2022) Uncovering pseudogenes and intergenic protein-coding sequences in TriTryps' genomes. Genome Biol Evol 14(10):evac142
18. Durante IM, Butenko A, Rašková V et al (2020) Large-scale phylogenetic analysis of trypanosomatid adenylate cyclases reveals associations with extracellular lifestyle and host-pathogen interplay. Genome Biol Evol 12(12):2403–2416
19. Santi AMM, Ribeiro JM, Reis-Cunha JL et al (2022) Disruption of multiple copies of the prostaglandin F2alpha synthase gene affects oxidative stress response and infectivity in *Trypanosoma cruzi*. PLoS Negl Trop Dis 16(10): e0010845
20. Reis-Cunha JL, Pimenta-Carvalho SA, Almeida LV et al (2024) Ancestral aneuploidy and stable chromosomal duplication resulting in differential genome structure and gene expression control in trypanosomatid parasites. Genome Res 34(3):441–453
21. Dumetz F, Imamura H, Sanders M et al (2017) Modulation of aneuploidy in *Leishmania donovani* during adaptation to different *in vitro* and *in vivo* environments and its impact on gene expression. mBio 8(3):e00599–e00517

22. Negreira GH, de Groote R, Van Giel D et al (2023) The adaptive roles of aneuploidy and polyclonality in *Leishmania* in response to environmental stress. EMBO Rep 24(9):e57413
23. Waithaka A, Maiakovska O, Grimm D et al (2022) Sequences and proteins that influence mRNA processing in *Trypanosoma brucei*: evolutionary conservation of SR-domain and PTB protein functions. PLoS Negl Trop Dis 16(10): e0010876
24. Novak EM, de Mello MP, Gomes HB et al (1993) Repetitive sequences in the ribosomal intergenic spacer of *Trypanosoma cruzi*. Mol Biochem Parasitol 60(2):273–280
25. Field H, Field MC (1997) Tandem duplication of *rab* genes followed by sequence divergence and acquisition of distinct functions in *Trypanosoma brucei*. J Biol Chem 272(16): 10498–10505
26. Zakharova A, Tashyreva D, Butenko A et al (2023) A neo-functionalized homolog of host transmembrane protein controls localization of bacterial endosymbionts in the trypanosomatid *Novymonas esmeraldas*. Curr Biol 33(13): 2690–2701
27. Záhonová K, Kostygov A, Ševčíková T et al (2016) An unprecedented non-canonical nuclear genetic code with all three termination codons reassigned as sense codons. Curr Biol 26(17):2364–2369
28. Opperdoes FR, Záhonová K, Škodová-Sveráková I et al (2024) *In silico* prediction of the metabolism of *Blastocrithidia nonstop*, a trypanosomatid with non-canonical genetic code. BMC Genomics 25(1):184
29. Kachale A, Pavlíková Z, Nenarokova A et al (2023) Short tRNA anticodon stem and mutant eRF1 allow stop codon reassignment. Nature 613(7945):751–758
30. Albanaz ATS, Carrington M, Frolov AO et al (2023) Shining the spotlight on the neglected: new high-quality genome assemblies as a gateway to understanding the evolution of Trypanosomatidae. BMC Genomics 24(1):471
31. Butenko A, Kostygov AY, Sádlová J et al (2019) Comparative genomics of *Leishmania* (*Mundinia*). BMC Genomics 20(1):726
32. Andrews S (2019) FastQC: a quality control tool for high throughput sequence data. http://www.bioinformatics.babraham.ac.uk/projects/fastqc. Accessed 25 Feb 2025
33. Bushnell B, Rood J, Singer E (2017) BBMerge – accurate paired shotgun read merging *via* overlap. PLoS One 12(10):e0185056
34. Allam A, Kalnis P, Solovyev V (2015) Karect: accurate correction of substitution, insertion and deletion errors for next-generation sequencing data. Bioinformatics 31(21): 3421–3428
35. Bankevich A, Nurk S, Antipov D et al (2012) SPAdes: a new genome assembly algorithm and its applications to single-cell sequencing. J Comput Biol 19(5):455–477
36. Kajitani R, Toshimoto K, Noguchi H et al (2014) Efficient *de novo* assembly of highly heterozygous genomes from whole-genome shotgun short reads. Genome Res 24(8): 1384–1395
37. Luo R, Liu B, Xie Y et al (2012) SOAPdenovo2: an empirically improved memory-efficient short-read *de novo* assembler. Gigascience 1(1):18
38. Wick RR (2017) Filtlong. https://github.com/rrwick/Filtlong. Accessed 25 Feb 2025
39. Hackl T, Hedrich R, Schultz J et al (2014) *Proovread*: large-scale high-accuracy PacBio correction through iterative short read consensus. Bioinformatics 30(21):3004–3011
40. Kolmogorov M, Yuan J, Lin Y et al (2019) Assembly of long, error-prone reads using repeat graphs. Nat Biotechnol 37(5):540–546
41. Camacho C, Coulouris G, Avagyan V et al (2009) BLAST+: architecture and applications. BMC Bioinformatics 10:421
42. Buchfink B, Reuter K, Drost HG (2021) Sensitive protein alignments at tree-of-life scale using DIAMOND. Nat Methods 18(4): 366–368
43. Gurevich A, Saveliev V, Vyahhi N et al (2013) QUAST: quality assessment tool for genome assemblies. Bioinformatics 29(8):1072–1075
44. Tegenfeldt F, Kuznetsov D, Manni M et al (2025) OrthoDB and BUSCO update: annotation of orthologs with wider sampling of genomes. Nucleic Acids Res 53(D1):D516–D522
45. Langmead B, Salzberg SL (2012) Fast gapped-read alignment with Bowtie 2. Nat Methods 9(4):357–359
46. Danecek P, Bonfield JK, Liddle J et al (2021) Twelve years of SAMtools and BCFtools. Gigascience 10(2):1–4
47. Kim D, Paggi JM, Park C et al (2019) Graph-based genome alignment and genotyping with HISAT2 and HISAT-genotype. Nat Biotechnol 37(8):907–915
48. Trapnell C, Roberts A, Goff L et al (2012) Differential gene and transcript expression

analysis of RNA-seq experiments with TopHat and cufflinks. Nat Protoc 7(3):562–578
49. Pertea G, Pertea M (2020) GFF utilities: GffRead and GffCompare. F1000Res 9:304
50. Chan PP, Lin BY, Mak AJ et al (2021) tRNAscan-SE 2.0: improved detection and functional classification of transfer RNA genes. Nucleic Acids Res 49(16):9077–9096
51. Laslett D, Canback B (2004) ARAGORN, a program to detect tRNA genes and tmRNA genes in nucleotide sequences. Nucleic Acids Res 32(1):11–16
52. Robinson JT, Thorvaldsdottir H, Winckler W et al (2011) Integrative genomics viewer. Nat Biotechnol 29(1):24–26
53. Carver T, Harris SR, Berriman M et al (2012) Artemis: an integrated platform for visualization and analysis of high-throughput sequence-based experimental data. Bioinformatics 28(4): 464–469
54. Steinbiss S, Silva-Franco F, Brunk B et al (2016) Companion: a web server for annotation and analysis of parasite genomes. Nucleic Acids Res 44(W1):W29–W34
55. Stanke M, Keller O, Gunduz I et al (2006) AUGUSTUS: *ab initio* prediction of alternative transcripts. Nucleic Acids Res 34: W435–W439
56. Valach M, Moreira S, Petitjean C et al (2023) Recent expansion of metabolic versatility in *Diplonema papillatum*, the model species of a highly speciose group of marine eukaryotes. BMC Biol 21(1):99
57. Tashyreva D, Simpson AGB, Prokopchuk G et al (2022) Diplonemids – a review on “new” flagellates on the oceanic block. Protist 173(2): 125868
58. Dlugosz M, Deorowicz S (2017) RECKONER: read error corrector based on KMC. Bioinformatics 33(7):1086–1089
59. Kallenborn F, Hildebrandt A, Schmidt B (2021) CARE: context-aware sequencing read error correction. Bioinformatics 37(7): 889–895
60. Zimin AV, Marcais G, Puiu D et al (2013) The MaSuRCA genome assembler. Bioinformatics 29(21):2669–2677
61. Di Genova A, Buena-Atienza E, Ossowski S et al (2021) Efficient hybrid *de novo* assembly of human genomes with WENGAN. Nat Biotechnol 39(4):422–430

Chapter 9

Assembly and Annotation of Kinetoplastid and Diplonemid Mitochondrial Genomes

Evgeny S. Gerasimov, Matus Valach, Gertraud Burger, Kristína Záhonová, and Vyacheslav Yurchenko

Abstract

This chapter outlines the assembly and annotation of mitochondrial genomes and transcriptomes of kinetoplastids and diplonemids using high-throughput sequencing data. The mitochondrial genomes of these protists are highly atypical, characterized by multipartition and extensive RNA editing. These complexities pose significant challenges for genome assembly, gene prediction, and transcript reconstruction. Here, we describe analytical strategies developed in our laboratories that address these issues.

Key words Kinetoplastea, Trypanosomatids, Diplonemids, Mitochondrion, RNA editing, Next-generation sequencing

1 Introduction

As all euglenozoans, kinetoplastids and diplonemids share in the mitochondrial (mt) DNA alternative genetic code, in which UGA stop codon codes for tryptophan [1]. Although they are sister clades in phylogenetic trees, their mitochondrial genomes differ drastically from each other, calling for dedicated approaches in bioinformatics genome analysis.

1.1 Kinetoplast Mitochondrial Genomes and Kinetoplast (K) DNA

The apomorphy for all kinetoplastid protists (Euglenozoa: Kinetoplastea) is the presence of a large mass of mtDNA called kinetoplast (k)DNA [2, 3], which is (1) compacted and localized in the vicinity of the flagellar pocket (= eukinetoplast) as in trypanosomatids, (2) distributed throughout the mitochondrion in several nearly identical clusters (= polykinetoplast) as in bodonids, or (3) unevenly dispersed as a diffuse mass (= pankinetoplast) as in prokinetoplastids [4–6].

In trypanosomatids, the by-far best studied kinetoplastids because of their medical and economical importance [7], kDNA

Paul A. M. Michels et al. (eds.), *Euglenozoa: Methods and Protocols, Volume 1*, Methods in Molecular Biology, vol. 3013, https://doi.org/10.1007/978-1-0716-5142-1_9,

is comprised of circular double-stranded DNA molecules of two functional types: maxicircles and minicircles [8, 9]. Of note, the situation in other kinetoplastid groups is less clear and warrants further investigation, as it has been recently speculated that parabodonids, for example, contain not circular but linear versions of minicircles [10]. The typical maxicircle of Trypanosomatidae is 20–50 kb in size and present in dozens of identical copies. It encodes mitochondrial genes and is, therefore, an analog of mtDNA of other organisms, but not without peculiarities. Maxicircle can be divided into two regions. The coding region (CR), which is tightly packed with compactly organized and typically very syntenic genes among trypanosomatid species, is typically defined as the sequence that extends from the 12S ribosomal RNA (*rRNA*) gene to the *ND5* gene [11–13]. These genes contain extremely short intergenic spacers or may even overlap in some species. For naming convention of kinetoplastids' mtDNA genes, *see* **Note 1**.

The remainder of the maxicircle sequence is termed the divergent region (DR) and comprises arrays of repeats of various length and composition, its sequence assembly and analysis being challenging [14]. The DR cannot be directly assembled from short sequencing reads, therefore most of the maxicircle sequences deposited in nucleotide databases contain only the CR with short flanks. However, recent advances in sequencing technologies (especially long-read platforms) have resolved this issue.

The bizarre molecular feature of trypanosomatid kinetoplast is U-insertion/deletion RNA editing that modifies many primary maxicircle transcripts [15–17]. Edited genes are termed cryptogenes. Some transcripts are edited so extensively that over a half of their translatable sequence is created by inserted Us. Examples of extensively (pan-) edited genes are *RPS12*, *ND3*, *ND8*, *ND9*, *G3*, and *G4* [18], but some genes tend to change the editing degree over the course of evolution [19]. Insertions and deletions carried out by the RNA editing machinery are guided by special small guide RNAs (gRNAs), which are typically encoded by minicircles [20–22]. Minicircles are typically small (ranging between 600 and 2000 bp in most species, but with some notable exceptions [9, 23]) and present in thousands of copies in the kDNA. The common characteristic of these molecules is the high heterogeneity within sequence classes, although minicircles often have characteristic common sequence motifs called conserved sequence blocks (CSBs) [24–28]. Sequence and copy-number heterogeneity of minicircles impose algorithmic difficulties on the kDNA repertoire assembly because many routinely used methods rely on uniformity of coverage and do not work properly when a common sequence block is present in different sequences.

U-insertion/deletion RNA editing also complicates the analysis of the mitochondrial transcriptome. From the algorithmic point of view, the edited RNA is very different from its genomic cryptogene sequence so that "general purpose" read alignment algorithms fail to build accurate read alignments. Another problem is that a variety of partially and alternatively edited transcripts (or isoforms) are found, which are very difficult to distinguish [29]. This led to the development of several dedicated analysis toolkits (i.e., T-Aligner [30] or TREAT [31]), which implement editing-aware algorithms and address problems that arise from extreme transcriptome complexity and the presence of multitude erroneous editing products. Hereafter, we will focus on usage of T-Aligner for RNA editing analysis of trypanosomatids.

1.2 Diplonemid Mitochondrial Genomes

The mtDNAs of diplonemids are also multi-partite but organized very differently compared to kinetoplastids. The mitochondrial genome of the type species, *Paradiplonema papillatum* (originally referred to as *Diplonema papillatum* [32]), comprises 81 distinct circular molecules that are 6 and 7 kb long. Both circle classes mostly contain repeats with ~80% sequence identity across the groups [33]. In addition, the copy numbers of the various classes can differ by more than 100-fold [34]. As with kinetoplastid mini-circles discussed above, the assembly of diplonemid mtDNA from shotgun and short-read data is virtually impossible. It can only be achieved in certain species on rare occasions when circle classes are unique and lack longer repeats [35]. The diplonemid mtDNA circles also include coding sequence, albeit extraordinarily short, as diplonemid mitochondrial genes are fragmented in an unprecedented way. For example, in *P. papillatum*, mitochondrial genes are split into up to 11 pieces, with sizes between ~40 and 540 bp, each encoded on a separate molecule. These gene pieces (termed modules) are transcribed separately and then assembled by RNA ligation into mature mRNAs and rRNAs. Due to their small size, identifying coding sequence in mtDNA can be difficult, but the major hurdle of gene identification is the massive RNA editing operating in diplonemid mitochondria. *P. papillatum* has two types of RNA editing. One is substitutions of As by Is and Cs by Us, i.e., deaminations, with densely packed clusters of up to 50% edited positions over a 100 bp stretch. The other type is U-appendage editing, where up to ~50 Us are added to the 3′-end of module transcripts, prior to their assemblage into mature RNAs [36]. Gene identification, therefore, requires transcriptome information. As mentioned in the case of kinetoplastids, the sometimes-considerable DNA-RNA differences complicate the alignment of RNA reads to genome sequence, and, thus, the assignment of transcripts to their coding regions.

The above-given examples of genome multipartition, gene fragmentation, and RNA editing come from *P. papillatum*, a species with relatively moderate deviations in mitochondrial genome architecture, gene structure, and encryption. Other diplonemids, especially hemistasiids, show far more extreme features in all three aspects, further complicating in silico analyses of their mtDNA [35].

The methodological approaches described below will be divided into two parts because of the fundamental differences between kinetoplastid and diplonemid mtDNA. Yet, we believe that it is beneficial to present them within one chapter devoted to the analysis of mt genome of Euglenozoa.

2 Materials

2.1 Software

BBTools (e.g., BBDuk, BBMap) [37], BEDtools [38], BLAST+ [39], Bowtie2 [40], BWA [41, 42], exonerate [43], FastQC [44], fastp [45], Flye [46], FreeBayes [47], Geneious [48], MAFFT [49], SAMtools [50], SPAdes [51], T-Aligner [30], Trinity [52], VCFtools [53]. Kinetoplastids' RNA editing pipeline is better run under native Linux with python 3.10+ and packages "numpy", "matplotlib", "pandas", and "scipy" installed.

2.2 Sequencing Data

For the analysis of kinetoplastid mtDNA, short sequencing Illumina paired-end HiSeq 150 nt or MiSeq 250 nt reads are convenient. Isolation of DNA material from the mitochondria-enriched fraction is highly recommended, but total DNA sequenced to a median nuclear genome coverage of 200× (assuming a size of 20–40 Mb) is usually sufficient to obtain a quasi-complete mitochondrial genome sequence. For species with large (longer than 2 kb) minicircles or for full-length maxicircle assembly, long sequencing reads are necessary. For RNA sequencing, Illumina paired-end HiSeq 150 nt reads or MiSeq 250 nt reads, poly(A)-enriched reads or reads from mitochondria-enriched fraction are preferred. Total RNA sequencing without enrichment and selection usually contains only a few edited reads, which limits the downstream analyses.

For the analysis of diplonemid mtDNA, we recommend Illumina MiSeq paired-end reads, but for systematic full-length chromosome assembly, long sequencing reads (Oxford Nanopore Technologies or alternatively, PacBio) are required. For RNA sequencing, we recommend Illumina HiSeq paired-end strand-specific reads produced from poly-A RNA fraction, but total RNA libraries may be necessary to identify mitochondrial rRNA genes.

3 Methods

3.1 Analysis of mtDNA of Kinetoplastids

3.1.1 Quality Control

1. Check read base quality and adapter presence with FastQC:

```
mkdir fastqc_reports
fastqc -t 4 -o fastqc_reports <read_files_*fastq>
```

2. Trim low-quality bases and technical adapter sequences with fastp.

```
fastp -a <adapter_seq> -l 100 -q 25 -o <trimmed_R1.fastq> -O
<trimmed_R2.fastq> -i <totrim_R1.fastq> -I <totrim_R2.fastq>
```

Use adapter sequence **<adapter_seq>** from the sequencing kit's manual. Alternatively, do not use **-a** option, instead, rely on fastp's feature of automatic adapter detection or provide adapters file with **--adapter_fasta <file_with_adapters.fasta>** option.

3. Estimate the relative abundance of mtDNA reads by counting CSB3 sequence (common for most minicircles, but can slightly diverge from consensus in different species) with:

```
grep -cP "GGGGTTGGTGTA" <read_R1.fastq>
```

Since DNA sequencing is not strand-specific, you can probe only forward sequence to get a rough estimate.

3.1.2 Maxicircle Assembly

1. Assemble a mitochondrial maxicircle using one of the following strategies depending on what type of input data you have.

1.1. Assemble the coding region of maxicircle using short sequencing (e.g., Illumina) reads with SPAdes:

```
spades.py -t 4 --pe1-1 <read_R1.fastq> --pe1-2 <read_R2.fastq>
-o <spades_out> --disable-gzip-output [--isolate]
```

Adding the **--isolate** option can usually give slightly better results. Full maxicircle assembly is only possible with long PacBio or Oxford Nanopore sequencing reads (see next step).

1.2. Assemble full-size maxicircle from long PacBio sequencing reads:

```
flye --genome-size <gs> --pacbio-raw <pacbio_reads.fastq> --
out-dir <flye_out> --threads 8
```

or long Oxford Nanopore reads:

```
flye --genome-size <gs> --nanopore <nanopore_reads.fastq> --
out-dir <flye_out> --threads 8
```

See **Note 2** on how to set the **<gs>** parameter.

1.3. Perform hybrid assembly of maxicircle using short and long sequencing reads in single command:

```
spades.py --threads 10 --disable-gzip-output --pacbio <pac-
bio_reads.fastq> --pe1-1 <illumina_R1.fastq> --pe1-2 <illumi-
na_R2.fastq> -o <spades_hybrid>
```

Further steps will be performed using the output of **step 1.1** but can be easily adjusted for two other cases.

2. Locate maxicircle contig using 12S and ND5 homologues from other species (*Leishmania tarentolae* or *Trypanosoma brucei*). Firstly, prepare nucleotide blast database from assembly scaffolds file:

```
makeblastdb -in <spades_out>/scaffolds.fasta -out <spades_-
out>/scaffolds.db -dbtype nucl
```

3. Search for marker genes in assembly scaffolds with BLASTn and figure out the scaffold entry name, which corresponds to a potential maxicircle (usually both searches hit the same contig/scaffold):

```
blastn -query <ND5_sequences.fasta> -db <spades_out>/scaf-
folds.db -evalue 1e-6 -word_size 5 -max_target_seqs 5 -outfmt
6
blastn -query <12S_sequences.fasta> -db <spades_out>/scaf-
folds.db -evalue 1e-6 -word_size 5 -max_target_seqs 5 -outfmt
6
```

4. Extract the assembled maxicircle scaffold:

```
samtools faidx <spades_out>/scaffolds.fasta <NODE_XXX_ful-
l_name> > <maxicircle.fasta>
```

5. (If needed) Reverse-complement the maxicircle sequence to put 12S rRNA gene on the forward strand:

```
seqtk seq -r <maxicircle.fasta> > <maxicircle_final.fasta>
```

3.1.3 Maxicircle Annotation

Use SnapGene or similar software to annotate the ~18 mitochondrial genes. SnapGene allows viewing the sequence and in silico translations of six frames.

1. Find open reading frames (ORFs) corresponding to never-edited genes. These genes are encoded as continuous ORFs so they can be detected after in silico translation of all possible ORFs longer than 100 amino acids using “Mold, protozoan” genetic code table (usually equivalent of NCBI’s translation table #4). At this step, *ND1*, *ND4*, *ND5*, *CO1*, *CO3*, *MURF1*, and *MURF2* will be annotated for most trypanosomatid genomes.

2. Detect rRNA gene (*12S* and *9S*) boundaries by best local alignment with known *12S* and *9S* sequences and translate detected coordinates on the maxicircle:

```
exonerate --bestn 1 <maxicircle_final.fasta> <12S.fasta>
```

Exonerate will output the best alignment, giving the target and query ranges included into the alignment. Usually, the middle part of ribosomal RNA will be included in the alignment. Expand this region by adding the unaligned nucleotides of the query to get full gene length.

3. Inspect 5′-truncated ORFs of partially edited genes. Use synteny with known sequences of *L. tarentolae* (GenBank accession M10126.1) or *T. brucei* (GenBank accession M94286.1) as guidance. Most of these genes (e.g., *ND7*, *A6*, *CYb*) will have short edited domains on the 5′. Mark these incomplete ORFs with flanks of ~100 nt, which must include the edited domain.
4. Check and mark G-rich regions corresponding to heavily edited cryptogenes (usually, *ND3*, *ND8*, *ND9*, *G3*, *G4*, and *RPS12*). The typical size of the marked region will be ~250 nt.
5. Prepare a tabulated (gff/gtf-like) file <**annotation.gff**>, that will include exact boundaries and strand information ("+" or "–") of genes annotated in **steps 1** and **2**, and the approximate boundaries of cryptogenes annotated in **steps 3** and **4**.
6. Get annotated sequences in FASTA format:

```
bedtools getfasta -s -fi <maxicircle_final.fasta> -bed <annotation.gff> > <annotated_references.fasta>
```

This file will be used to reconstruct edited mRNAs and refine gene boundaries with T-Aligner.

3.1.4 Minicircles Assembly

1. Assemble an initial set of DNA contigs with SPAdes:

```
spades.py -t 4 --pe1-1 <read_R1.fastq> --pe1-2 <read_R2.fastq> -o <spades_minicircles> --disable-gzip-output --isolate -k 25,41,81,101,127
```

2. Assemble an initial set of DNA contigs with megahit:

```
megahit -1 <read_R1.fastq> -2 <read_R2.fastq> -t 8 -o <megahit_minicircles>
```

3. Extract minicircle contigs using marker sequence. Usually, the best is to use CSB3 sequence as marker:

```
cat <spades_minicircles>/scaffolds.fasta <megahit_minicircles>/final.contigs.fa | awk 'BEGIN{data=""}{if(substr($1,1,1) == ">") {print data; data=$1 "\n"} else {data=data $1;}}END{print data}' | grep -B 1 -P "GGGGTTGGTGTA|TACACCAACCCC" | grep -v -P "--" - > <possible_minicircles.fasta>
```

4. Filter **<possible_minicircles.fasta>** file by expected sequence length and/or by the presence of other marker minicircle sequences (for example, a typical size range of 600–800 and CSB1/2 block sequences can be used in case of *Leishmania* spp.).
5. (Alternative) Follow the KOMICS [54] package tutorial at https://github.com/FreBio/komics. KOMICS is a specialized package for minicircle repertoire assembly from short sequencing reads using megahit as an internal assembler. It also offers the rKOMICS R package for subsequent data analysis [55], which works particularly well with *Leishmania*-like monomeric minicircles (*see* **Note 3** about the assembly of large minicircles).

3.1.5 Transcriptome Assembly

1. Join all sequencing reads in a single file:

```
cat <read_R1.fastq> <read_R2.fastq> > <joined_rnaseq.fastq>
```

If you have two or more RNAseq libraries, join all of them.

2. Extract reference cryptogene sequence with flanks using coordinates obtained in Subheading 3.1.3:

```
echo -e"<Maxicircle_contig_name>\t<gene_start-50>\t<gene_end+50>\t.\t.\t<gene_strand+/->" | bedtools getfasta -s -fi <maxicircle_final.fasta> -bed - > <cryptogene_ref.fasta>
```

3. Align RNAseq data on the reference sequence with T-Aligner's `alignlib` tool:

```
~/t-aligner/alignlib --in_ref <cryptogene_ref.fasta> --in_lib <joined_rnaseq.fastq> --out_prefix <t_aligner_out_>
```

The output of this step will include mapped reads in ". fastq" format and in T-Aligner-specific ".taf" format, which can be used to visualize cryptogene coverage and editing states with auxiliary scripts and perform other downstream analyses.

4. Visualize editing states and read coverage depth for the reference cryptogene:

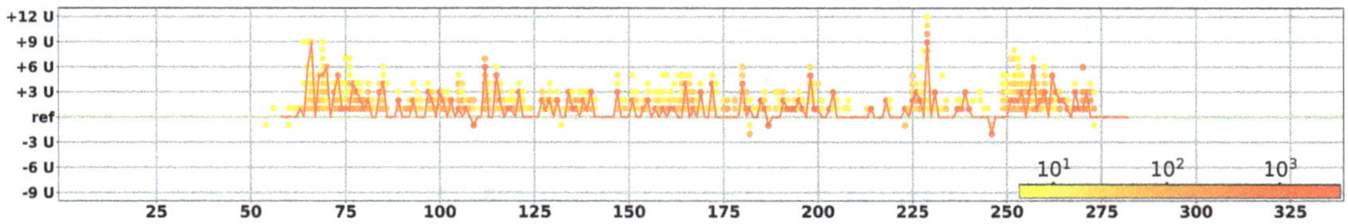

Fig. 1 A dot matrix plot for the *Vickermania spadyakhi ND8* cryptogene produced by T-Aligner's coverage.py script. The dot matrix is a heatmap, where dot intensities reflect read support and dots reflect editing events (termed "editing states"). The unedited cryptogene is drawn by green baseline ($Y = 0$), the editing states connected with U insertion are depicted by dots above this line ($Y > 0$), and deletions of U are shown below the line ($Y < 0$). The Y scale is the number of inserted/deleted Us, the X scale is the number of A/G/C nucleotide of analyzed reference sequence. As editing can modify the number of Us in each position of the reference, the default T-Aligner's approach is to ignore each T in a given DNA reference sequence and operate with a so-called T-less profile; the same approach is good for plotting the results. The solid red line depicts the path that reflects the editing events needed to produce a canonical fully edited *ND8* mRNA

```
python ~/t-aligner/coverage.py --r <cryptogene_ref.fasta> --t
<t_aligner_out_mapped_reads.taf> --o <dotmatrix_plot.png> --g
dots
python ~/t-aligner/coverage.py --r <cryptogene_ref.fasta> --t
<t_aligner_out_mapped_reads.taf> --o <coverage_plot.png> --g
bars
```

These plots give an overall view of the transcription and editing events for the studied cryptogene. Most plotting tools from T-Aligner package rely on ".taf" format mappings. The dot matrix visualization of editing events (Fig. 1) is a powerful feature of the package that gives an overview of all editing events (with their support and interrelations), which is extremely useful for studies of complex transcripts pools generated by RNA editing.

5. Assemble transcripts with arbitrary long ORFs with `findorfs` command:

```
~/t-aligner/findorfs --in_lib <t_aligner_out_mapped_reads.
fastq> --in_ref <cryptogene_ref.fasta> --out_prefix <assemble-
d_orfs_> --orf_tracing_mode extension --orf_min_orf_aa 80 --
orf_filter_esd 10
```

This step will assemble all transcripts for a given cryptogene, including multitude of isoforms and partially edited products. Canonical edited mRNA will be usually assembled too. *See* **Notes 4** and **5** for more details on `findorfs` command and how to locate canonical edited mRNA sequence.

3.1.6 gRNA Annotation

1. Predict gRNA-coding loci by alignment using `find_grna` tool from T-Aligner package and edited mRNA sequence(s) from Subheading 3.1.5.

```
find_grna --mrna <canonical_mRNA.fasta> --grna <minicircles.
fasta> --seed_score 20 --seed_length 16 --length 21 --score
25 --gu 15 --mm 4 --anchor 5 > <fow.out> 2> /dev/null
```

The command predicts gRNAs by alignment on the given strand of minicircles. Allowed number of mismatches (--**mm**), G:U pairs (--**gu**), minimal alignment length (--**length**), anchoring region of gRNA (--**anchor**), minimal alignment score (--**score**), seed alignment length (--**seed_length**), and score (--**seed_score**) are controlled in this step. *See* **Notes 6** and **7** for more information.

2. Validate alignments with small RNA sequencing data (if available). Trim and merge small RNA sequencing reads:

```
fastp -i <small_rna1.fastq> -I <small_rna2.fastq> -q 25 -l
15 --merge --merged_out <merged.fq> --out1 <unmerged_1.fq> --
out2 <unmerged_2.fq> --thread 10
```

3. Index minicircles assembly:

```
bwa index -p <minicircles.bwa_index> <minicircles.fasta>
```

4. Map trimmed and merged reads onto a minicircle assembly:

```
bwa mem -t 10 <minicircles.bwa_index> <merged.fq> | samtools
sort -@ 4 -o <mapped.bam>
```

5. Process alignments and convert to FASTA format:

```
samtools index <mapped.bam>
samtools view <mapped.bam> | awk '{print ">" $1 "\n" $10; }' >
<mapped.fasta>
```

6. Cut polyA/U tails from the mapped reads:

```
for i in {0..10}; do echo Trimming gRNA ends cycle ${i}; sed -i
-e 's/^AAAA/A/1' <mapped.fasta>; sed -i -e 's/TTTT$/T/1'
<mapped.fasta>; done
```

7. Cluster sequences with `cd-hit` to produce clusters, which correspond to gRNAs:

```
cd-hit-est -i <mapped.fasta> -o <clustered.out> -c 0.90 -s
0.80 -d 40 -M 90000 -T 30 -n 8
```

From the clustered output file <**clustered.out**> pick gRNA sequences as clusters with 10+ sequences and length of 20–70 nt.

8. Compare resulting set of gRNAs predicted from alignment (**step 1**) with set of gRNAs obtained from small RNA sequencing data (**step 7**). Note that not all gRNAs predicted by alignment must be necessarily supported by small RNA sequencing for a variety of reasons (e.g., coverage and representation problems of small RNA library).

3.2 Analysis of mtDNA of Diplonemids

3.2.1 Read Processing

1. Check the quality of DNA- and RNA-Seq reads by FastQC as in **step 1** of Subheading 3.1.1.
2. Trim the reads with BBDuk to remove adaptors and low-quality reads. It is not necessary to unpack the read files beforehand. The typical command for trimming of paired-end short reads is:

```
bbduk in1=<forward_1.fastq.gz> in2=<reverse_2.fastq.gz>
out1=<trimmed_1.fastq.gz> out2=<trimmed_2.fastq.gz> ref=a-
dapters.fa usejni=t qtrim=rl trimq=20 ktrim=r k=22 mink=11
hdist=2 tpe tbo t=10 2> <bbduk_report.log>
```

The exact description of parameters can be found in the BBDuk manual.

3. Check the quality of trimmed reads by FastQC using the command from **step 1** of Subheading 3.1.1.

3.2.2 Genome and Transcriptome Assembly

1. Assemble DNA-Seq trimmed reads using SPAdes:

```
spades.py --pe1-1 <trimmed_DNA_1.fastq.gz> --pe1-2 <trimmed_D-
NA_2.fastq.gz> --careful -t 40 -o <spades_output_directory>
```

If using Nanopore or Pacbio data, refer to **step 1.2** of Subheading 3.1.2 (and also *see* **Note 2**).

2. Assemble RNA-Seq trimmed reads by Trinity:

```
Trinity --seqType fq --left <trimmed_RNA_1.fastq.gz> --right
<trimmed_RNA_1.fastq.gz> --output <trinity_output_directory>
--max_memory 100G --CPU 30
```

3.2.3 Identification of Mature Transcripts and Genomic Modules

1. Prepare nucleotide databases from the assembled genome and transcriptome:

```
makeblastdb -in <spades_output_directory>/scaffolds.fasta
-dbtype nucl
makeblastdb -in <trinity_output_directory>/Trinity.fasta
-dbtype nucl
```

2. Identify mature transcripts by BLASTx searches using previously identified diplonemids' proteins as queries against the prepared transcriptomic database:

```
blastx -query <diplonemids_proteins.fasta> -db <trinity_output_directory>/Trinity.fasta -out <transcripts.tsv> -outfmt 6
```

3. Extract identified transcripts from the assembled transcriptome to a separate file (**<identified_transcripts.fasta>**).
4. Translate found transcripts into protein sequences using "Mold, protozoan" genetic code (NCBI's translation table #4).
5. Identify gene modules by BLASTn searches using identified transcripts as queries against the prepared genomic database:

```
blastn -query <identified_transcripts.fasta> -db <spades_output_directory>/scaffolds.fasta -out <modules.tsv> -outfmt 6
```

6. Extract identified gene modules from the assembled genome to a separate file (**<identified_modules.fasta>**).
7. Map identified modules to a corresponding transcript to corroborate module assignments (*see* **Note 8**).

3.2.4 Chromosome Classification

1. Extend the module-containing contigs by read mapping using BBMap or Geneious built-in mapper (*see* **Note 8**).
2. Compare module-containing contigs among each other using BLASTn:

```
blastn -query <identified_modules.fasta> -db <spades_output_directory>/scaffolds.fasta
```

   This identifies cassettes and constant regions. Cassettes represent unique sequences up- and downstream of modules, which are flanked by constant region that exhibit over 90% identity over 100 bp adjacent to the cassettes. Extract cassette sequences into a separate file (**<identified_cassettes.fasta>**).
3. Assign contigs having the same cassette-flanking regions to the same chromosome class.
4. Order classes from the highest to lowest count of chromosomes and name them A, B, etc.
5. Align contigs of one class using MAFFT:

```
mafft --auto --adjustdirection <one_class_contigs.fasta> > <one_class_contigs_aligned.fasta>
```

6. Place cassette's boundaries at positions where sequence identities drop below 90% over 100 bp.

3.2.5 Identification of DNA Polymorphic and RNA Editing Sites

1. Preliminary RNA editing site identification can be achieved by comparing mature transcripts with aligned genomic modules from **step 7** of Subheading 3.2.3. However, this is only possible when the mitochondrial genome (and consequently transcriptome) does not exhibit a high degree of DNA polymorphism, which is particularly prevalent in hemistasiids [34, 35, 56]. For this reason, we recommend performing variant calling on DNA-Seq and RNA-Seq reads to identify DNA-DNA differences and RNA-DNA differences in a systematic manner as detailed in the following.
2. Map trimmed DNA-Seq reads using Geneious built-in mapper or BBMap (*see* **Note 8**) onto mitochondrial genomic contigs containing annotated cassettes (i.e., the file <**identified_cassettes.fasta**>).

   For Geneious, use the following parameters:

```
Save assembly report; Save list of used reads (include mates);
Save contigs; Do not trim; Map multiple best matches to all;
Only map paired reads that both map; Allow at maximum 2% read
gaps; Maximum gap size 4; Word length 48; Index word length 14;
Minimum overlap identity 95%; Allow at maximum 2% mismatches
per read; Maximum ambiguity 4
```

   For BBMap (e.g., 100 bp reads with library insert sizes <700 bp, "very sensitive" preset):

```
java -Xmx4g -cp <path_to_directory> align2.BBMap pairlen=500
rescuedist=1000 maxindel=2 ambiguous=all k=8 saa=f vslow=t
ref=<identified_cassettes.fasta> out=<DNA_reads_to_identi-
fied_cassettes>.sam nodisk in1=<reads1.fastq> in2=<reads2.
fastq> qin=33
```

3. Generate a sorted BAM from the SAM read-alignment file:

```
samtools view -S -b <DNA_reads_to_identified_cassettes.sam> >
<DNA_reads_to_identified_cassettes.bam>
samtools sort -o <DNA_reads_to_identified_cassettes-sorted.
bam> <DNA_reads_to_identified_cassettes.bam>
samtools index <DNA_reads_to_identified_cassettes-sorted.bam>
```

4. Call DNA-DNA differences using FreeBayes. This can be also done within Geneious, which includes a FreeBayes distribution, or using a stand-alone FreeBayes installation:

```
freebayes -f <identified_cassettes.fasta> --ploidy 100 --min-mapping-quality 30 --min-alternate-count 2 --min-alternate-fraction 0.01 --use-duplicate-reads <DNA_reads_to_identified_cassettes-sorted.bam> > <dna_dna_differences.vcf>
```

Assign as DNA polymorphic sites those that possess at least two different nucleotides in ≥10% of the DNA-Seq reads (*see* also **Note 9**).

5. Map trimmed RNA-Seq reads using the Geneious built-in mapper or BBMap (*see* **Note 8**) onto mitochondrial identified transcripts (*see* **Note 10**).

   When using Geneious:

```
Save assembly report; Save list of used reads (include mates); Save contigs; Do not trim; Map multiple best matches to all; Allow at maximum 10% read gaps; Maximum gap size 4; Word length 20; Index word length 14; Allow at maximum 10% mismatches per read; Maximum ambiguity 8
```

When using BBMap (e.g., 100 bp reads with library insert sizes <400 bp, "very sensitive" preset):

```
java -Xmx4g -cp /path/to/directory align2.BBMap pairlen=200 rescuedist=400 maxindel=2 ambiguous=all k=8 saa=f vslow=t ref=<identified_transcripts.fasta> out=<RNA_reads_to_identified_transcripts.sam> nodisk in1=<reads1.fastq> in2=<reads2.fastq> qin=33
```

6. Generate a sorted BAM from the SAM read-alignment file:

```
samtools view -S -b <RNA_reads_to_identified_transcripts.sam> > <RNA_reads_to_identified_transcripts.bam>
samtools sort -o <RNA_reads_to_identified_transcripts-sorted.bam> <RNA_reads_to_identified_transcripts.bam>
samtools index <RNA_reads_to_identified_transcripts-sorted.bam>
```

7. Call RNA-DNA differences using FreeBayes:

```
freebayes -f <identified_transcripts.fasta> --ploidy 100 --min-mapping-quality 30 --min-alternate-count 2 --min-alternate-fraction 0.01 --use-duplicate-reads <RNA_reads_to_identified_transcripts-sorted.bam> > <rna_dna_differences.vcf>
```

8. To identify all genuine editing sites, i.e., variant positions present among the RNA-DNA differences, but absent from among the genomic polymorphisms, intersect the VCF files of RNA-DNA and DNA-DNA differences with the tool vcf-isec of the VCFtools suite:

```
vcf-isec -c <rna_dna_differences.vcf> <dna_dna_differences.vcf> > <rna_editing_sites.vcf>
```

Assign as the default RNA editing sites those are present in ≥50% of the RNA-Seq reads.

4 Notes

1. For historical reasons, the kinetoplastid mtDNA genes still use slightly different naming than in other eukaryotes although representing homologous sequences. Genes encoding NAD$^+$ subunits of NADH dehydrogenase (complex I) are abbreviated as *ND*, cytochrome *b* (complex III) as *CYb* instead of *cob*, *cox* subunits of cytochrome *c* oxidase (complex IV) as *CO*, and the ATP6 subunit of the ATP synthase (complex V) as *A6*. Moreover, several genes were reported as having unknown function and were named MURF (mitochondrial unassigned reading frame) and CR (C-rich region) [57, 58]. With more advanced HMM analysis and structural modeling, MURF1 was assigned to *nad2* (or ND2 in the kinetoplastid terminology), MURF5 to *rps3*, CR3 to *nad4L* (or G3), CR4 to *nad6* (or G4), and CR5 to *nad3* (or ND3) [59–61].
2. When assembling the genome with **flye**, the tool requires an estimate of genome size, i.e., the expected total assembly size. If you use mitochondrial fraction enrichment, set the **--genome-size** parameter to the expected maxicircle size (e.g., "***35k***" for *Leishmania*). If you use a total-cell DNA library, set this parameter to "***32m***", which is a typical nuclear genome size of trypanosomatid. For diplonemids, the typical values are "***500k***" when using enriched mitochondrial DNA and "***250m***" when starting from total cell DNA.
3. The assembly of large minicircles from species like *Vickermania ingenoplastis* is usually possible only with long sequencing reads. Flye assembler can be used with default settings, though genome length should be set to the value around "***50m***".
4. The key option of the **find_orfs** tool is --**orf_tracing_mode**, which determines the preferrable way of reads overlap. It can take the value "extension" (prefers best extension of ORF, default, usually gives best results), "overlap" (more confident, best read overlap preferred), "editing" (prefers extension with

read sequence that has maximal number of edited sites out of all possible, can work best for pan-edited cryptogenes), or "coverage" (prefers extension passing through most supported editing states). The option --**orf_filter_esd** filters out unedited reads (reads with 10+ edits will be used by example command above), for partially edited cryptogenes (with 5′-short edited domains, such as in the case of *Leishmania* CYb), this option should not be used! Other useful options are --**orf_min_overlap** and --**orf_min_extension**, which control minimal overlap and extension and are set to 10 by default. It is recommended to increase these values to 15 or 20 for longer reads (MiSeq platform) or for very deeply covered genes. Option --**orf_output_orf_alignments true** will output alignment of assembled transcript against cryptogene sequence and generate a separate. fasta file for each assembled transcript.

5. The output of the `find_orfs` tool usually includes various partially edited and alternatively edited mRNA sequences assembled from sequencing data. The priority task is to identify the canonically edited mRNA. The most straightforward approach is to use the predicted translated proteins output file of `find_orfs` tool as a BLAST database (use `makeblastdb` to create it) and a set of known homologous sequences for model species (*L. tarentolae*, *T. brucei*, and others) as queries.
6. Note that `find_grna` models mRNA:gRNA alignments only for a given strand of assembled minicircles. That means that if you expect that both strands of minicircles can encode gRNAs, you should also run the tool with reverse-complemented minicircles assembly file as input.
7. Alignment gRNA:mRNA modelling is controlled by command line options of `find_grna` tool. Allow less mismatches and G:U pairs, and set longer seed with higher seed score to produce more strict alignments, which will reflect longer gRNAs (suits well for *Leishmania*). In contrast, for species with short and divergent gRNAs use short seed and decrease the alignment length and maximal score, which will allow to model alignments with higher mismatch/G:U per nucleotide density.
8. Our best experience has been with the built-in aligner and mapper of the Geneious software. However, since this is not an open-source software, other software (e.g., BBMap [37] or Bowtie2 [40]) may be used to essentially the same effect.
9. Some apparent polymorphisms can arise from mitochondrial sequences that were transferred to the nuclear genome (a.k.a. NUMTs). These can even be expressed from the nuclear genome because many localize to 3′ UTRs, as documented extensively for *P. papillatum* [62]. This confounding effect is

especially important for low-copy-number mitochondrial chromosomes.

10. The identified mitochondrial transcript sequences (**identified_transcripts.fasta**) represent the edited transcriptome reference. While the Geneious built-in mapper is usually able to properly map essentially all RNA-Seq reads whether they originate from fully edited, partially edited, or pre-edited transcripts, other tools (e.g., BBMap, Bowtie2, minimap2) require a virtual sequence representing the pre-edited transcriptome reference, especially when using a short-read technology. This virtual reference is created by concatenating genomic module sequences of individual genes in the same order as in the identified transcripts. Thus, when using Bowtie2 or similar tools, map RNA-Seq reads to both references and then merge the resulting BAM files using the edited transcripts as the default reference. When interested in examining substitution RNA editing intermediates, the most practical approach is to map reads onto a virtual sequence, in which known sites of, e.g., A-to-I and C-to-U editing have been replaced by R and Y, respectively; most tools accept such "degenerate" references, even if read mapping generally will not be as efficient as when using the virtual pre-edited reference or the edited transcriptome reference, especially for editing clusters that are longer than half the read size.

Acknowledgments

The work on kinetoplastids was primarily supported by the European Union's Operational Program "Just Transition" (LERCO CZ.10.03.01/00/22_003/0000003 to V.Y.) and the Czech Ministry of Education, Youth and Sports (INTER-EXCELLENCE-LUASK22033 to V.Y.). The work on diplonemids was supported by the Fonds de Recherche du Québec—Nature et Technologies (FRQNT; grant 2018-PR-206806 to G.B.) and the Natural Sciences and Engineering Research Council of Canada (NSERC; grants RGPIN-2014–05286 and RGPIN-2019–04024 to G.B.). Computational resources were provided by the e-INFRA CZ project (ID:90254) supported by the Czech Ministry of Education, Youth and Sports.

References

1. Záhonová K, Lax G, Leonard G et al (2021) Single-cell genomics unveils a canonical origin of the diverse mitochondrial genomes of euglenozoan. BMC Biol 19:103
2. Vickerman K (1976) Comparative cell biology of the kinetoplastid flagellates. In: Vickerman K, Preston TM (eds) Biology of Kinetoplastida, vol 1. Academic, London, pp 35–130

3. Kostygov AY, Karnkowska A, Votýpka J et al (2021) Euglenozoa: taxonomy, diversity and ecology, symbioses and viruses. Open Biol 11: 200407
4. Lukeš J, Guilbride DL, Votýpka J et al (2002) Kinetoplast DNA network: evolution of an improbable structure. Eukaryot Cell 1(4): 495–502
5. d'Avila-Levy CM, Boucinha C, Kostygov A et al (2015) Exploring the environmental diversity of kinetoplastid flagellates in the high-throughput DNA sequencing era. Mem Inst Oswaldo Cruz 110(8):956–965
6. Moreira D, López-García P, Vickerman K (2004) An updated view of kinetoplastid phylogeny using environmental sequences and a closer outgroup: proposal for a new classification of the class Kinetoplastea. Int J Syst Evol Microbiol 54:1861–1875
7. Stuart K, Brun R, Croft S et al (2008) Kinetoplastids: related protozoan pathogens, different diseases. J Clin Invest 118(4):1301–1310
8. Jensen RE, Englund PT (2012) Network news: the replication of kinetoplast DNA. Ann Rev Microbiol 66:473–491
9. Yurchenko V, Hobza R, Benada O et al (1999) *Trypanosoma avium*: large minicircles in the kinetoplast DNA. Exp Parasitol 92(3): 215–218
10. Gerasimov ES, Afonin DA, Korzhavina OA et al (2022) Mitochondrial RNA editing in *Trypanoplasma borreli*: new tools, new revelations. Comput Struct Biotechnol J 20:6388–6402
11. Camacho E, Rastrojo A, Sanchiz A et al (2019) *Leishmania* mitochondrial genomes: maxicircle structure and heterogeneity of minicircles. Genes 10(10):758
12. Afonin DA, Gerasimov ES, Škodová-Sveráková I et al (2024) *Blastocrithidia nonstop* mitochondrial genome and its expression are remarkably insulated from nuclear codon reassignment. Nucleic Acids Res 52:3870–3885
13. Berná L, Greif G, Pita S et al (2021) Maxicircle architecture and evolutionary insights into *Trypanosoma cruzi* complex. PLoS Negl Trop Dis 15(8):e0009719
14. Gerasimov ES, Zamyatnina KA, Matveeva NS et al (2020) Common structural patterns in the maxicircle divergent region of Trypanosomatidae. Pathogens 9(2):100
15. Benne R, Van den Burg J, Brakenhoff JP et al (1986) Major transcript of the frameshifted *coxII* gene from trypanosome mitochondria contains four nucleotides that are not encoded in the DNA. Cell 46(6):819–826
16. Read LK, Lukeš J, Hashimi H (2016) Trypanosome RNA editing: the complexity of getting U in and taking U out. Wiley Interdiscip Rev RNA 7(1):33–51
17. Maslov DA, Opperdoes FR, Kostygov AY et al (2019) Recent advances in trypanosomatid research: genome organization, expression, metabolism, taxonomy and evolution. Parasitology 146(1):1–27
18. Koslowsky D, Sun Y, Hindenach J et al (2014) The insect-phase gRNA transcriptome in *Trypanosoma brucei*. Nucleic Acids Res 42(3): 1873–1886
19. Gerasimov ES, Kostygov AY, Yan S et al (2012) From cryptogene to gene? *ND8* editing domain reduction in insect trypanosomatids. Eur J Protistol 48(3):185–193
20. Aphasizheva I, Alfonzo J, Carnes J et al (2020) Lexis and grammar of mitochondrial RNA processing in trypanosomes. Trends Parasitol 36(4):337–355
21. Gerasimov ES, Gasparyan AA, Afonin DA et al (2021) Complete minicircle genome of *Leptomonas pyrrhocoris* reveals sources of its non-canonical mitochondrial RNA editing events. Nucleic Acids Res 49(6):3354–3370
22. Seiwert SD, Stuart K (1994) RNA editing: transfer of genetic information from gRNA to precursor mRNA *in vitro*. Science 266(5182): 114–117
23. Gerasimov ES, Afonin DA, Škodová-Sveráková I et al (2025) Evolutionary divergent kinetoplast genome structure and RNA editing patterns in the trypanosomatid *Vickermania*. Proc Natl Acad Sci USA 122(15):e2426887122
24. Hong M, Simpson L (2003) Genomic organization of *Trypanosoma brucei* kinetoplast DNA minicircles. Protist 154(2):265–279
25. Yurchenko V, Kolesnikov AA (2001) Minicircular kinetoplast DNA of Trypanosomatidae. Mol Biol (Mosk) 35(1):1–10
26. Thomas S, Martinez LL, Westenberger SJ et al (2007) A population study of the minicircles in *Trypanosoma cruzi*: predicting guide RNAs in the absence of empirical RNA editing. BMC Genomics 8:133
27. Cooper S, Wadsworth ES, Ochsenreiter T et al (2019) Assembly and annotation of the mitochondrial minicircle genome of a differentiation-competent strain of *Trypanosoma brucei*. Nucleic Acids Res 47(21): 11304–11325
28. Cooper S, Wadsworth ES, Schnaufer A et al (2022) Organization of minicircle cassettes and guide RNA genes in *Trypanosoma brucei*. RNA 28(7):972–992

29. Zimmer SL, Simpson RM, Read LK (2018) High throughput sequencing revolution reveals conserved fundamentals of U-indel editing. Wiley Interdiscip Rev RNA 9(5): e1487
30. Gerasimov ES, Gasparyan AA, Kaurov I et al (2018) Trypanosomatid mitochondrial RNA editing: dramatically complex transcript repertoires revealed with a dedicated mapping tool. Nucleic Acids Res 46(2):765–781
31. Simpson RM, Bruno AE, Bard JE et al (2016) High-throughput sequencing of partially edited trypanosome mRNAs reveals barriers to editing progression and evidence for alternative editing. RNA 22(5):677–695
32. Tashyreva D, Simpson AGB, Prokopchuk G et al (2022) Diplonemids – a review on "new" flagellates on the oceanic block. Protist 173(2): 125868
33. Vlcek C, Marande W, Teijeiro S et al (2011) Systematically fragmented genes in a multipartite mitochondrial genome. Nucleic Acids Res 39(3):979–988
34. Valach M, Moreira S, Hoffmann S et al (2017) Keeping it complicated: mitochondrial genome plasticity across diplonemids. Sci Rep 7(1): 14166
35. Kaur B, Záhonová K, Valach M et al (2020) Gene fragmentation and RNA editing without borders: eccentric mitochondrial genomes of diplonemids. Nucleic Acids Res 48(5): 2694–2708
36. Moreira S, Valach M, Aoulad-Aissa M et al (2016) Novel modes of RNA editing in mitochondria. Nucleic Acids Res 44(10): 4907–4919
37. Bushnell B, Rood J, Singer E (2017) BBMerge – accurate paired shotgun read merging *via* overlap. PLoS One 12(10):e0185056
38. Quinlan AR (2014) BEDTools: the swiss-army tool for genome feature analysis. Curr Protoc Bioinformatics 47:11.12.11–11.12.34
39. Camacho C, Coulouris G, Avagyan V et al (2009) BLAST+: architecture and applications. BMC Bioinformatics 10:421
40. Langmead B, Salzberg SL (2012) Fast gapped-read alignment with Bowtie 2. Nat Methods 9(4):357–359
41. Li H, Durbin R (2010) Fast and accurate long-read alignment with Burrows-Wheeler transform. Bioinformatics 26(5):589–595
42. Li H, Durbin R (2009) Fast and accurate short read alignment with Burrows-Wheeler transform. Bioinformatics 25(14):1754–1760
43. Slater GS, Birney E (2005) Automated generation of heuristics for biological sequence comparison. BMC Bioinf 6:31
44. Andrews S (2019) FastQC: a quality control tool for high throughput sequence data. http://www.bioinformatics.babraham.ac.uk/projects/fastqc. Accessed 27 Apr 2025
45. Chen S (2023) Ultrafast one-pass FASTQ data preprocessing, quality control, and deduplication using fastp. iMeta 2(2):e107
46. Kolmogorov M, Yuan J, Lin Y, Pevzner PA (2019) Assembly of long, error-prone reads using repeat graphs. Nat Biotechnol 37(5): 540–546
47. Garrison E, Marth G (2012) Haplotype-based variant detection from short-read sequencing. arXiv 1207:3907
48. Kearse M, Moir R, Wilson A et al (2012) Geneious basic: an integrated and extendable desktop software platform for the organization and analysis of sequence data. Bioinformatics 28(12):1647–1649
49. Katoh K, Standley DM (2013) MAFFT multiple sequence alignment software version 7: improvements in performance and usability. Mol Biol Evol 30(4):772–780
50. Danecek P, Bonfield JK, Liddle J et al (2021) Twelve years of SAMtools and BCFtools. Gigascience 10(2):1–4
51. Prjibelski A, Antipov D, Meleshko D et al (2020) Using SPAdes *de novo* assembler. Curr Protoc Bioinformatics 70(1):e102
52. Grabherr MG, Haas BJ, Yassour M et al (2011) Full-length transcriptome assembly from RNA-Seq data without a reference genome. Nat Biotechnol 29(7):644–652
53. Danecek P, Auton A, Abecasis G et al (2011) The variant call format and VCFtools. Bioinformatics 27(15):2156–2158
54. Van den Broeck F, Savill NJ, Imamura H et al (2020) Ecological divergence and hybridization of Neotropical *Leishmania* parasites. Proc Natl Acad Sci USA 117(40):25159–25168
55. Geerts M, Schnaufer A, Van den Broeck F (2021) rKOMICS: an R package for processing mitochondrial minicircle assemblies in population-scale genome projects. BMC Bioinf 22(1):468
56. Yabuki A, Tanifuji G, Kusaka C et al (2016) Hyper-eccentric structural genes in the mitochondrial genome of the algal parasite *Hemistasia phaeocysticola*. Genome Biol Evol 8(9): 2870–2878
57. Duarte M, Tomás AM (2014) The mitochondrial complex I of trypanosomatids–an overview of current knowledge. J Bioenerg Biomembr 46(4):299–311
58. Opperdoes FR, Michels PA (2008) Complex I of Trypanosomatidae: does it exist? Trends Parasitol 24(7):310–317

59. Kannan S, Burger G (2008) Unassigned *MURF1* of kinetoplastids codes for NADH dehydrogenase subunit 2. BMC Genomics 9: 455
60. Valach M, Leveille-Kunst A, Gray MW, Burger G (2018) Respiratory chain Complex I of unparalleled divergence in diplonemids. J Biol Chem 293(41):16043–16056
61. Ramrath DJF, Niemann M, Leibundgut M et al (2018) Evolutionary shift toward protein-based architecture in trypanosomal mitochondrial ribosomes. Science 362(6413): eaau7735
62. Valach M, Moreira S, Petitjean C et al (2023) Recent expansion of metabolic versatility in *Diplonema papillatum*, the model species of a highly speciose group of marine eukaryotes. BMC Biol 21(1):99

Chapter 10

Cleavage Under Targets and Release Using Nuclease (CUT&RUN) for *Trypanosoma brucei* Parasites

Danae Schulz

Abstract

Many gene regulatory processes depend on proteins that interact with DNA. Characterizing these interactions can shed light on the molecular mechanisms that allow cells to control which RNA is made, when it is made, and how much is made. Additionally, DNA protein interactions are essential for cell division, DNA replication, and repair. Chromatin ImmunoPrecipitation followed by sequencing (ChIP-seq) has been an invaluable tool for understanding gene regulatory processes in the eukaryotic kinetoplastid parasite *Trypanosoma brucei* by mapping genomic binding sites for a protein of interest. We recently sought to expand the repertoire of available techniques to interrogate protein-DNA interactions by optimizing Cleavage Under Targets and Release Using Nuclease (CUT&RUN) for use in *Trypanosoma brucei*. The protocol presented here details a CUT&RUN protocol suitable for proteins in small complexes for which there is an available antibody.

Key words *Trypanosoma brucei*, Chromatin, Epigenetics, DNA-protein interactions

1 Introduction

The kinetoplastid, *Trypanosoma brucei*, is an intriguing model system for exploring gene regulation in an early branching eukaryote. *T. brucei* gene regulation differs from better studied model organisms in a number of important ways. While many eukaryotic cells, such as yeast, worms, and flies, utilize a dedicated promoter for each gene, *T. brucei* transcription occurs polycistronically, with one Pol II promoter initiating transcription of up to 100 genes [1]. Unlike bacterial systems, the genes transcribed from a given promoter in *T. brucei* are not necessarily functionally related (reviewed in [2–4]). Maturation of mRNAs in *T. brucei* includes co-transcriptional polyadenylation and trans-splicing (reviewed in [2]). Unlike other eukaryotic model systems, *T. brucei* lack obvious DNA sequence-specific transcription factor binding sites; thus post-transcriptional processes are important for regulating

Paul A. M. Michels et al. (eds.), *Euglenozoa: Methods and Protocols, Volume 1*, Methods in Molecular Biology, vol. 3013, https://doi.org/10.1007/978-1-0716-5142-1_10,

transcript levels [5, 6]. Trypanosomes have histone proteins that form structures similar to those found in other model systems [7], but the amino acid sequences of the histones and the histone variants have diverged quite substantially [8, 9]. Consequently, the histone tails of *T. brucei* have alternate post-translational modifications. A number of outstanding papers have characterized the location of *T. brucei* histone variants and modifications and begun to explore their function [8, 10–13]. However, how each variant and histone modification contributes to gene regulation is still not well understood.

Given the presence of numerous post-translational modifications on the histones of *T. brucei*, it is perhaps not surprising that the parasites also contain many proteins that read, write, and erase those modifications [9, 10, 13–20]. We are starting to build a more complete picture for where chromatin interacting proteins localize and for the composition of their complexes [14]. However, there is still much to be learned about the role for epigenetic processes in *T. brucei* gene regulation, particularly as it relates to transcriptional reprogramming that occurs during life cycle transitions.

The optimization of Chromatin ImmunoPrecipitiation (ChIP-seq) in *T. brucei*, first published in 2009 [10], has been instrumental to the study of gene regulatory processes in this organism. Since then, additional optimization of methods that reveal protein-DNA interactions has been enormously helpful for elucidating mechanisms of gene regulation in this early branching eukaryote. Examples of these additional methods include fragment ion patchwork quantification for histone acetylation [21], high-resolution MNase-ChIP-seq [11], chromosome conformation capture (Hi-C) [22] and high-resolution DNA-DNA contact mapping (Micro-C) [23].

While ChIP-seq protocols have led to seminal discoveries on genome organization and gene regulation in *T. brucei* [10, 14, 19], these protocols are time consuming, often requiring multiple days to process samples. In addition, they require specialized equipment to sonicate DNA and are subject to epitope masking and false positive binding sites that can sometimes result from formaldehyde crosslinking [24–28]. To avoid these issues, we adapted a method developed in Steven Henikoff's lab called Cleavage Under Targets and Release Using Nuclease (CUT&RUN) [29] for use in *T. brucei* [20, 30]. We present the method here with some additional notes. The method can be performed in 4–5 h using 50–75 million parasites as input. Parasites are first permeabilized using saponin, then incubated with an antibody to a DNA-binding target of interest. A protein A/protein G micrococcal nuclease fusion protein is then added to the permeabilized parasites, where it binds the antibody for the protein of interest. Calcium is then added to activate cleavage of the DNA surrounding the protein of interest, and these small segments of DNA leak out of the nucleus, while the larger unbound fragments remain. After centrifugation, smaller micrococcal nuclease cut DNA samples from the supernatant are harvested and spike-in DNA from a non-trypanosome species is added so that samples can be normalized following the preparation of high-throughput sequencing libraries (Fig. 1). If harvesting DNA from the supernatant, DNA can be purified with either a phenol/chloroform protocol or using AMPure XP beads. After the DNA is isolated, high-throughput sequencing libraries can be prepared using the lab's preferred method. The data from high-throughput sequencing can be processed in a similar way as data generated from ChIP-seq using peak calling algorithms [31] and/or peak calling algorithms combined with measures of occupancy using DiffBind [32] or another similar method.

If the proteins under study are in large complexes that are too big to leak out of the nucleus, then in theory nuclei can be ruptured using SDS and the small DNA fragments can be recovered following proteinase K treatment and size selection with Beckman Coulter AMPure XP beads. However, our lab has not performed experiments on large protein complexes and this procedure may require optimization.

To ensure that parasites are properly permeabilized, we recommend preparing control samples using a histone antibody. Following permeabilization and incubation with the histone antibody, parasites can be stained using a fluorescently labeled anti-IgG antibody and analyzed by flow cytometry to verify that permeabilization has been successful. Examples of expected results and processing workflows can be found in Ashby et al. [20] and Miller et al. [30]. If control samples processed with histone antibodies are taken all the way to completion of the CUT&RUN protocol, the resulting DNA can be run on an agarose gel and a characteristic ladder of 150 bp spaced bands should result [20, 29, 30]. Below we present a detailed protocol for CUT&RUN that we optimized for *T. brucei* bloodstream parasites [30]. The protocol may have to be optimized further for use in procyclic parasites.

2 Materials

2.1 Chemicals and Biologics

1. Spermidine trihydrochloride.
2. Saponin.
3. Sodium chloride.
4. Calcium chloride.
5. Tris hydrochloride (Tris–HCl).
6. cOmplete™ EDTA-free Protease Inhibitor Cocktail (Sigma-Aldrich).
7. 0.5 M pH 8.0 UltraPure EDTA Buffer.
8. 0.5 M pH 8.0 EGTA Buffer.

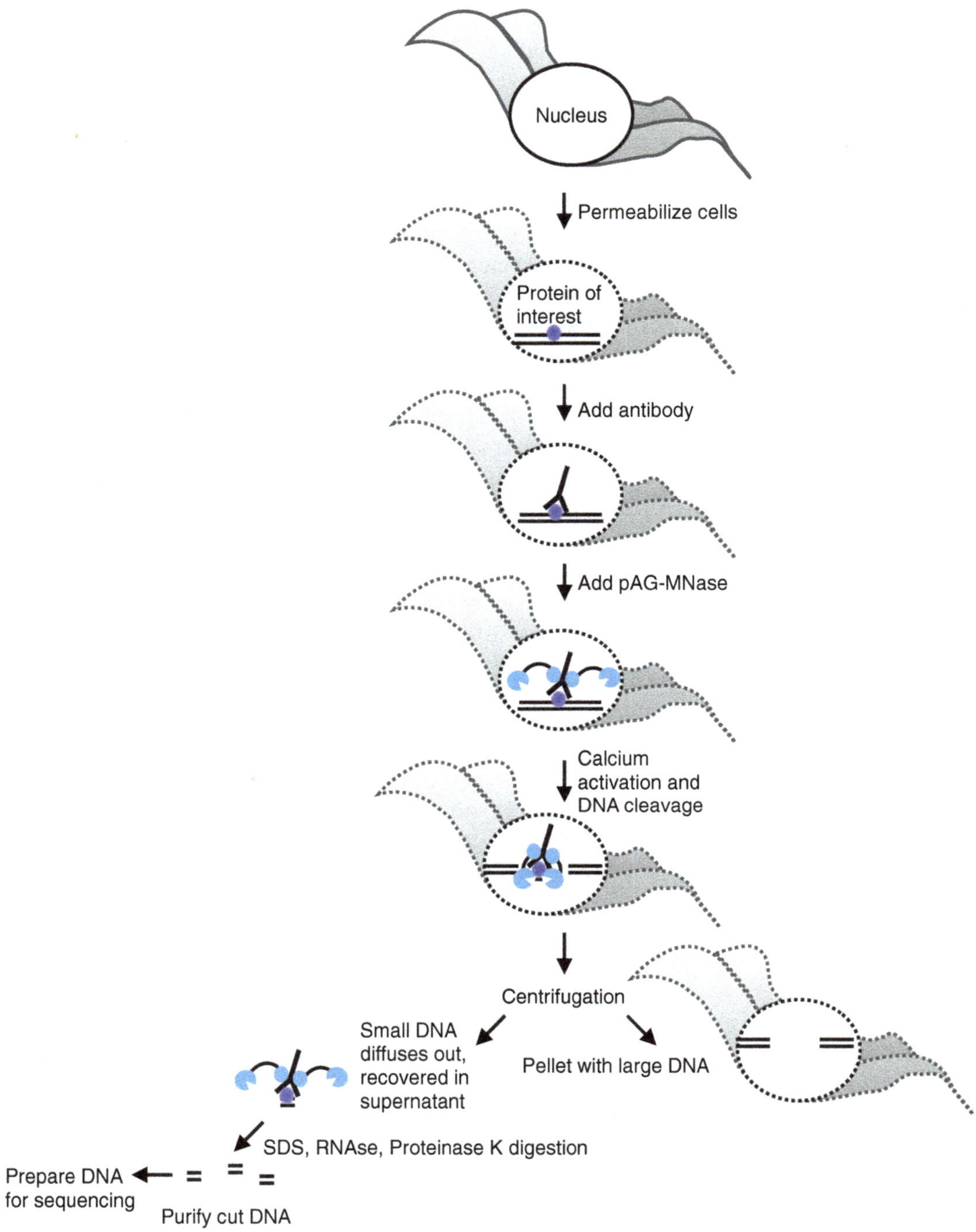

Fig. 1 Overview of CUT&RUN protocol [29]. Trypanosomes are permeabilized using saponin, and antibodies are added to the protein of interest. pAG-MNase binds to the antibody and cleaves the DNA bound to the protein of interest. Small fragments of DNA diffuse out of the nuclease and are harvested for high-throughput sequencing

9. Mouse anti-rabbit IgG-PE.
10. Primary antibody, this will vary depending on your system.
11. Rabbit anti-H3 (kind gift from Christian Janzen).
12. Sodium dodecyl sulfate solution.
13. Proteinase K.
14. RNase A.
15. Ampure XP beads (Beckman Coulter).
16. CUT&RUN pAG-MNase and Spike-In DNA (Cell Signaling Technology).
17. HMI-9 medium [33].
18. 80% ethanol.

2.2 Equipment and Supplies

1. Microcentrifuge.
2. Swinging bucket centrifuge.
3. 1.7 mL Olympus microtubes.
4. Olympus 50 mL centrifuge tubes.
5. Rotator with tube holders.
6. Magnetic stand with Eppendorf tube adaptor.
7. Flow cytometer (Novocyte 2000 or equivalent).

2.3 Buffers

1. NP-S Buffer A with detergent: 0.5 mM spermidine, 0.1% Saponin (vol/vol), 50 mM NaCl, 10 mM Tris–HCl (pH 7–8). Add protease inhibitors just before use. Store at 4 °C.
2. NP-S Buffer B no detergent: 0.5 mM spermidine, 50 mM NaCl, 10 mM Tris–Cl (pH 7–8). Add protease inhibitors just before use. Store at 4 °C.
3. 2× STOP Buffer: 20 mM EDTA, 40 mM EGTA, add 50 pg of yeast spike-in DNA to each reaction or 50 pg/100 μL of 2× STOP Buffer.

   TE Buffer: 10 mM Tris–HCl (pH 7–8), 1 mM EDTA.

3 Methods

Parasites are cultured in HMI-9 medium with incubation at 37 °C and 5% CO_2. Cultures should be prepared in advance so that 50–75 million bloodstream parasites are available for each sample.

3.1 Prepare Cells

1. Parasite cultures should be counted with a hemocytometer or other preferred counting method.

2. Centrifuge parasites at 2800× *g* for 10 min. This spin step can be performed at room temperature or at 10 °C. Remove supernatant and resuspend in the remaining medium (~100 μL).
3. Samples should be combined into an Eppendorf tube such that each tube has 50–75 million cells. Samples should be centrifuged again at 2800× *g* for 4 min at 10 °C in a microcentrifuge. Remove supernatant.

3.2 Permeabilize Cells (See Note 1)

1. Samples should be washed with 1 mL NP-S Buffer A (*see* **Note 2**).
2. Centrifuge samples at 4600× *g* for 4 min at 10 °C. Remove the supernatant.

3.3 Primary Antibody Binding

1. Each sample should be resuspended in 100 μL NP-S Buffer A.
2. EDTA should be added for a final concentration of 2 mM (4 μL of 0.05 M EDTA/tube).
3. 5 μg of antibody toward the target of interest should be added (the amount of antibody may vary depending on the epitope) **OR**

   5 μg α-H3 antibody should be added (for control sample, this might require titration, *see* **Note 3**) **OR**

   5 μg α-IgG antibody should be added (for control sample).
4. Samples should be incubated for 45 min on tube rotator at room temperature.
5. 1 mL NP-S Buffer B should be added to each sample.
6. Centrifuge samples at 4600× *g* for 4 min at 10 °C. Remove supernatant.
7. 1 mL NP-S Buffer B should be added to each sample.
8. An aliquot of one million cells should be removed for flow cytometry analysis (~14 μL if starting with 75 million cells). The aliquot should be set aside on ice.
9. The remainder of the sample should be centrifuged at 4600× *g* for 4 min at 10 °C. Remove supernatant (total two washes).

3.4 Protein AG-MNase Binding

1. Each sample should be resuspended in 100 μL of NP-S Buffer B.
2. 1.5 μL of protein AG-MNase enzyme should be added to each sample. This amount may need to be titrated based on the manufacturer (*see* **Note 3**).
3. Samples should be incubated for 45 min at room temperature on a tube rotator.

3.5 Flow Cytometry Sample Preparation

1. 1 mL NP-S Buffer B should be added to the aliquot of one million cells prepared above (alternatively, we routinely use HMI-9 medium for the flow cytometry antibody incubation and washes). In some cases, it may be appropriate to prepare a second aliquot of cells to use as a negative control (either permeabilized cells unstained with antibody during the Primary Antibody Binding step or cultured cells that have not been permeabilized).
2. Samples should be centrifuged at 4600× *g* for 4 min at 10 °C.
3. Samples should be resuspended in 100 μL NP-S Buffer B.
4. Samples should be incubated with α-rabbit IgG PE at 1:200 for 15 min at room temperature on a tube rotator.
5. Samples should be washed in 1 mL NP-S Buffer B (or HMI-9), then centrifuged at 4600× *g* for 4 min at 10 °C. Remove supernatant.
6. The wash should be repeated by resuspending samples in 1 mL NP-S Buffer B (or HMI-9) and centrifuging at 4600× *g* for 4 min at 10 °C. Remove supernatant.
7. Samples should be resuspended in 300 μL NP-S Buffer B (or HMI-9).
8. Samples should be transferred into a flow cytometry tube.
9. Flow cytometry should be used to analyze whether cells have been successfully permeabilized (*see* **Note 4**).

3.6 Protein AG-MNase Wash (Continued After 45 min Incubation Above)

1. 1 mL NP-S Buffer B should be added to each sample.
2. Centrifuge samples at 4600× *g* for 4 min at 10 °C. Remove supernatant.
3. 1 mL NP-S Buffer B should be added to each sample.
4. Centrifuge samples at 4600× *g* for 4 min at 10 °C. Remove supernatant.

3.7 Targeted Digestion

1. Make sure to prepare enough 2× Stop Buffer with spike in control (*see* **Notes 5** and **6**). Samples should be resuspended in 100 μL buffer NP-S Buffer B.
2. Samples should be incubated at room temperature for 5 min.
3. 2 μL of 100 mM $CaCl_2$ should be added to all samples (final concentration = 2 mM), flick samples to mix (*see* **Note 7**).
4. Samples should be incubated at 25 °C for 5 min. We recommend using a heat block as ambient temperature in the lab can vary.
5. 100 μL 2× Stop buffer should be added to each sample. Samples should be mixed by flicking.

3.8 Chromatin Release

1. Samples should be incubated for 10 min at 37 °C to release CUT&RUN fragments from the insoluble nuclear chromatin.
2. Centrifuge samples at 4600× *g* for 4 min at 10 °C. Remove supernatant into new tube. *Save the supernatant* (*see* **Note 8**).

3.9 DNA Extraction

1. 2 μL of 10% SDS (final concentration = 0.1% SDS), 3.3 μL of 10 mg/mL proteinase K (final concentration = 165 μg/mL), and 1.33 μL of 1 mg/mL RNAse A (final concentration = 6.5 μg/mL) should be added to each sample.
2. Samples should be mixed by gentle flicking and incubated for 10 min at 70 °C.
3. Add 1.8× volume of AMPure XP beads (360 μL) to the solution (*see* **Note 9**).
4. Incubate at room temperature for 5 min.
5. Add samples to magnetic stand for 5 min.
6. Wash samples with 200 μL 80% ethanol while on the magnetic stand.
7. Remove supernatant.
8. Wash samples again with 200 μL 80% ethanol.
9. Remove supernatant.
10. Dry samples for 5 min.
11. Resuspend beads with 12 μL TE buffer to elute DNA.
12. Incubate at room temperature for 5 min.
13. Add samples to magnetic stand for 5 min.
14. Remove supernatant new tube.
15. Following DNA purification, high-throughput sequencing libraries can be prepared using the preferred method of the research lab (*see* **Note 10**). The analysis pipeline will also depend on the particular experimental question. In general, analysis pipelines for CUT&RUN are similar to those used for ChIP-seq (*see* **Note 11**).

4 Notes

1. The protocol works best when everything is kept cold prior to cutting with the protein AG-MNase fusion protein. We recommend keeping buffers chilled on ice and pre-cooling centrifuges to 4–10 °C.
2. Protease inhibitors should be added to the NP-S buffer just before commencing the experiment.
3. If titration of pAG MNase is necessary, we suggest using an anti-histone antibody. When taken through to completion

using a histone antibody, the CUT&RUN supernatant should contain a characteristic ladder of bands in 150 bp increments, which can be checked on an agarose gel by running the supernatant fraction following DNA purification [20, 29, 30].

4. Examples of flow cytometry plots showing successful permeabilization can be found in [20, 30]. Our parasites are analyzed in the PE channel with a laser voltage of 489 and filtered for FSC-H >50,000 using a Novocyte 2000. The parameters will vary according to the individual lab instrument.
5. The amount of 2× Stop Buffer required for the experiment should be calculated and yeast spike-in DNA should be added prior to starting this step (*see* 2× Stop Buffer recipe above).
6. The timing is very important in this section to avoid untargeted cleavage.
7. During troubleshooting, a sample can be prepared that is not activated with calcium to ensure that there is no non-specific cutting.
8. If troubleshooting, it is possible to also analyze the pellet to ensure that small pieces of DNA are not present in this fraction. To process the DNA from the pellet, resuspend it in 200 μL NP-S Buffer B and process in parallel with the supernatant samples.
9. Phenol/chloroform extraction is an alternate option if Ampure XP beads are not available.
10. To prepare libraries for high-throughput sequencing we use the New England Biolabs NEBNext Ultra II DNA Library Prep Kit for Illumina combined with New England Biolabs NEBNext Multiplex Oligos for Illumina following the manufacturer's instructions. The preferred library kit will depend on the preferences of the individual lab.
11. Examples of expected results using α-H3 antibodies and analysis workflows can be found in ref. [20, 30]. We sequenced reads on a HISEQ3000 with 50 bp single end reads. We trimmed FASTQ reads with TrimGalore and aligned them to the Tb927v5.1 genome using bowtie [34]. MACS [35] was used for peak calling and Diffbind [32] was used to analyze called peaks for differential occupancy over time. The individual analysis pipeline will vary according to the experimental question.

References

1. Berriman M (2005) The genome of the African trypanosome Trypanosoma brucei. Science 309:416–422. https://doi.org/10.1126/science.1112642
2. Clayton CE (2016) Gene expression in Kinetoplastids. Curr Opin Microbiol 32:46–51. https://doi.org/10.1016/j.mib.2016.04.018
3. Clayton CE (2014) Networks of gene expression regulation in Trypanosoma brucei. Mol Biochem Parasitol 195:96–106. https://doi.org/10.1016/j.molbiopara.2014.06.005
4. Siegel TN, Gunasekera K, Cross GAM, Ochsenreiter T (2011) Gene expression in Trypanosoma brucei: lessons from high-throughput

RNA sequencing. Trends Parasitol 27:434–441. https://doi.org/10.1016/j.pt.2011.05.006

5. Clayton C (2013) The regulation of trypanosome gene expression by RNA-binding proteins. PLoS Pathog 9:e1003680. https://doi.org/10.1371/journal.ppat.1003680
6. Luzak V, Osses E, Danese A et al (2025) SLAM-seq reveals independent contributions of RNA processing and stability to gene expression in African trypanosomes. Nucleic Acids Res 53:gkae1203. https://doi.org/10.1093/nar/gkae1203
7. Deák G, Wapenaar H, Sandoval G et al (2023) Histone divergence in trypanosomes results in unique alterations to nucleosome structure. Nucleic Acids Res 51:7882–7899. https://doi.org/10.1093/nar/gkad577
8. Maree JP, Tvardovskiy A, Ravnsborg T et al (2022) *Trypanosoma brucei* histones are heavily modified with combinatorial post-translational modifications and mark Pol II transcription start regions with hyperacetylated H2A. Nucleic Acids Res 50:9705–9723. https://doi.org/10.1093/nar/gkac759
9. Figueiredo LM, Cross GAM, Janzen CJ (2009) Epigenetic regulation in African trypanosomes: a new kid on the block. Nat Publ Group 7:504–513. https://doi.org/10.1038/nrmicro2149
10. Siegel TN, Hekstra DR, Kemp LE et al (2009) Four histone variants mark the boundaries of polycistronic transcription units in Trypanosoma brucei. Genes Dev 23:1063–1076. https://doi.org/10.1101/gad.1790409
11. Wedel C, Siegel TN (2017) Genome-wide analysis of chromatin structures in Trypanosoma brucei using high-resolution MNase-ChIP-seq. Exp Parasitol. https://doi.org/10.1016/j.exppara.2017.03.003
12. Wright JR, Siegel TN, Cross GAM (2010) Histone H3 trimethylated at lysine 4 is enriched at probable transcription start sites in Trypanosoma brucei. Mol Biochem Parasitol 172:141–144. https://doi.org/10.1016/j.molbiopara.2010.03.013
13. Kraus AJ, Vanselow JT, Lamer S et al (2020) Distinct roles for H4 and H2A.Z acetylation in RNA transcription in African trypanosomes. Nat Commun 11:1498. https://doi.org/10.1038/s41467-020-15274-0
14. Staneva DP, Carloni R, Auchynnikava T et al (2021) A systematic analysis of *Trypanosoma brucei* chromatin factors identifies novel protein interaction networks associated with sites of transcription initiation and termination. Genome Res 31:2138–2154. https://doi.org/10.1101/gr.275368.121
15. Janzen CJ, Hake SB, Lowell JE, Cross GAM (2006) Selective Di- or Trimethylation of histone H3 lysine 76 by two DOT1 homologs is important for cell cycle regulation in Trypanosoma brucei. Mol Cell 23:497–507. https://doi.org/10.1016/j.molcel.2006.06.027
16. Figueiredo LM, Janzen CJ, Cross GAM (2008) A histone methyltransferase modulates antigenic variation in African trypanosomes. PLoS Biol 6:e161. https://doi.org/10.1371/journal.pbio.0060161
17. Siegel TN, Kawahara T, DeGrasse JA et al (2007) Acetylation of histone H4K4 is cell cycle regulated and mediated by HAT3 in Trypanosoma brucei. Mol Microbiol 67:762–771. https://doi.org/10.1111/j.1365-2958.2007.06079.x
18. Kawahara T, Siegel TN, Ingram AK et al (2008) Two essential MYST-family proteins display distinct roles in histone H4K10 acetylation and telomeric silencing in trypanosomes. Mol Microbiol 69:1054–1068. https://doi.org/10.1111/j.1365-2958.2008.06346.x
19. Schulz D, Mugnier MR, Paulsen E-M et al (2015) Bromodomain proteins contribute to maintenance of bloodstream form stage identity in the African trypanosome. PLoS Biol 13:e1002316–e1002338. https://doi.org/10.1371/journal.pbio.1002316
20. Ashby E, Paddock L, Betts HL et al (2022) Genomic occupancy of the Bromodomain protein Bdf3 is dynamic during differentiation of African trypanosomes from bloodstream to Procyclic forms. mSphere 7:e0002322. https://doi.org/10.1128/msphere.00023-22
21. ElBashir R, Vanselow JT, Kraus A et al (2015) Fragment ion patchwork quantification for measuring site-specific acetylation degrees. Anal Chem 87:9939–9945. https://doi.org/10.1021/acs.analchem.5b02517
22. Müller LSM, Cosentino RO, Förstner KU et al (2018) Genome organization and DNA accessibility control antigenic variation in trypanosomes. Nature 563:121–125. https://doi.org/10.1038/s41586-018-0619-8
23. Rabuffo C, Schmidt MR, Yadav P et al (2024) Inter-chromosomal transcription hubs shape the 3D genome architecture of African trypanosomes. Nat Commun 15:10716. https://doi.org/10.1038/s41467-024-55285-9
24. Teytelman L, Thurtle DM, Rine J, van Oudenaarden A (2013) Highly expressed loci are vulnerable to misleading ChIP localization of multiple unrelated proteins. Proc Natl Acad Sci

110:18602–18607. https://doi.org/10.1073/pnas.1316064110

25. Park SH, Nguyen TN, Kirkham JK et al (2011) Transcription by the multifunctional RNA polymerase I in Trypanosoma brucei functions independently of RPB7. Mol Biochem Parasitol:1–8. https://doi.org/10.1016/j.molbiopara.2011.06.008
26. Jain D, Baldi S, Zabel A et al (2015) Active promoters give rise to false positive 'phantom peaks' in ChIP-seq experiments. Nucleic Acids Res 43:6959–6968. https://doi.org/10.1093/nar/gkv637
27. Baranello L, Kouzine F, Sanford S, Levens D (2016) ChIP bias as a function of cross-linking time. Chromosom Res 24:175–181. https://doi.org/10.1007/s10577-015-9509-1
28. Meyer CA, Liu XS (2014) Identifying and mitigating bias in next-generation sequencing methods for chromatin biology. Nat Rev Genet 15:709–721. https://doi.org/10.1038/nrg3788
29. Skene PJ, Henikoff S (2017) An efficient targeted nuclease strategy for high-resolution mapping of DNA binding sites. eLife 6: e21856. https://doi.org/10.7554/eLife.21856
30. Miller G, Rollosson LM, Saada C et al (2023) Adaptation of CUT&RUN for use in African trypanosomes. PLoS One 18:e0292784. https://doi.org/10.1371/journal.pone.0292784
31. Thomas R, Thomas S, Holloway AK, Pollard KS (2016) Features that define the best ChIP-seq peak calling algorithms. Brief Bioinform bbw035. https://doi.org/10.1093/bib/bbw035
32. Stark R, Gord B (2011) DiffBind: differential binding analysis of ChIP-Seq peak data
33. Hirumi H, Hirumi K (1989) Continuous cultivation of Trypanosoma brucei blood stream forms in a medium containing a low concentration of serum protein without feeder cell layers. J Parasitol 75:985–989
34. Langmead B, Trapnell C, Pop M, Salzberg SL (2009) Ultrafast and memory-efficient alignment of short DNA sequences to the human genome. Genome Biol 10:R25. https://doi.org/10.1186/gb-2009-10-3-r25
35. Zhang Y, Liu T, Meyer CA et al (2008) Model-based analysis of ChIP-Seq (MACS). Genome Biol 9:R137. https://doi.org/10.1186/gb-2008-9-9-r137

Chapter 11

Mapping Open Chromatin in *Trypanosoma brucei* Using ATAC-Seq

Ruth Shelton and Keith Matthews

Abstract

Post-transcriptional control is well established as a key mechanism of gene regulation in trypanosomatids. However, recent studies suggest that transcriptional regulation may also play a role, challenging long-standing dogma. The Assay for Transposase Accessible Chromatin with sequencing (ATAC-seq) provides a genome-wide overview of chromatin accessibility (i.e., whether chromatin is more open or closed), without requiring prior knowledge of chromatin markers which may be absent or poorly understood in trypanosomatids. Here, we present an optimized ATAC-seq protocol for use in *Trypanosoma brucei*, which has been used in both bloodstream and procyclic forms, and can be used to inform application of the method to other Euglenozoa. We also provide guidance on bioinformatic analysis, including integration of output files with established differential accessibility and RNA-seq data analysis pipelines.

Key words Epigenomics, Chromatin accessibility, ATAC-seq, Transcriptional regulation, RNA-seq, Differential accessibility analysis, Next-generation sequencing

1 Introduction

In eukaryotes, epigenetic regulation—inheritable changes in gene expression that occur without altering the underlying DNA sequence—shapes genome function in ways that extend beyond the genetic code itself. However, while gene expression is regulated at the transcriptional level in the overwhelming majority of prokaryotes and eukaryotes, Euglenozoa predominantly regulate gene expression at the post-transcriptional level [1]. In *Trypanosoma brucei*, for instance, the majority of the parasite's ~8000 genes are organized into polycistronic transcription units (PTUs) comprising functionally unrelated genes arranged in a strand-specific manner, separated by either divergent or convergent strand switch regions. Transcription is initiated from a limited number of divergent Pol II transcription sites (dTSSs), and terminates at convergent regions (cTTSs) [2]. Transcription from these dTSSs has traditionally been

Paul A. M. Michels et al. (eds.), *Euglenozoa: Methods and Protocols, Volume 1*, Methods in Molecular Biology, vol. 3013, https://doi.org/10.1007/978-1-0716-5142-1_11,

considered constitutive, lacking gene-specific control [3]. Instead, regulation occurs post-transcriptionally, through mechanisms such as *trans*-splicing, polyadenylation, nuclear degradation and export, cytosolic mRNA decay and regulation of translation initiation or elongation [4]. As a result, transcription was not thought to be subject to epigenetic regulation. Known exceptions to this include the major surface protein genes—procyclins and variant surface glycoproteins (*VSGs*)—which are instead transcribed by Pol I, rather than Pol II, and therefore subject to transcriptional control [5]. However, recent studies suggest that epigenetic control of Pol II does in fact occur: the SPARC complex, comprising several proteins with homology to histone readers and writers, has been identified as a regulator of Pol II transcription initiation [6]; additionally, selection for TLF resistance has been shown to promote deposition of base J at TSSs, inhibiting Pol II binding and leading to reversible silencing of PTUs [7]. These findings challenge the long-standing assumption that transcription in *T. brucei* is entirely constitutive and post-transcriptionally regulated, instead pointing to a more nuanced model that includes both transcriptional and epigenetic mechanisms. As such, tools that can systematically investigate the role of epigenetic regulation in gene expression are increasingly important for understanding gene regulation in trypanosomes and other Euglenozoa.

The Assay for Transposase-Accessible Chromatin with sequencing (ATAC-seq) has become a foundational technique for investigating genome-wide chromatin accessibility across eukaryotes. It utilizes a mutant Tn5 transposase, preloaded with sequencing adaptors, that preferentially cleaves and inserts adaptors into regions of open chromatin [8, 9]. This allows for high-resolution mapping of accessible DNA and enables the detection of dynamic changes in chromatin structure across different developmental stages or cellular states. ATAC-seq offers two main advantages for examining epigenetic changes in Euglenozoa, including *T. brucei*: first, it does not require any a priori knowledge of chromatin-associated markers or histone modifications typically linked to open or closed chromatin states where conventional regulatory features may be absent, divergent, or poorly characterized; second, ATAC-seq requires significantly less starting material than comparable methods such as DNase-seq or FAIRE-seq [10], making it particularly well-suited for species with compact genomes or limited biological material.

In this chapter, we present the complete workflow for performing ATAC-seq in *T. brucei*, from sample preparation and bioinformatic analysis through to integration with established differential accessibility and RNA-seq analysis pipelines. The protocol is compatible with a range of *T. brucei* strains and life cycle stages and can be used to inform application of the method to other euglenozoans. The approach provides a valuable tool for dissecting the contribution of chromatin structure to gene regulation in a system with non-canonical transcriptional organization.

2 Materials

2.1 ATAC-Seq Library Preparation and Analysis

2.1.1 Sample Preparation, Tagmentation, and Purification

1. Media as appropriate for study organism.
2. 1 × 10^6 cells (if doing ATAC-seq only) or 1–5 × 10^7 cells (if combining with RNA-seq) per sample/replicate.
3. ATAC-seq Kit (Active Motif).
4. 1.5 mL Eppendorf tubes.
5. 1× PBS.
6. Centrifuge.
7. PCR tubes.
8. Mini-centrifuge able to hold PCR tubes.
9. Eppendorf ThermoMixer C with SmartBlock 1.5 mL (Eppendorf).
10. If combining with RNA-seq data: RNAlater (Invitrogen).
11. pH indicator strips.

2.1.2 PCR Amplification of Tagmented DNA

1. PCR tubes.
2. If preparing >16 samples: Nextera™-Compatible Multiplex Primers (ActiveMotif).
3. PCR thermocycler.

2.1.3 Purification of Amplified Libraries

1. 80% ethanol.
2. Vortex.
3. 1.5 mL Eppendorf tubes.
4. Magnetic rack.

2.1.4 Quantification and QC

1. Qubit fluorometer and Qubit 1× dsDNA high sensitivity assay (Invitrogen).
2. TapeStation and high-sensitivity D1000 ScreenTape (Agilent).

2.1.5 Data Analysis Overview

All analyses can be carried out on a computer running UNIX or Linux. Most packages can be installed using Anaconda or installed into R using BiocManager.

2.1.6 Datasets

1. reference_genome.fa: the relevent reference genome to your study, to which sequencing data will be aligned. This can be downloaded from TriTrypDB.org→ Data → Download data sets.

2. annotation_file.gff: gff file corresponding to reference genome, containing genome annotation information. This can be downloaded from TriTrypDB.org→ Data → Download data sets.

2.1.7 Preprocessing, Alignment, and Filtering

1. FastQC (https://www.bioinformatics.babraham.ac.uk/projects/fastqc/).
2. TrimGalore! (https://www.bioinformatics.babraham.ac.uk/projects/trim_galore/).
3. Cutadapt (https://cutadapt.readthedocs.io/en/stable/).
4. Bowtie2 (https://bowtie-bio.sourceforge.net/bowtie2/index.shtml).
5. Samtools (http://www.htslib.org).
6. GATK (https://gatk.broadinstitute.org/hc/en-us).

2.1.8 Visualization

1. IGV (http://www.htslib.org).
2. Samtools (http://www.htslib.org).
3. Khmer (https://khmer.readthedocs.io/en/latest/).
4. Deeptools (https://deeptools.readthedocs.io/en/latest/).

2.1.9 Peak Calling

1. Samtools (http://www.htslib.org).
2. Genrich (https://github.com/jsh58/Genrich).
3. Bedtools (https://bedtools.readthedocs.io).

2.1.10 Quality Control

1. GATK (https://gatk.broadinstitute.org/hc/en-us).
2. R Studio (https://posit.co/download/rstudio-desktop/).
3. R Library "tidyverse" (https://www.tidyverse.org).
4. Deeptools (https://deeptools.readthedocs.io/en/latest/).
5. Subread (https://subread.sourceforge.net).

2.1.11 ATAC-Seq Peak Annotation

1. R Studio (https://posit.co/download/rstudio-desktop/).
2. R Library "GenomicRanges" (https://posit.co/download/rstudio-desktop/).
3. R Library "rtracklayer" (https://bioconductor.org/packages/release/bioc/html/rtracklayer.html).
4. R Library "tidyverse" (https://www.tidyverse.org).

2.2 RNA-Seq Library Preparation and Analysis

2.2.1 RNA Extraction

1. RNeasy Mini Kit (Qiagen).
2. β-mercaptoethanol.
3. Centrifuge.
4. Vortex.
5. 70% ethanol.
6. Nuclease-free water.

2.2.2 DNase Treatment

1. TURBO DNA-free Kit (Invitrogen).
2. 1.5 mL Eppendorf tubes.
3. Heatblock (can use ThermoMixer from ATAC-seq tagmentation step with no agitation).

2.2.3 Quantification and QC

1. Qubit fluorometer and Qubit RNA Broad Range assay kit.
2. Nanodrop spectrophotometer.
3. TapeStation and RNA ScreenTape (Agilent).

2.2.4 Data Analysis Overview

All analyses can be carried out on a computer running UNIX or Linux. Most packages can be installed using Anaconda or installed into R using BiocManager.

2.2.5 Datasets

1. annotated_transcripts.fa: the relevant annotated transcript file to your study, to which will be used as the reference for quantifying transcript counts with Salmon. This can be downloaded from TriTrypDB.org→ Data → Download data sets.
2. annotation_file.gff: gff file corresponding to cell line being studied, containing genome annotation information. This can be downloaded from TriTrypDB.org→ Data → Download data sets.

2.2.6 RNA-Seq Preprocessing

1. FastQC (https://www.bioinformatics.babraham.ac.uk/projects/fastqc/).
2. TrimGalore! (https://www.bioinformatics.babraham.ac.uk/projects/trim_galore/).
3. Cutadapt (https://cutadapt.readthedocs.io/en/stable/).

2.2.7 Transcript Quantification

1. GFFread (https://ccb.jhu.edu/software/stringtie/gff.shtml).
2. Salmon (https://combine-lab.github.io/salmon/).
3. R Studio (https://posit.co/download/rstudio-desktop/).
4. R library "readr" (https://readr.tidyverse.org).
5. R library "rtracklayer" (https://bioconductor.org/packages/release/bioc/html/rtracklayer.html).
6. R library "GenomicRanges" (https://bioconductor.org/packages/release/bioc/html/GenomicRanges.html).
7. R library "tximport" (https://bioconductor.org/packages/release/bioc/html/tximport.html).
8. R library "edgeR" (https://bioconductor.org/packages/release/bioc/html/edgeR.html).

2.2.8 Integrating ATAC-Seq and RNA-Seq Data

1. R Studio (https://posit.co/download/rstudio-desktop/).
2. R library "readr" (https://readr.tidyverse.org).

3 Methods

3.1 ATAC-Seq Library Preparation

All steps are done at room temperature, unless stated otherwise. The Active-Motif ATAC-seq protocol has been optimized for 1× 10^6 *T. brucei* bloodstream form (BSF) cells (*see* **Note 1**).

3.1.1 Sample Preparation, Tagmentation, and Purification

1. From log-phase *T. brucei* culture, collect 1 × 10^6 cells by centrifuging for 10 min at 1000 × *g*.

   If cells will be collected for RNA-seq (*see* **step 8**), add enough medium to each flask to ensure cells will not overgrow (i.e., exceed 1 × 10^6 cells/mL) during ATAC-seq sample preparation.
2. Wash pellet in 100 μL of ice-cold 1× PBS and centrifuge at 500 × *g* for 5 min.
3. Remove supernatant and resuspend cells in 100 μL of ice-cold ATAC Lysis Buffer.
4. Transfer to a PCR tube (*see* **Note 2**) and spin for a further 10 min at 500 × *g*. At this point, the tagmentation master mix can be prepared:

Reagents	Volume
2× tagmentation buffer	25 μL
10× PBS	2 μL
1.0% digitonin	0.5 μL
10% Tween 20	0.5 μL
H_2O	12 μL
Assembled transposomes	10 μL

5. Remove the supernatant carefully (*see* **Note 3**).
6. Gently resuspend the lysis pellet in 50 μL of the tagmentation master mix. Transfer suspension to a 1.5 mL Eppendorf tube.
7. Incubate for 1 h at 37 °C in a ThermoMixer set at 800 RPM (*see* **Note 4**).
8. If integrating ATAC-seq and RNA-seq, cells can also be collected during incubation and stored in RNAlater for subsequent RNA extraction (*see* **Note 5**).
   (a) Pellet 1–5 × 10^7 cells by centrifuging at 1200 × *g* for 10 min.

(b) Discard supernatant and transfer cells to a 1.5 mL Eppendorf tube by dissolving pellet in ~100 μL of PBS.

(c) Pellet cells by centrifuging at 1200 × *g* for 10 min.

(d) Remove supernatant.

(e) Resuspend cell pellet in a small volume of PBS (e.g., 20 μL) and add 10 volumes of RNAlater. Cells can be kept at 2–8 °C for up to a month, or at −20 °C for longer term storage.

9. Add 250 μL DNA Purification Binding Buffer and 5 μL 3 M sodium acetate to each sample. Check that the pH is less than 7.5 using pH indicator strips (*see* **Note 6**). If the pH is over 7.5, add additional sodium acetate in 5 μL aliquots until the correct pH is achieved.
10. Transfer the sample to a labelled DNA purification column.
11. Spin at 17,000 × *g* for 1 min.
12. Discard the flow-through and return the collection tube to the column.
13. Add 750 μL of Wash Buffer to the column. Prior to first use, 40 mL of 100% ethanol will need to be added to the bottle to bring it to a final concentration of 80%.
14. Centrifuge at 17,000 × *g* for 1 min.
15. Discard flow through and return column to collection tube.
16. Centrifuge empty columns for 2 min at 17,000 × *g* to remove residual Wash Buffer.
17. Transfer columns to a clean Eppendorf tube.
18. Add 35 μL of DNA Purification Elution Buffer to the center of the column and incubate for 1 min.
19. Centrifuge for 1 min at 17,000 × *g*.

    To increase yield, add the eluted DNA Purification Elution Buffer to the same column and centrifuge again.
20. The purified DNA can be stored at −20 °C or taken directly into the PCR amplification steps.

3.1.2 PCR Amplification of Tagmented DNA

1. Set up PCR reactions below. If pooling libraries for sequencing, each sample will need a unique i5/i7 primer combination (*see* **Note 7**).

Reagent	Volume
Tagmented DNA	33.5 μL
i7 Indexed Primer (25 μM)	2.5 μL
i5 Indexed Primer (25 μM)	2.5 μL

(continued)

Reagent	Volume
dNTPs (10 mM)	1 μL
5× Q5 Reaction Buffer	10 μL
Q5 Polymerase (2 U/μL)	0.5 μL

2. Perform PCR with the following cycling conditions. *See* **Note 8** for information about optimizing the number of PCR cycles.

Temperature	Time	Number of cycles
72 °C	For 5 min	1
98 °C	For 30 s	1
98 °C 63 °C 72 °C	For 10 s For 30 s For 1 min	10
10 °C	∞	1

3. Remove SPRI beads from fridge to allow them to reach room temperature prior to bead clean up step.

3.1.3 Purification of Amplified Libraries

1. Prepare 400 μL of 80% ethanol per sample.
2. Vortex the room temperature SPRI beads thoroughly to ensure they are well mixed.
3. Transfer each PCR product to an Eppendorf tube and add 60 μL of SPRI beads.
4. Vortex briefly and incubate for 5 min at room temperature.
5. Apply magnet to samples.
6. Remove supernatant once solution is clear.
7. While samples are still attached to the magnet, carefully add 180 μL of 80% ethanol to each sample.
8. Incubate at room temperature for 30 s.
9. Aspirate the supernatant.
10. Repeat **steps 7–9**.
11. Allow residual ethanol to evaporate for 5 min, being careful not to over dry the beads; under-dried beads will slightly shiny, while over-dried beads will start to separate/crack on the sides of the PCR tube. Adequately dried beads will be matte in appearance.
12. Remove samples from magnet.
13. Add 20 μL of DNA Purification Elution Buffer and vortex to mix.

14. Incubate samples for 5 min at room temperature.
15. Apply magnet to samples.
16. Transfer supernatant to a fresh Eppendorf tube once the solution is clear.

3.1.4 Quantification and QC

1. Quantify libraries using a Qubit 1× dsDNA HS assay (*see* **Note 9**).
2. Prepare 3 μL of ~1 ng/μL of each sample to assess size distribution on a High Sensitivity D1000 TapeStation (Fig. 1; *see* **Note 10**).
3. Libraries are ready for sequencing (*see* **Note 11**).

3.2 RNA-Seq Library Preparation

All steps are done according to manufacturer's instructions, unless otherwise noted.

3.2.1 RNA Extraction

If integrating ATAC-seq and RNA-seq results, cells collected in RNAlater need to be pelleted and RNA extracted prior to sequencing. RNA can be extracted using the RNeasy Mini Kit (Qiagen) or other comparable methods, depending on laboratory preference.

1. Add 10 μL of β-mercaptoethanol per 1 mL of Buffer RLT; protocol requires 600 μL of Buffer RLT per sample.
2. Pellet cells through centrifugation at 3000 × *g* and aspirate RNAlater (*see* **Note 12**).
3. Loosen the cell pellet thoroughly by flicking the tube.
4. Add 600 μL of Buffer RLT + β-mercaptoethanol.
5. Homogenize cells by vortexing each tube for 1 min (*see* **Note 13**).
6. Add 600 μL of 70% ethanol and mix well through pipetting.
7. Transfer up to 700 μL of sample to an RNeasy spin column placed in a 2 mL collection tube. Centrifuge for 15 s at >8000 × *g* and discard flow-through. As the sample volume exceeds the maximum volume of the column, this will have to be done in two stages. The collection tube can be reused in next step.
8. Add 700 μL of Buffer RW1 to spin column and centrifuge for 15 s at >8000 × *g*. Discard flow through, ensuring that the tip of the column does not come into contact with the flow-through.
9. Add 500 μL of Buffer RPE to the column. Centrifuge for 2 min at >8000 × *g*.
10. Place spin column into a new 2 mL collection tube, once again ensuring that the column does not come into contact with the flow-through to prevent carryover of ethanol that may

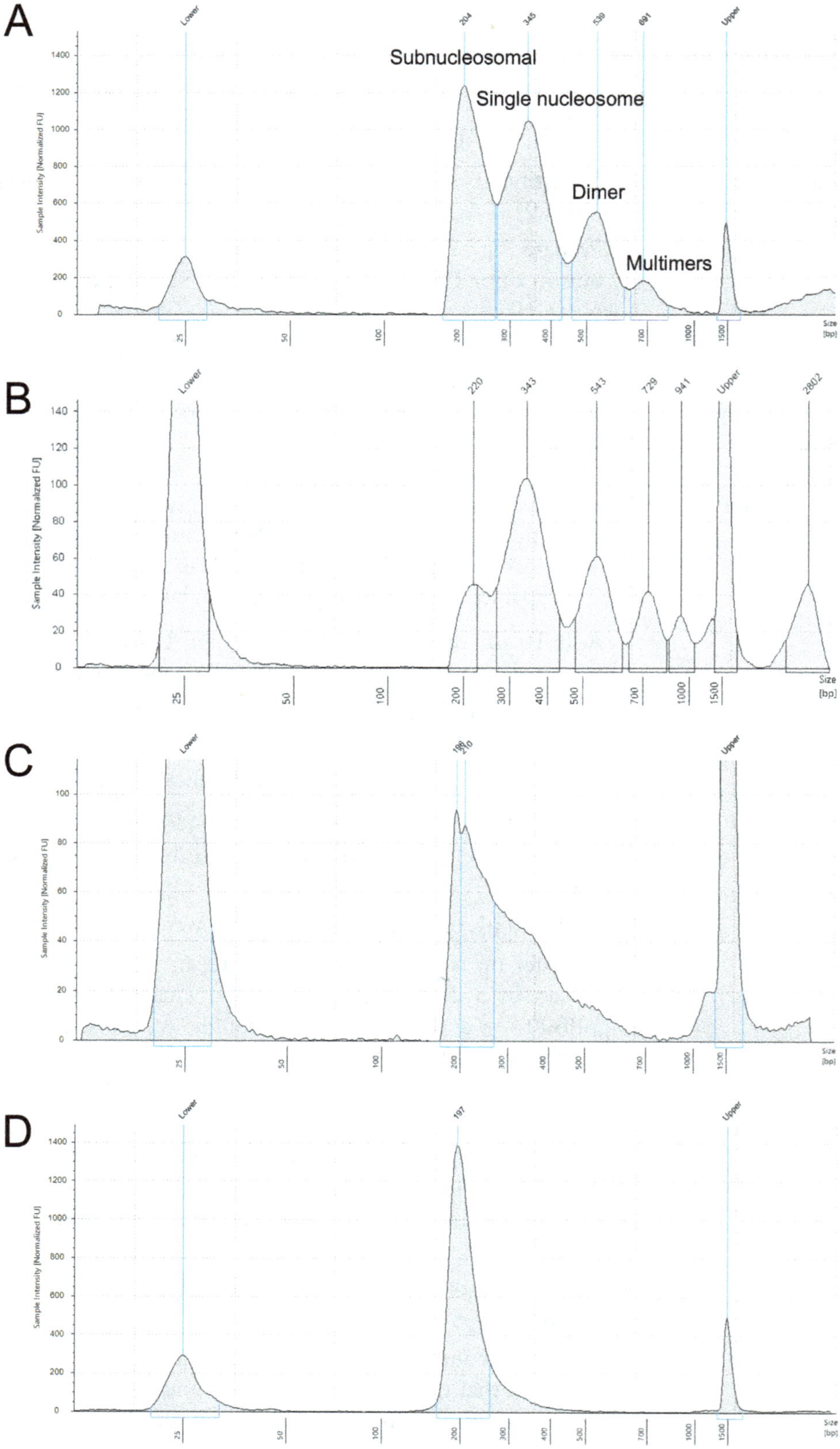

Fig. 1 Example TapeStation traces from successful and unsuccessful ATAC-seq library preparations. (**a**) Annotated TapeStation trace from a successful ATAC-seq library generation. The 150–180 bp periodicity in the

interfere with downstream sequencing. Centrifuge at full speed for 1 min.

11. Place spin column in a new 1.5 mL collection tube and add 30–50 μL of RNAse-free water to the spin column membrane and incubate for 1 min at room temperature. Centrifuge for 1 min at >8000 × *g* to elute.
12. To increase final RNA concentration, reuse the eluate from the previous step by applying it back onto the spin column membrane, then centrifuge again for 1 min at >8000 × *g*.

3.2.2 DNase Treatment

To prevent any contaminating genomic DNA from interfering with downstream analysis, samples should be treated with DNase. The RNeasy mini kit allows for on-column DNase treatment, or after extraction in solution. We recommend using the TURBO DNA-free Kit (Invitrogen), after RNA extraction, according to manufacturer's instructions.

3.2.3 Quantification and QC

1. Quantify libraries using a Qubit and a Nanodrop.
2. Prepare 3 μL of 25–500 ng/μL of each sample to assess size distribution on an RNA ScreenTape for TapeStation (*see* **Note 14**).
3. Libraries are ready for sequencing.

3.3 ATAC-Seq Data Analysis Overview

Analysis of ATAC-seq data can be broadly divided into four key stages: preprocessing and alignment, read filtering, peak-calling, and quality control steps. The resulting output can subsequently be visualized, loaded into a differential ATAC-seq workflow and integrated with RNA-seq data (Fig. 2).

3.4 Preprocessing and Alignment

Initial pre-processing of raw ATAC-seq data follows a standard protocol, whereby read quality is checked and sequencing adaptors are removed. Reads are then aligned to a reference genome before proceeding to the post-alignment filtering steps.

1. In the command line, download the relevant reference genome from TriTrypDB.org using wget.

```
wget <link-to-reference-genome-fa>
```

2. FastQC should be run to do some simple quality control on the raw sequence data; the resulting.html file contains a summary

Fig. 1 (continued) trace indicates successful library generation: the initial peak represents nucleosome free DNA, with each subsequent peak reflecting the addition of another nucleosome. (**b** and **c**) Other example of TapeStation traces from successful ATAC-seq library preparations. (**d**) TapeStation trace from an unsuccessful ATAC-seq library generation lacking nucleosomal periodicity, attributed to a faulty thermomixer

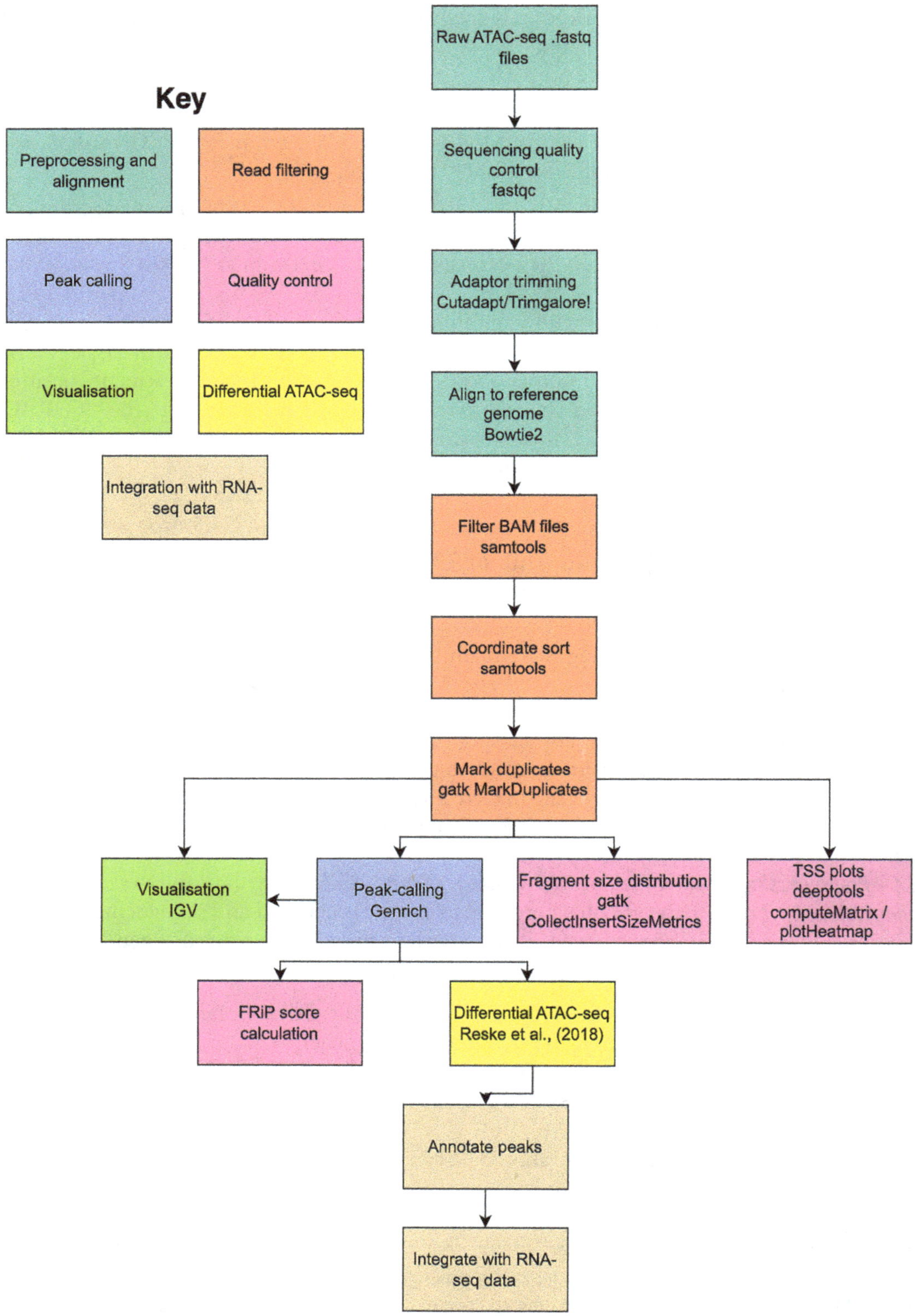

Fig. 2 Overview of ATAC-seq data analysis pipeline. The pipeline can be split up into the following steps: preprocessing and alignment, read filtering, peak-calling, and quality control steps. This involves running fastqc on raw sequencing reads, trimming read adaptors from raw reads using Cutadapt and TrimGalore!, aligning to a reference genome with Bowtie2, before filtering low-quality or unmapped reads with samtools

including base quality, adaptor contamination, GC content and overrepresented sequences. Examples of some FastQC modules that may fail for ATAC-seq samples are shown in Fig. 3.

Run fastqc on all gzipped files using a wildcard (*).

```
fastqc *.gz
```

3. Use Cutadapt and Trim Galore! to trim low-quality bases and adaptor sequences from paired-end reads, as well as run FastQC on the trimmed files to assess data quality after trimming.

   # TrimGalore! is a wrapper for Cutadapt, providing extra functionality such as adaptor detection and integration with FastQC, and thus requires Cutadapt to be installed.

   # --phred33 specifies the quality encoding of input fastq files; --paired indicates that sequencing data is paired-end; --fastqc instructs TrimGalore! to run FastQC on the output data; there will be two files for each sample if paired-end sequencing was performed.

```
trim_galore --phred33 --paired --fastqc sam-
ple1_S1_L001_R1_001.fastq.gz sample1_S1_L001_R2_001.fastq.gz
```

   # The fastqc output should now show no adaptor contamination, although there will be a warning for sequence length distribution.

4. Align trimmed reads to the reference genome using Bowtie2

   # Bowtie2 is a fast, memory efficient tool ideal for aligning paired-end, short-read sequences to a reference genome.

   # Bowtie2 requires that the reference genome be indexed prior to alignment. This should result in six files ending in.bt2

   # <reference_genome.fa> will be the input reference genome downloaded from TriTrypDB; <indexed_reference_-genome> will be the basename for the resulting index files – e.g. 'antat1.1'.

```
bowtie2-build <reference_genome.fa> <indexed_reference_gen-
ome>
```

Fig. 2 (continued) and removing PCR duplicates with GATK MarkDuplicates. Peaks (regions of significantly enriched reads) can then be called using a peak-caller such as Genrich. Output BAM/bigwig files can be viewed in IGV, which can also be used to inform peak-calling parameter optimization. Overall quality of ATAC-seq reads can be judged using nucleosomal periodicity plots, enrichment at TSSs seen in TSS plots, and the fraction of reads in peaks (FRiP) score. Output files can subsequently be loaded into a differential ATAC-seq workflow and integrated with RNA-seq data

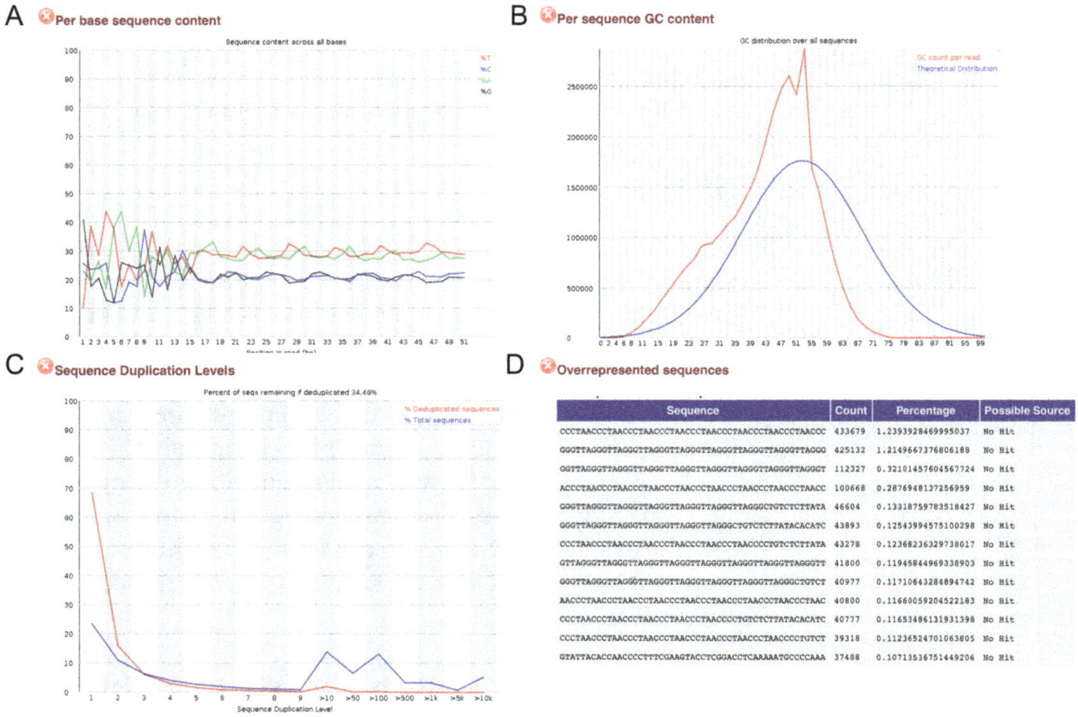

Fig. 3 FastQC modules that may fail when running on ATAC-seq samples. (**a**) Per base sequence content may fail due to the pattern observed in the first ~15 bp. This is likely due to the Tn5 insertion site sequence bias. (**b**) The per sequence GC content has failed in all ATAC-seq experiment we have performed. This may be linked to the GC content of regions of open chromatin. (**c**) The sequence duplication module may fail due to the repeated sampling of open chromatin by ATAC-seq, exacerbated by *T. brucei*'s compact genome as well as the presence of PCR duplicates. (**d**) Highly repetitive sequences, indicative of telomeric repeats, may be found in the overrepresented sequences module, causing it to fail

Now align trimmed reads to the indexed genome.

--phred33 specifies the quality encoding of input fastq files; --mm enables memory mapping, reducing RAM usage – this can be useful if the genome is large, or if working on a server where memory is limited; --maxins sets the maximum fragment size; --very-sensitive prioritises sensitivity over speed, improving accuracy for reads that would otherwise be more difficult to map due to sequencing errors or mismatches etc.; --threads determines the maximum number of cores to be used; -x specifies the indexed reference genome. This should only include the prefix used during indexing – e.g., 'indexed_reference_genome.1.bt2' would be 'indexed_reference_genome', or 'antat1.1.1.bt2' would be 'antat1.1; -1 and -2 specify the fastq files for forward (-1) and reverse (-2) reads from paired-end data; 2> redirects the Bowtie2 standard error to a log file which reports mapping rates. The Bowtie2 output file is then piped (|) to samtools to convert it to a BAM file: -h includes the SAM header in the output; -b outputs the file in BAM format; -

reads from the standard output of Bowtie; > writes the output BAM file.

```
bowtie2 --phred33 --mm --maxins 2000 \
--very-sensitive --threads 10 \
-x <indexed_reference_genome> \
-1 sample1_S1_L001_R1_001_val_1.fq.gz \
-2 sample1_S1_L001_R2_001_val_2.fq.gz \
2> sample1.log | samtools view -h -b - > sample1.bam
```

3.5 Post-alignment Read Filtering

After aligning, most ATAC-seq protocols will perform three key filtering steps: removal of low-quality or unmapped reads; removal of mitochondrial reads (*see* **Note 15**) and removal of PCR duplicates.

1. Filter BAM files to remove non-aligned/poorly mapped reads using samtools view, leaving high-confidence alignment and properly paired reads for downstream analysis.

   # -h includes the header in the output file which includes information such as sample name, reference sequences, sequencing platform etc.; -b outputs the file in BAM format; -F excludes reads with specific flags specified by the subsequent integer. In this case, -F 1804 excludes unmapped reads, reads where the mate is unmapped, secondary alignments, and reads failing quality checks; -f includes reads with specific flags specified by the subsequent integer. -f 2 only includes reads that are properly paired; -q sets the minimum mapping quality (MAPQ)score threshold; -@ specifies the number of threads to be use for multithreaded processing; -o determines the output file name and location.

   ```
   samtools view -h -b -F 1804 -f 2 -q 30 -@ 10 sample1.bam -o sample1_filtered.bam
   ```

2. Use samtools sort to sort reads by coordinate before marking or removing duplicates that arise due to PCR amplification.

   # PCR duplicates do not provide additional information as they are a technical artefact of the PCR amplification step. These can be filtered out using GATK MarkDuplicates as they will have identical start and end coordinates for both read pairs. Prior to duplicate removal, BAM files need to be sorted by coordinate.

   # samtools sort sorts by coordinate by default; -@ specifies the number of threads to be use for multithreaded processing; -o determines the output file name and location.

   ```
   samtools sort -@ 10 sample1_filtered.bam -o sample1_filtered_sorted.bam
   ```

   # GATK also requires read group information. As we didn't specify this during alignment, we will need to add it now using AddOrReplaceReadGroups.

--INPUT specifies the input bam file to which read groups will be added; --OUTPUT determines the output file name and location; RGID is a unique identifier and must unique to each sample; RGLB specifies the library that the sample came from – samples sequenced on the same run will have the same RGLB; RGPL is the platform used for sequencing; RGPU is the platform unit, and can be the same for all sample if prepared on the same platform and under the same conditions; RGSM determines the sample name.

```
gatk AddOrReplaceReadGroups \
--INPUT sample1_filtered_sorted.bam \
--OUTPUT sample1_filtered_sorted_RG.bam \
--RGID 1 \
--RGLB lib1 \
--RGPL Illumina \
--RGPU unit1 \
--RGSM sample1
```

Finally, duplicates are removed using GATK MarkDuplicates.

-I should be the BAM file with poorly mapped reads removed, with read group information added and alignments sorted by coordinate; -O determines the output file name and location; -M determines the name and location of a.txt file containing information on the duplication rate; --ASSUME_-SORT_ORDER specifies which order the file has been sorted in (in this case, by coordinate).

Ideally, duplicates should be as low as possible. Depending on sequencing depth, we get a duplication rate of ~15-25%.

```
gatk MarkDuplicates \
--INPUT sample1_filtered_sorted.bam \
--OUTPUT sample1_filtered_sorted_marked.bam \
--METRICS_FILE sample1_dups_metrics.txt \
--ASSUME_SORT_ORDER coordinate
```

3.5.1 Visualization

The processed ATAC-seq BAM files can now be visualized directly using a genome browser such as IGV or UCSC. However, conversion of BAM files to bigwig allows for more efficient visualization; the bigWig file format is used to store continuous genomic data and is also much smaller than the BAM file from which they were generated. Converting the BAM files to bigWig will also be useful for plotting TSS enrichment during subsequent quality control. The deepTools command line package provides a tool, bamCoverage, that can be used to generate the bigWig files. Their compact, indexed format allows for much faster loading and smoother navigation compared to using BAM files, and normalization using RPGC accounts for differences in read depth, enabling

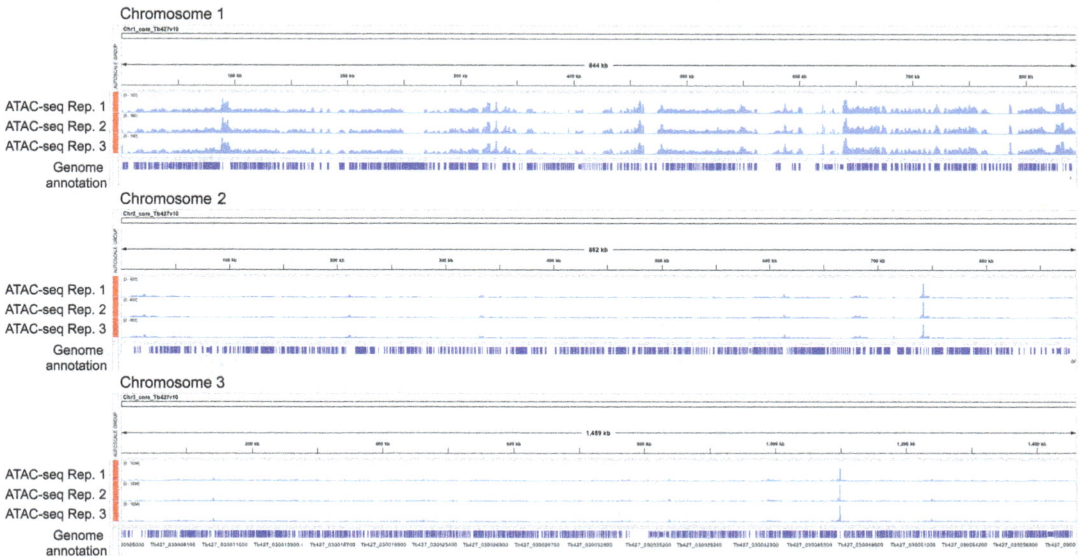

Fig. 4 Visualization of 3 ATAC-seq replicates across chromosomes 1, 2, and 3. Pale blue tracks represent the normalized ATAC-seq signal bigwig files. The dark blue tracks represent the genome annotation file (GFF), containing information about gene names and locations

comparisons across samples. Visualization in this example will be performed using IGV (Fig. 4).

1. Prior to conversion to bigwig, BAM files need to be indexed using samtools.

   # samtools index requires the input file as the first argument, and output file as the second.

   ```
   samtools index sample1_filtered_sorted_marked.bam sample1_filtered_sorted_marked.bam.bai
   ```

2. Calculate the effective genome size with khmer unique-kmers.py.

   # bamCoverage has several normalisation options. For ATAC-seq, we will use the reads per genomic content (RPGC), which is calculated using the total number of mapped reads * fragment length / effective genome size. To do this, we will therefore need to calculate the effective genome size, which represents the mappable genome, using khmerEnv's unique-kmers.py.

   # -k determines the k-mer size; -R indicates the name and location that the report generated should be written to; the final parameter should be the fasta file for the genome that we are calculating the effective genome size of.

   ```
   unique-kmers.py -k 100 -R effective_genome_size.txt <reference_genome.fa>
   ```

3. Convert BAM files to bigwig using featureCounts.

   # --bam defines input file; --outFileName defines the output file name/location; --outFileFormat determines the format

of the output file – in this case, bigwig; --normalizeUsing determines which normalisation method to use out of RPKM, CPM, BPM or RPGC; --ignoreDuplicates removes reads with the same orientation and start position. This shouldn't be an issue if using the marked bam file; --centerReads shifts the position of reads to the midpoint of fragments to better represent where transposition occurred; --effectiveGenomeSize is required for RPGC normalisation and should be the number calculated earlier; --binsize sets the resolution of the bigwig files in basepairs. A binsize of 1 creates a bigwig file to basepair resolution, although will be slower to computer. For TSS plots, ideally this value should be 10 or below.

```
bamCoverage \
--bam sample1_filtered_sorted_marked.bam \
--outFileName sample1_filtered_sorted_marked.bw \
--outFileFormat bigwig \
--normalizeUsing RPGC \
--ignoreDuplicates --centerReads \
--effectiveGenomeSize <effective-genome-size> \
--binSize 1
```

4. Open the IGV desktop application.
5. Import genome used for alignment into IGV using Genomes > Load genome from file.
6. Load ATAC-seq.bw files into IGV using File > Load from file.
   (a) To make comparisons between multiple samples, highlight all.bw files in IGV, right click > Group Autoscale.
7. Load.narrowPeak files using File > Load from file.
   (a) This is useful for comparing peak calling software/parameters. If performing differential ATAC-seq, it also makes it much easier to identify regions specified in the output file of significantly more enriched regions.
8. Download relevant.gff genome annotation file from TriTrypDB. The GFF file contains information about gene names and locations.
9. Load.gff file into IGV using File > Load from file.

3.5.2 Peak Calling

For most downstream applications, ATAC-seq peaks—corresponding to regions where there is significant read-pile up compared to a background—need to be called using a peak caller such as MACS2/3 or Genrich. Which peak caller, and which parameters to use, will depend on the samples being analyzed, and should be optimized by comparing results visually in IGV. Although the ENCODE pipeline recommends MACS2 for peak calling [11], we have found that *T. brucei* chromatin accessibility is characterized by diffuse, broad regions rather than sharp, discrete

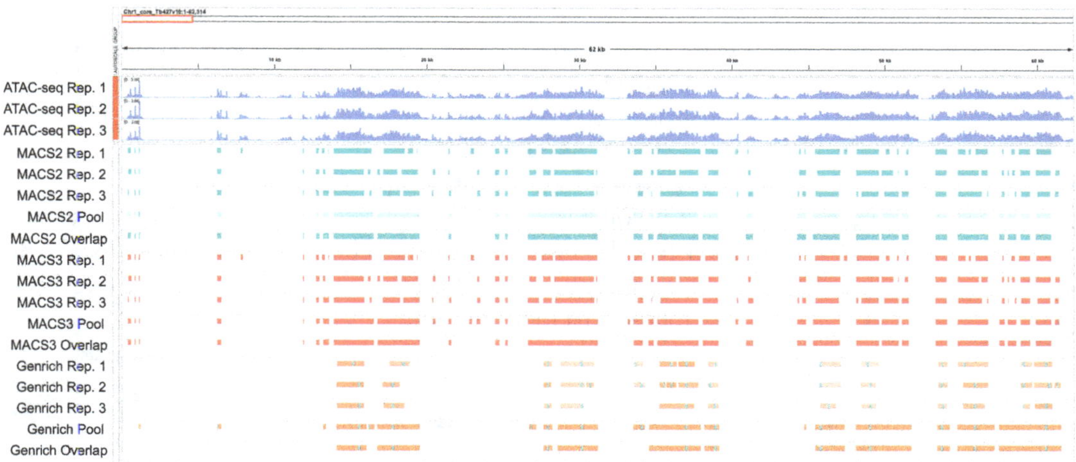

Fig. 5 Comparing MACS2, MACS3, and Genrich peak calling software. The first three tracks are the bigwigs for three replicates of AnTat1.1 BSFs, with the peaks called for MACS2 (green), MACS3 (red), and Genrich (orange) underneath. For each of the peak calling software, the first three tracks are the individual replicates, followed by the pooled replicates and finally the overlap peaks, where for peaks called on pooled replicates, only peaks that overlap at least 50% with a peak from each of the individual replicates are kept. On the whole, Genrich is more conservative with the number of peaks it calls and is more likely to merge multiple smaller peaks into one larger peak. For all samples, using the overlap peak calling is a more conservative approach

peaks, likely due to its polycistronic transcription. Genrich captures these broad accessibility regions more effectively than MACS2 (Fig. 5) and, as it has an "ATAC-seq mode" (whereas MACS2/3 are designed for ChIP-seq datesets), it also performs Tn5 shifting, adjusting peak positions to account for transposase insertion bias and more accurately reflect true Tn5 binding sites.

1. Sort BAM files by name using samtools sort.

   # Although MACS2/3 can use the coordinate sorted alignments, Genrich requires BAM files to be sorted by name – if opting to use MACS, this initial step can be skipped.

   # The parameters are very similar to the previous coordinate sort. -n instructs samtools to sort by name.

   ```
   samtools sort -n -@ 10 sample1_filtered_sorted_marked.bam -o sample1_filtered_sorted_marked_namesort.bam
   ```

2. Call peaks using Genrich on individual replicates.

   # Genrich peaks for each sample can now be called.

   # -t specifies the input BAM file; -o determines the output peak file name and location; -q sets the maximum FDR threshold for peak calling—0.05 means only peaks with a FDR of <5% will be kept; -j sets Genrich into ATAC-seq mode; -v enables verbose mode, printing status updates, including the number of peaks found and any warnings or errors, to stderr; -b produces a bed file of the reads/fragments/intervals used to call peaks and can be useful when trying to refine parameters.

```
Genrich -t sample1_filtered_sorted_marked_namesort.bam \
 -o sample1_genrich.narrowPeak \
 -q 0.05 -j -v \
 -b sample1_genrich.bed
```

3. Use Genrich to call peaks on pooled replicates.

   # If you have multiple replicates in a group, Genrich can be used to call peaks on pooled replicates. It first calls peaks on individual replicates and then combines the results using Fisher's method to identify significant peaks across replicates.

```
Genrich -t sample1_filtered_sorted_marked_namesort.bam,sample2_filtered_sorted_marked_namesort.bam,sample3_filtered_sorted_marked_namesort.bam \
 -o bsf_genrich_pool.narrowPeak \
 -q 0.05 -j -v \
 -b bsf_genrich_pool.bed
```

4. Calculate "overlap peaks" from pooled peaks.

   # The ENCODE pipeline recommends that, from peaks called on pooled replicates, only peaks that overlap at least 50% with a peak from each of the individual replicates are kept. Reske et al. [12] published a bash script to execute this on two replicates where peaks were called with MACS2. The script below is a modified version of the Reske et al. [12] script, adapted for three replicates where peaks were called using Genrich: the Genrich narrowPeak file contains additional columns compared to a MACS2 narrowPeak file requiring adjustments to column references in the script. We will use intersectBed from bedtools.

   # intersectBed -wo looks for overlaps between input files and reports the overlapping regions, including all columns from both input files with the final column (column 21) indicating the number of base pairs that they overlap by; -a indicates the first input file (initially the pooled peaks file); -b indicates the second input file (initially the first replicate peak file); awk filters the intersectBed output according to the parameters supplied within single quotes; BEGIN{FS="\t"; OFS="\t"} instructs awk that the input (FS) and output (OFS) are tab-separated; s1 = $3–$2 calculates the width of the peak in the first input file by subtracting the start of the peak (column 2) from the end of the peak (column 3); s2 = $13–$12 calculates the width of the peak in the second input file by subtracting the start of the peak (column 12) from the end of the peak (column 13); if (($21/s1 > = 0.5) || ($21/s2 > = 0.5)) {print $0}} ensures that a peak is retained if at least 50% of its length in either input file is covered by the overlap (column 21) reported by intersectBed; cut -f 1–10 keeps only columns 1–10, which contain the peak information from the first input file; sort alphabetizes the output; uniq

removes duplicate lines ensuring only unique peaks are kept; the entire process is then repeated with the remaining replicates, using the previous output as the first input (-a stdin).

```
#!/bin/bash
REP1=$1
REP2=$2
REP3=$3
POOL=$4
# First intersection: ${POOL} with ${REP1}
intersectBed -wo -a ${POOL} -b ${REP1} | \
awk 'BEGIN{FS="\t";OFS="\t"} {s1=$3-$2; s2=$13-$12; if
(($21/s1 >= 0.5) || ($21/s2 >= 0.5)) {print $0}}' | \
cut -f 1-10 | sort | uniq | \
# Second intersection: result with ${REP2}
intersectBed -wo -a stdin -b ${REP2} | \
awk 'BEGIN{FS="\t";OFS="\t"} {s1=$3-$2; s2=$13-$12; if
(($21/s1 >= 0.5) || ($21/s2 >= 0.5)) {print $0}}' | \
cut -f 1-10 | sort | uniq | \
# Third intersection: result with ${REP3}
intersectBed -wo -a stdin -b ${REP3} | \
awk 'BEGIN{FS="\t";OFS="\t"} {s1=$3-$2; s2=$13-$12; if
(($21/s1 >= 0.5) || ($21/s2 >= 0.5)) {print $0}}' | \
cut -f 1-10 | sort | uniq > overlap_peaks.narrowPeak
```

This code can be submitted as a bash script; create an overlap_peaks.sh file containing the code and submit using bash overlap_peaks.sh sample1_genrich.narrowPeak sample2_-genrich.narrowPeak sample3_genrich.narrowPeak bsf_gen-rich_pool.narrowPeak.

3.5.3 Quality Control Overview

There are several QC metrics that can be used to confirm whether ATAC-seq library generation was successful. Several of them, including mapping statistics, mitochondrial contamination and duplication rate have been discussed previously. In this next section, fragment size distribution, transcription start site (TSS) enrichment, and the fraction of reads in peaks (FRiP) score shall be explained.

3.5.4 Fragment Size Distribution Plots

A successfully generated ATAC-seq library should show clear periodicity of around 200 bp in fragment size distribution plots, reflecting a nucleosome-free peak (~35–100 bp), mononucleosome peak (~150–250 bp), and dinucleosome and multinucleosome peaks (~350–450 bp and 600 bp+, respectively; Fig. 6). These plots can be generated via the R package ATACseqQC, or in the command line using Picard CollectInsertSizeMetrics. The code to generate plots using CollectInsertSizeMetrics will be presented here, as it is

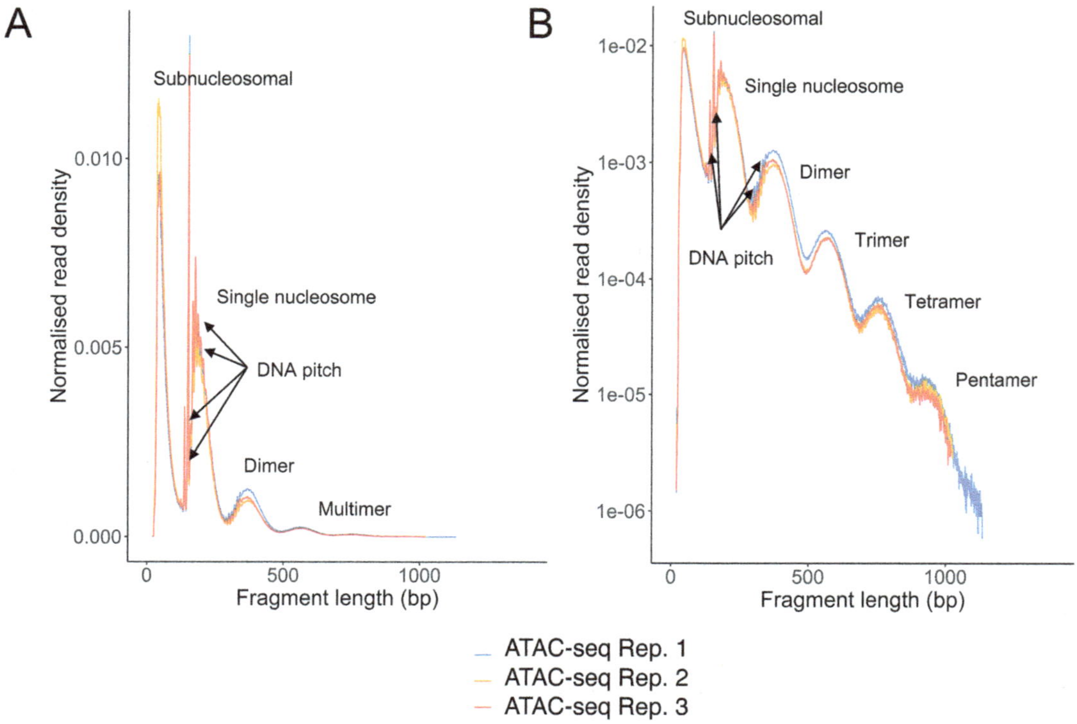

Fig. 6 Nucleosomal periodicity plots from three *T. brucei* BSF replicates on a linear (**a**) and logarithmic (**b**) scale. The clear periodicity within the insert size distribution corresponds to fragments protected by an increasing number of nucleosomes, with the first peak being subnucleosomal (i.e., no nucleosomes attached). The helical pitch of DNA is also apparent in these samples, evidenced by high-frequency oscillations between DNA fragments shorter than ~300 bp

easier to combine the results into one fragment size distribution plot for all samples in R.

1. Calculate the insert size distributions on the BAM files prior to duplicate removal using CollectInsertSizeMetrics. This will also produce an insert size histogram that can be used to assess the fragment size distribution.

   # gatk also includes some Picard tools, including CollectInsertSizeMetrics.

   # I = specifies the input BAM file; O = specifies the name and location of the output txt file which contains the insert size distribution and can be used for plotting in R; H = specifies the file to write the insert size histogram plot to.

```
gatk CollectInsertSizeMetrics \
I=sample1_filtered_sorted.bam \
O=sample1_fragment_size_dist.txt \
H=sample1_fragment_size_hist.pdf
```

2. To generate one plot containing the fragment size distributions for all samples, the CollectInsertSizeMetrics output.txt files need to be loaded into R, and the fragment counts normalized by dividing the count per insert size by the total number of paired reads for each sample.

```
library(tidyverse)
```

Make a list including the insert size distribution.txt file, the sample name, and the number of read pairs for each sample. The number of read pairs can be found in the insert size distribution txt file under "READ_PAIRS".

```
file_info <- list(
c("sample1_fragment_size_dist.txt", "Antat BSF Replicate 1", 15340662),
c("sample2_fragment_size_dist.txt", "Antat BSF Replicate 2", 16296420),
c("sample3_fragment_size_dist.txt", "Antat BSF Replicate 3", 17326892)
```

Define the process_file function, which will add a sample_name column and calculate the normalized read counts for each insert size, and then apply it to samples.

```
process_file <- function(file_info) {
file_name <- file_info[1]
sample_name <- file_info[2]
total_reads <- as.numeric(file_info[3])
```

Read file, skipping the first 9 rows containing file information and summary statistics.

```
df <- read_delim(file_name, delim = "\t", skip = 10)
```

Add a "sample_name" column.

```
df <- mutate(df, sample_name = sample_name)
```

Coerce "All_Reads.fr_count" column to numeric type.

```
df$All_Reads.fr_count <- as.numeric(df$All_Reads.fr_count)
```

Coerce "insert_sizes" column to numeric type.

```
df$insert_sizes <- as.numeric(df$insert_size)
```

Calculate the normalized fragment count by dividing the raw read counts by the total read count.

```
df <- mutate(df, normalised = 'All_Reads.fr_count' / total_-
reads)

return(df)
}
```

Apply the process_file function to each file.

```
data_frames <- map(file_info, process_file)
```

Combine the individual data frames into one.

```
all_counts <- bind_rows(data_frames)
```

Check that the combined data frame looks as expected.

```
View(all_counts)
```

Use ggplot to make the combined fragment size distribution plot for all samples.

```
p<-ggplot(data=all_counts, aes(x=insert_size, y=nor-
malised, colour=sample_name)) +
geom_line() +
theme
(legend.title = element_blank()) +
labs(x="Fragment length (bp)", y="Normalised read den-
sity", colour= "Sample") +
theme_classic()
p
```

The periodicity becomes more apparent when the y-axis is plotted on a logarithmic scale.

```
plog <- p + scale_y_log10()
plog
```

3.5.5 TSS Enrichment

A common QC metric used in assessing ATAC-seq library quality is to calculate the enrichment of reads near TSSs; in eukaryotes, various studies have shown that TSSs are enriched in accessible chromatin [8]. It is therefore common to plot the enrichment around TSSs as QC metric, comparing enrichment at TSSs to the regions 2–3000 bp upstream and downstream of the TSS. Although *T. brucei* transcription is polycistronic, *T. brucei* divergent TSSs (dTSSs) are enriched for histone variants H2AV and H2BV, which are also associated with regions of open chromatin [13]. Here, we present a method of plotting TSS enrichment for dTSSs using the computeMatrix and plotHeatmap packages from deeptools (Fig. 7).

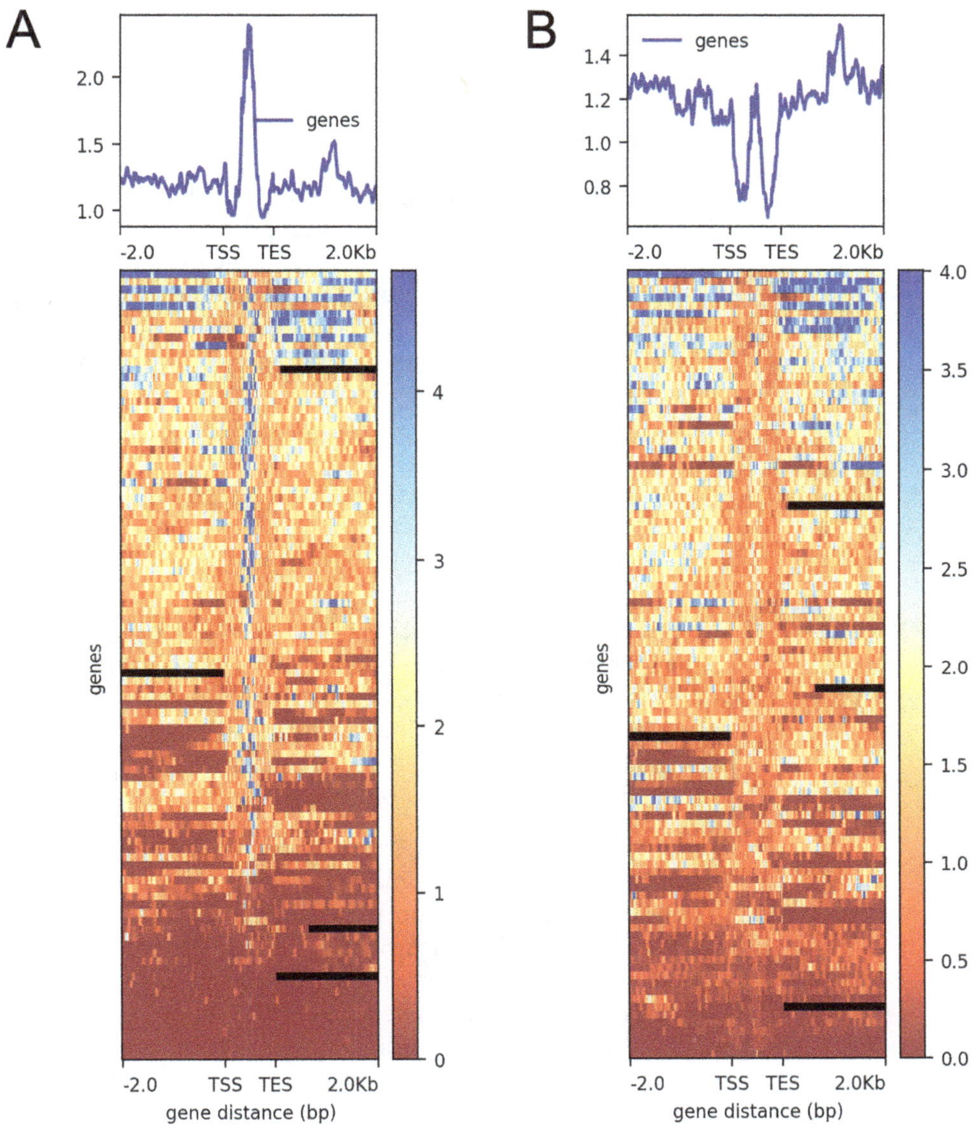

Fig. 7 TSS plots are used as a quality control metric for ATAC-seq libraries. Successful libraries should show a short, sharp peak of reads enriched around the TSS (shown in the center of the plot) relative to regions more distal from it. (**a**) Example of a successful library generation, with a clear peak in the center of the plot. (**b**) A less successful library generation, with a much smaller peak around the TSS

1. Identify dTSSs from genome annotation file and save in.bed format.
2. Prepare data for plotting using deeptools computeMatrix.

   # scale-region; -S determines input bigwig file; -R determines the bed file containing the TSS coordinates; -b 2000 sets the distance upstream of the dTSS to 2000 bp; -a sets the distance downstream of the dTSS; --outFileName determines the name of the ouput matrix file used in the plotHeatmap step.

```
computeMatrix \
scale-regions \
-S sample1_filtered_sorted_marked.bw \
-R <link-to-coord-file> \
-b 2000 -a 2000 \
--outFileName sample1_dtss.dat
```

3. Create TSS plot and heatmap using deeptools plotHeatmap.

   # -m determines the input matrix file from computeMatrix; -out determines the name of the plot image that will be generated; --heatmapHeight sets the height of the plot in cm; --sortUsing orders the rows of the heatmap by the mean signal across all samples.

```
plotHeatmap \
-m sample1_dtss.dat \
-out sample1_dtss.png \
--heatmapHeight 15 \
--sortUsing mean
```

3.5.6 FRiP Score

The Fraction of Reads in Peaks (FRiP) score is a common quality control metric that measures the proportion of total mapped reads that are within called peaks. It is used as a measure of signal-to-noise, and, generally speaking, a high FRiP score (>0.3) reflects high-quality data. This can be calculated using the featureCounts package from Subread.

1. Convert the Genrich narrowPeak file to SAF.

   # featureCounts requires the genomic features of interest (in this case, peaks) to be in the SAF file format, containing ID, Chr, Start, End, and Strand columns. We will combine the Chromosome name, start and end for an ID column, and as the peak data is unstranded, put "." in the strand column.

   # awk is a command-line tool used for text editing; -F tells awk the input is tab-delimited; BEGIN {OFS = FS} ensures the output is also tab-delimited; $2 = $2 + 1 shifts the start coordinate from 0-based (as seen in BED format) ti 1-based, required for SAF files; peakid = "narrowPeak_" ++nr generates sequential peak IDs, starting from 1 and will increase with each line; print peakid,$1,$2,$3,"." prints columns containing the peak ID, a column containing the chromosome name, a column containing the peak start coordinate, a column containing the peak end coordinate and a column containing strand information, in this case, "." for all as ATAC-seq data is unstranded; this is followed by the input peak file, > which takes the awk output and writes it to the specified output.saf file.

```
awk -F $'\t' 'BEGIN {OFS = FS} { $2=$2+1; peakid="-
narrowPeak_"++nr; print peakid,$1,$2,$3,"."}' sample1_gen-
rich.narrowPeak > sample1_genrich_peaks.saf
```

2. Calculate FRiP score using featureCounts.

   # -p treat input as a paired-end BAM file; -a defines the saf annotation file (i.e., peak information); -F SAF specifies that the annotation file is in SAF format; -o defines the output.txt file containing count data; 2 > writes the STDERR to a.txt file, which will contain the calculated FriP score.

   ```
   featureCounts -p -a sample1_genrich_peaks.saf -F SAF
   -o sample1_genrich_FRiP.txt sample1_filtered_sorted_mar-
   ked_namesort.bam 2> sample1_genrich_FRiP_summary.txt
   ```

3. Find the calculated FRiP score from the output summary.txt file under "successfully assigned alignments."

3.6 Differential ATAC-Seq

A valuable resource for performing differential ATAC-seq is Reske et al. [12], which offers a comprehensive, step-by-step guide to normalization and differential accessibility analysis using *csaw* and EdgeR. These steps will not be repeated in detail here. An important part of the normalization process requires, prior to duplicate removal, downsampling reads so samples have the same complexity, before continuing with the pipeline. To note, if performing differential ATAC-seq using EdgeR in combination with peaks called by Genrich, conversion of BAM files to BEDPE format and Tn5 shifting of reads are not required, as Genrich performs this shifting internally when run in ATAC-seq mode.

For the analysis, coordinate-sorted BAM files (i.e., prior to name-sorting for Genrich peak calling) will be used, as indexing with **samtools** requires BAM files to be sorted by genomic coordinates (i.e., the BAM files from **step 6** from Subheading 3.3.1 and the index BAM files from Subheading 3.3.2, **step 1**).

3.7 ATAC-Seq Peak Annotation

Linking peaks to genes after differential ATAC-seq is an important step for subsequent downstream analysis and for integration with RNA-seq data. When working with mammalian cells, it is common to link a peak to a gene when it is within the TSS of that gene, however, for euglenozoans such as *T. brucei*, the polycistronic nature of transcription prevents this. One option is to instead link peaks to genes where they overlap. As Genrich tends to call fairly broad peaks, this will mean that the same peak will likely contain multiple genes. The following R code uses the GRanges table/ object of regions of significantly different chromatin accessibility generated during the differential ATAC-seq analysis.

```
library(GenomicRanges)
library(rtracklayer)
library(readr)
```

```
#####
# Preparing gene information
```

Read in relevant gff file as a Granges object

```
gff_file <- "path/to/your/annotation_file.gff"
gff_data <- import.gff(gff_file)
```

Limit gff to just protein coding genes

```
genes <- gff_data[gff_data$type == "protein_coding_gene",]
```

Subset genes to include only relevant columns

```
genes_gr <- GRanges(
 seqnames = seqnames(genes), # Extract seqnames from the
GRanges object
 ranges = IRanges(start = start(genes), end = end(genes)),
 strand = strand(genes), # Extract strand from the GRanges
object
 gene_id = mcols(genes)$ID, # Use mcols to access metadata
columns
 description = mcols(genes)$description # Access the descrip-
tion from metadata
)
```

#####

Preparing differential peaks file

If necessary, read in the significant peaks file from differential ATAC-seq analysis as a Granges object.

```
final.merged.peaks <- read_tsv("differential_peaks.txt")


final.merged.peaks <- GRanges(
seqnames = final.merged.peaks$seqnames,
 ranges = IRanges(start = final.merged.peaks$start, end =
final.merged.peaks$end),
strand = "*", # or use final.merged.peaks$strand if you want
strand info
 mcols = final.merged.peaks[, setdiff(names(final.merged.
peaks), c("seqnames", "start", "end"))]
)
```

#####

ATAC-seq peak annotation

Find overlaps between peaks and genes

```
overlaps.genes <- findOverlaps(final.merged.peaks, genes_gr)
```

Set up empty gene_id and description columns with NA

```
mcols(final.merged.peaks)$gene_id <- NA
mcols(final.merged.peaks)$description <- NA
```

Assign gene_id and description based on overlap.genes indices

```
mcols(final.merged.peaks)$gene_id[queryHits(overlaps.genes)]
<- mcols(genes_gr)$gene_id[subjectHits(overlaps.genes)]
mcols(final.merged.peaks)$description[queryHits(overlaps.
genes)] <- mcols(genes_gr)$description[subjectHits(overlaps.
genes)]
```

Create a data frame with all metadata from peaks and gene information

```
annotated.peaks <- as.data.frame(final.merged.peaks)
```

Save the annotated peaks data frame as a.csv file

```
write.csv(annotated.peaks, file = "annotated_peaks.csv")
```

3.8 Differential RNA-Seq

If RNA samples were also collected, differential RNA-seq can be performed to identify genes that are more highly expressed in one group compared to the other. The main steps in a typical RNA-seq differential expression workflow include trimming reads, quantifying expression using tools such as Salmon or FeatureCounts, and building a statistical model in R to identify differentially expressed genes. As the differential ATAC-seq analysis was performed using EdgeR, it is best practice to use EdgeR for differential RNA-seq analysis too. Here, we will describe how to quantify transcript counts with Salmon, before converting from transcript-level counts to gene-level counts with tximport, ready for analysis with EdgeR. Differential gene expression analysis with EdgeR can then be performed following tutorials from the EdgeR User's Guide or Larinov [13] (http://larionov.co.uk/deg_ebi_tutorial_2020/index.html).

1. In the command line, download annotated transcript file from TriTrypDB.org using wget.

```
wget <link-to-annotated-transcripts-fa>
```

2. Perform FastQC on raw fastq files.
3. Remove adaptor sequences and low quality bases using TrimGalore!

```
trim_galore --phred33 --paired --fastqc sam-
ple1_RNA_1.fastq.gz sample1_RNA_2.fastq.gz
```

4. Quantify transcript counts using Salmon.

Salmon is programme that produces accurate, transcript level quantification information from RNA-seq data. It is also able to handle multi-mapping reads.

First, much like when aligning using Bowtie2, we need to index the transcriptome file using Salmon.

-t specifies the transcriptome fasta file; -i specifies the directory that the index files will be located in.

```
salmon index -t annotated_transcripts.fa -i
transcriptome_index
```

Quantify transcriptome counts using salmon quant. This will produce a folder containing a "quant.sf" file for each sample, containing information on the length, effective length (transcripts per million), and read count, which will be used when building the differential RNA-seq model.

-i denotes the folder containing the annotated transcripts index file; -l specifies the library type, with A indicating the program should do this automatically; −1 first fastq file for sample of interest; −2 the second fastq file for sample of interest; —validateMappings enables more selective mapping, which can be useful for more complex transcriptomes or multimapping-prone datasets.

```
salmon quant -i transcriptome_index -l A -1 sample1_R-
NA_1_val_1.fastq.gz -2 sample1_RNA_2_val_2.fastq.gz -o
sample_1_RNA --validateMappings
```

5. Convert the gff file to gtf using gffread.

gffread can be used to read, convert and filter gff and gtf files. We will be using it to convert the gff file from TriTrypDB to gtf.

-T -o converts file from gff to gtf.

```
gffread annotation_file.gff -T -o annotation_file.gtf
```

6. Now in R, annotate quant files using tximport, converting counts from transcript-level to gene-level.

```
library(GenomicRanges)
library(rtracklayer)
library(readr)
library(tximport)
library(edgeR)
```

#####

Prepare a sample information.csv file. This should contain a "group" column, specifying the groups to be compared, and an "ID" column including sample identifiers. Other relevant information that might be included, if applicable, would be "batch," "replicate," "treatment," "line," etc. The first column should include the name of the folder containing the quant file for the relevant sample.

```
sample_info <- read.csv("sample_info.csv", row.names=1, stringsAsFactors = FALSE)
samples <-rownames(sample_info)
```

quantfiles_path <- "path/to/salmon/output/files" quantfiles <- file.path(samples, "quant.sf")

Read in converted gtf file.

```
gtf_file<-"path/to/your/annotation_file.gtf"
head(gtf_file)
```

#####

Convert transcript-level counts to gene-level counts with tximport.

Extract only the CDS information from the GFF file.

```
annotation_transcript <- elementMetadata(import(gtf_file, feature.type = 'CDS'))
```

Prepare a transcript-to-gene mapping table.

```
tx2geneid <- annotation_transcript[,c('transcript_id', 'gene_id')]
```

Use the mapping table to integrate the transcript-level quants to gene-level.

```
txi_gene <- tximport(quantfiles, type = 'salmon', tx2gene = tx2geneid, ignoreTxVersion = FALSE, txOut=FALSE)
```

Extract the raw gene-level counts and assign sample names.

```
salmon_counts <- txi_gene$counts
colnames(salmon_counts)<-samples
```

7. Create a DGEList object of read counts.

```
dgl <- DGEList(
counts = salmon_counts,
samples = sample_info,
group = sample_info$group
)
```

8. Perform differential RNA-seq analysis using EdgeR. Tutorials on how to do this can be found in the EdgeR User's guide, in particular the case studies, or the EdgeR tutorial in ref. 13.

3.9 Integrating ATAC-Seq and RNA-Seq

Once differential ATAC-seq and RNA-seq analyses have been performed and peaks have been annotated with gene names, the two datasets can be integrated. This is a relatively straightforward process that involves matching gene names between the two outputs in R.

```
library(readr)
```

If necessary, load differential (annotated) ATAC-seq and RNA-seq outputs.

```
annotated.peaks <- read_csv("annotated_ATAC_seq_output.csv",
col_names = TRUE)
```

Depending on how the RNA-seq data was saved (i.e., as a.csv or. txt), you may need to use read_csv or read_delim.

```
RNA_seq <- read_csv("RNA_seq_output.csv", col_names = TRUE)
```

or

```
RNA_seq <- read_delim("RNA_seq_output.txt", delim = "\t",
col_names = TRUE)
```

Rename RNA expression data columns.

```
col_names <- c("gene_id", "logFC_RNA", "logCPM_RNA", "F_RNA",
"PValue_RNA", "FDR_RNA")

colnames(RNA_seq) <- col_names
```

Merge the gene expression data with the annotated peaks data.

```
annotated_peaks_with_rna <- merge(annotated.peaks, RNA_seq,
by = "gene_id", all.x = TRUE)
```

Save the integrated ATAC-seq and RNA-seq object as a.csv.

```
write_csv(annotated_peaks_with_rna, file = "annotated_-
peaks_with_rna.csv")
```

4 Notes

1. The ATAC-seq protocol was optimized for AnTat1.1 BSFs by testing 10^5, 10^6, and 10^7 cells, with varying concentrations of Tn5 and length of time for the tagmentation reaction. 10^6

cells, 20% Tn5, and a 1 h tagmentation time was deemed to be the most optimal method on the basis of the TapeStation plots, fragment size distribution, and TSS plots (Fig. 8).

2. The pellet from the lysis buffer is very small and is only visible when spun down in a PCR tube. The pellet will need to be resuspended in the tagmentation buffer and transferred to an Eppendorf tube prior to incubation on a ThermoMixer.
3. As the lysis pellet is very small, removing the supernatant can be challenging. Swapping to a smaller pipette can make aspirating the supernatant easier, however, in practice, a small amount of lysis buffer can be left and tagmentation be successful.
4. Although the manufacturers have found that tagmentation can proceed with simple incubation with no agitation, we have found that issues with the ThermoMixer significantly reduced library quality. We have had good results using the Eppendorf ThermoMixer F1.5.

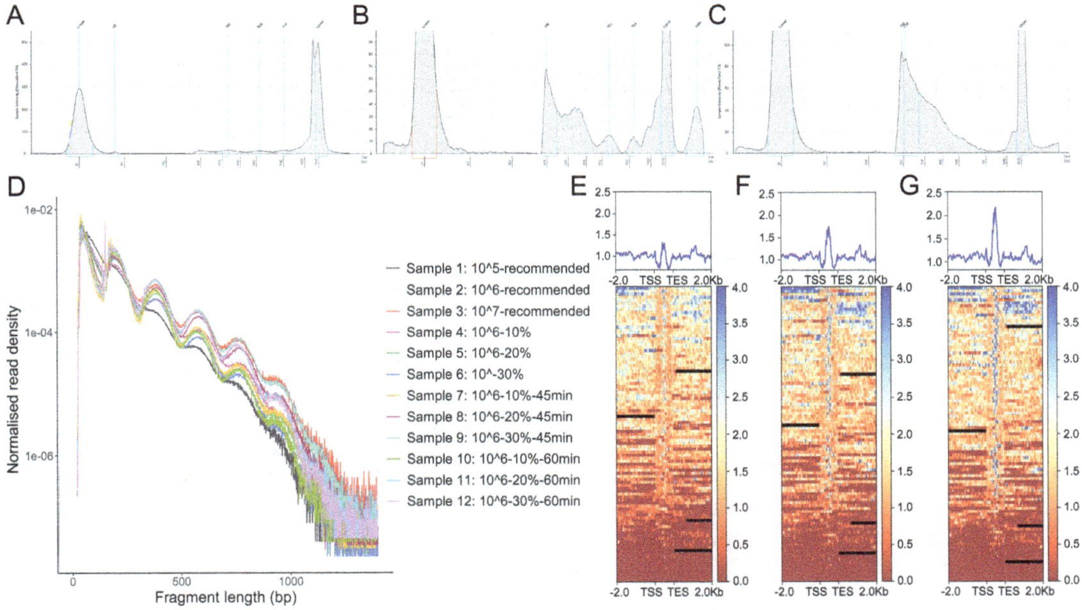

Fig. 8 Results from initial optimization, comparing 10^5, 10^6, and 10^7 cells, and 10^6 cells with different tagmentation times (30, 45, or 60 min) and concentrations of Tn5 (10%, 20% [manufacturer's recommendation], or 30%). (**a–c**) TapeStation traces generated using a 30 min tagmentation time and 10^5 (**a**) 10^6 (**b**), and 10^7 (**c**) cells. On the basis of these results, 10^6 cells were used for future optimization steps. (**d**) Nucleosomal periodicity plots for each of the optimization samples. While periodicity was generally consistent, signal was weaker in the 10^5 cell sample. Fewer peaks were also detected for this sample using MACS2/3 and Genrich (data not shown). (**e–g**) TSS plots for 10^6 cells with manufacturer's recommended concentration of Tn5 and 30-min (**e**), 45-min (**f**), and 60-min (**g**) tagmentation time. TSS signal improved as tagmentation time increased, and on this basis a 60-min tagmentation time was selected. Changing concentration of Tn5 had minimal impact, and 10^7 cells with 30 min tagmentation time also had a weaker TSS signal (data not shown)

5. Storing cells in RNAlater allows samples to be stored for longer time periods without loss of RNA, meaning ATAC-seq samples can be prepared and analyzed to confirm library generation was a success prior to going forward with RNA extraction and sequencing. If the number of ATAC-seq libraries is small, or you intend to proceed with RNA extraction and sequencing within a short period of time, cells can be pelleted and stored at –80 °C without the need for RNAlater.
6. The purification buffer contains a pH indicator to assess pH of the solution. However, the color change from pH 7.5–8.5 to <7.5 is very subtle and in practice it is much more straightforward to check the pH with indicator strips. In most cases, the pH is slightly above 7.5 and an additional 5 μL of sodium acetate lowers the pH to ~6.5. We have found that yields are higher when the pH is closer to 6.5 than to 7.0/7.5.
7. If planning on creating >16 libraries, additional i5/i7 primers can be purchased from ActiveMotif which allow multiplexing of up to 96 samples.
8. Too many PCR cycles can lead to excess PCR duplicates that have to be removed during data analysis. The optimal number of PCR cycles per sample can be calculated by qPCR or Qubit after three PCR cycles. An explanation for how to do this can be found in Grandi et al. [10]. However, in practice, we found that 9–11 PCR cycles were required and used the 10 cycles recommended by the kit when preparing multiple libraries at once.
9. The concentration of ATAC-seq libraries can be variable, averaging 20–50 ng/uL, although can be as low as 5 ng/μL and as high as 300 ng/μL. On the whole, concentration is a poor indicator of library quality, but reduced yields tended to occur when the pH was between 7.0 and 7.5 (*see* **Note 6**).
10. Running ATAC-seq samples on a Bioanalyzer or TapeStation is recommended to check for periodicity in fragment size (due to the addition of subsequent nucleosomes), which indicate a successful library generation. We have found, however, that the number of bands/level of periodicity is not necessarily a good indicator of sample quality, although some evidence of nucleosomal periodicity is required (*see* Fig. 9).
11. For sequencing, we aim for at least 30 million 100 bp paired-end reads per sample due to the number of reads lost due to mitochondrial contamination and other filtering steps.
12. RNAlater has a higher density than most cell culture media. The manufacturers therefore recommend pelleting cells at higher centrifugation speeds. However, when processing multiple samples (e.g., more than six), cell pellets can easily become resuspended during the time it takes to aspirate the supernatant

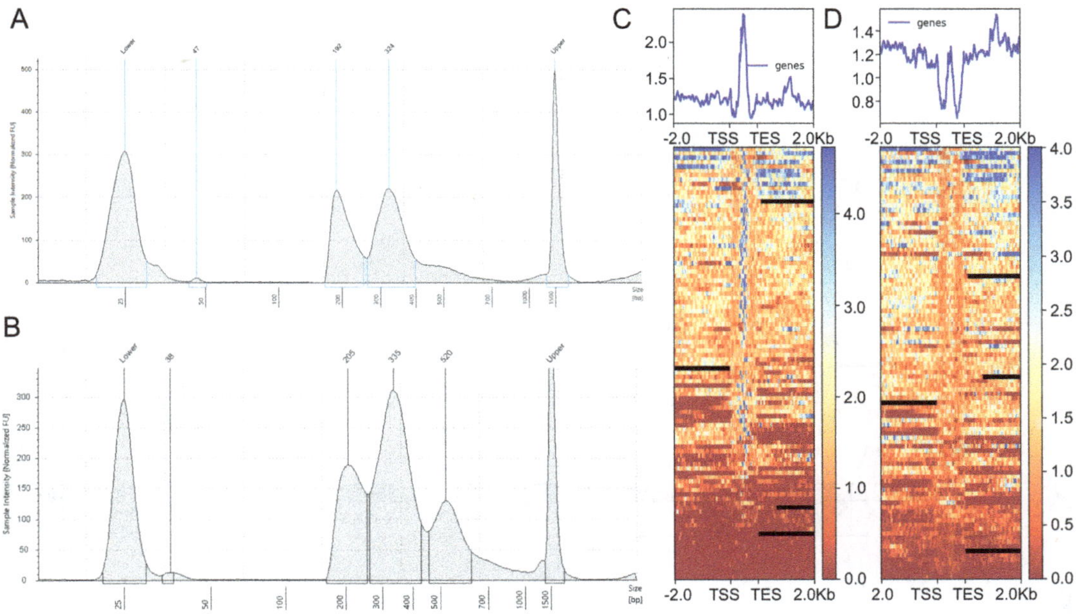

Fig. 9 The number of peaks in the TapeStation traces is not necessarily a good indicator of library quality. (**a**) TapeStation trace with fewer peaks. (**b**) TapeStation trace with stronger periodicity. (**c**) TSS plot with strong enrichment from same sample as seen in panel **a**. (**d**) TSS plot with very poor enrichment from same sample as seen in panel **b**

from previous tubes. To improve pelleting efficiency and reduce resuspension, RNAlater can be diluted 1:1 with BSA solution prior to centrifugation.

13. The kit provides multiple options for homogenization. Vortexing each sample for 1 min is by far the most straight-forward option, and resulting yields are sufficient for sequencing.

14. For *T. brucei*, successful RNA extraction should produce three peaks on the TapeStation trace, rather than the two expected for typical eukaryotes. However, this also means that the RNA integrity number (RIN) calculated by the TapeStation software can be very low, despite RNA being intact (*see* Fig. 10). As long as there are clear peaks/bands in the traces (i.e., no smearing), samples should be ready to sequence. For other Euglenozoans, the number of expected peaks may be different.

15. Most ATAC-seq analyses will include a step to remove mitochondrial reads; the original ATAC-seq protocol saw up to 80% of reads aligning to the mitochondrial genome [10], as the mitochondrial DNA is not chromatinized and so is easily accessible to the Tn5 transposase. However, the *T. brucei* reference genomes do not contain the kinetoplast genome and so do not need to be removed. If interested, the kinetoplast genome can be added into the reference fasta files. On average, we observed ~30% of reads aligning to the mitochondrial genome.

A

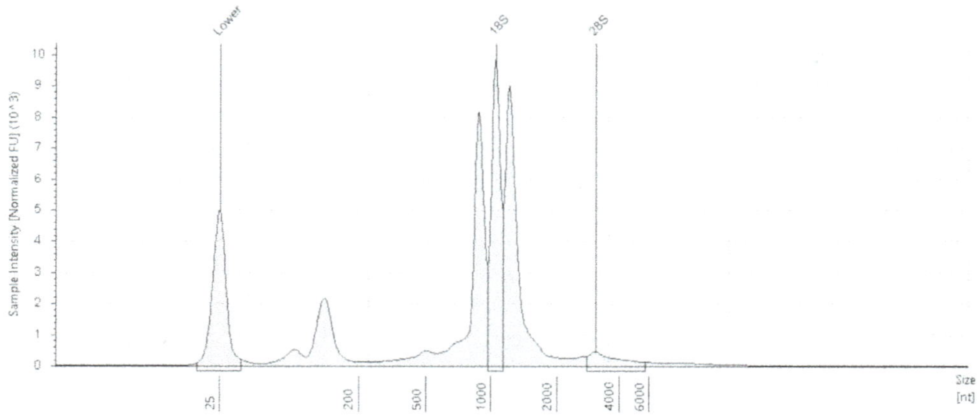

Sample Table

Well	RINe	28S/18S (Area)	Conc. [ng/µl]	Sample Description	Alert	Observations
A3	9.5	0.1	129	A6p30rep3		The upper ribosomal fragment has degraded

B

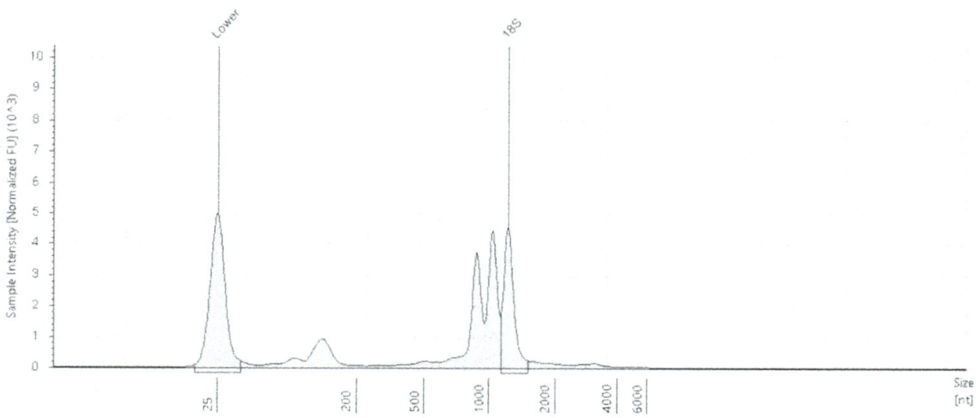

Sample Table

Well	RINe	28S/18S (Area)	Conc. [ng/µl]	Sample Description	Alert	Observations
F1	5.9	-	52.1	A7p30rep2		The upper ribosomal fragment has degraded

Fig. 10 RNA TapeStation from *T. brucei* RNA extractions. Successful RNA extraction from Trypanosoma samples should generate three peaks, as opposed to the two peaks observed in most other eukaryotes. Because of this, the RNA integrity number (RIN) score, which is used as a measure of RNA quality, can appear as low, depending on which of the three peaks is used for the calculation. (**a**) High quality RNA extraction with high RIN score. (**b**) High quality RNA extraction with low RIN score

Acknowledgments

This work was supported by the Wellcome trust [218492/Z/19/Z, 221717/Z/20/Z]. We would also like to thank past and present members of the Matthews lab for their advice, particularly Dr. Kirsty McWilliam for her support in the lab, and Dr. Emma Briggs and Dr. Guy Oldrieve for their assistance with bioinformatic analysis.

References

1. Kostygov AY, Karnkowska A, Votýpa J et al (2021) Euglenozoa: taxonomy, diversity and ecology, symbioses and viruses. Open Biol 11(3). https://doi.org/10.1098/rsob.200407
2. Siegel TN, Gunasekera K, Cross GAM et al (2011) Gene expression in Trypanosoma brucei: lessons from high-throughput RNA sequencing. Trends Parasitol 27(10):434–441. https://doi.org/10.1016/j.pt.2011.05.006
3. Clayton CE (2016) Gene expression in kinetoplastids. Curr Opin Microbiol 32:46–51. https://doi.org/10.1016/j.mib.2016.04.018
4. Biebinger S, Rettenmaier S, Flaspohler J et al (1996) The PARP promoter of Trypanosoma brucei is developmentally regulated in a chromosomal context. Nucleic Acids Res 24(7):1202–1211. https://doi.org/10.1093/nar/24.7.1202
5. Staneva DP, Bresson S, Auchynnikava T et al (2022) The SPARC complex defines RNAPII promoters in Trypanosoma brucei. elife 11:e83135. https://doi.org/10.7554/elife.83135
6. Kieft R, Cliffe L, Yan H et al (2025) Monoallelic epigenetic regulation of polycistronic transcription initiation by RNA polymerase II in Trypanosoma brucei. MBio 16(2):e02328–e02324. https://doi.org/10.1128/mbio.02328-24
7. Buenrostro JD, Giresi PG, Zaba LC et al (2013) Transposition of native chromatin for fast and sensitive epigenomic profiling of open chromatin, DNA-binding proteins and nucleosome position. Nat Methods 10(12):1213–1218. https://doi.org/10.1038/nmeth.2688
8. Buenrostro JD, Wu B, Chang HY, Greenleaf WG (2015) ATAC-Seq: a method for assaying chromatin accessibility genome-wide. Curr Protoc Mol Biol 109(1):21.29.1–21.29.9. https://doi.org/10.1002/0471142727.mb2129s109
9. Grandi FC, Modi H, Kampman L, Corces MR (2022) Chromatin accessibility profiling by ATAC-Seq. Nat Protoc 17(6):1518–1552. https://doi.org/10.1038/s41596-022-00692-9
10. Landt SG, Marinov GK, Kundaje A et al (2012) ChIP-seq guidelines and practices of the ENCODE and modENCODE Consortia. Genome Res 22(9):1813–1831. https://doi.org/10.1101/gr.136184.111
11. Reske JJ, Wilson MR, Chandler RL (2020) ATAC-seq normalization method can significantly affect differential accessibility analysis and interpretation. Epigenetics Chromatin 13(1):42. https://doi.org/10.1186/s13072-020-00342-y
12. Wright JR, Siegel TN, Cross GAM (2010) Histone H3 trimethylated at lysine 4 is enriched at probable transcription start sites in Trypanosoma brucei. Mol Biochem Parasitol 172(2):141–144. https://doi.org/10.1016/j.molbiopara.2010.03.013
13. Larionov A (2020) Differential gene expression analysis. RNA-seq Data Analysis course, EBI, April 2020. http://larionov.co.uk/deg_ebi_tutorial_2020/index.html. Accessed 4 Jul 2025

Chapter 12

Transcriptomic Analysis of Trypanosomatids

Carlos Mario Ospina, Tatiana M. Cáceres, Luz Helena Patiño, and Juan David Ramírez

Abstract

The study of transcriptomics in trypanosomatids is essential for understanding their unique gene expression mechanisms, which are predominantly regulated post-transcriptionally due to polycistronic transcription and a lack of specific promoters. These parasites, responsible for diseases such as Chagas disease and leishmaniasis, pose significant global health challenges. To improve transcriptomic data analysis, a structured bioinformatics workflow has been developed. This method integrates various tools to efficiently process RNA sequencing data. The workflow begins with raw data preprocessing, including sequence cleaning and filtering. Differential expression analysis is then performed to identify genes with altered activity under different conditions. Functional annotation follows, utilizing databases such as Gene Ontology (GO) and KEGG Orthology (KO) to assign gene functions. Finally, data visualization techniques are applied to present findings clearly. This approach accounts for the complexity of trypanosomatid transcriptomes, optimizing research by providing a standardized framework. It enables precise gene expression profiling, facilitating insights into parasite survival, adaptation, immune evasion, and drug resistance. The methodology is also applicable to other parasites and fields like molecular epidemiology and targeted therapy development, making it a crucial tool for advancing research and treatment strategies against these pathogens.

Key words Transcriptomics, Trypanosomatids, RNA-Seq, Differential expression analysis, Gene Ontology

1 Introduction

Transcriptomics is the study of the complete RNA transcript set in a biological entity at a specific time [1]. It is crucial for understanding gene expression, biological processes, and disease mechanisms [2]. This is particularly relevant for trypanosomatids, organisms with unique genomic and transcriptional features [3].

These organisms have compact genomes (18.0–32.6 Mb) with variable gene numbers [3–5]. Their genomes feature high gene density due to the absence of introns and short intergenic regions [3, 6]. Genes are arranged in tandem arrays on the same DNA

Paul A. M. Michels et al. (eds.), *Euglenozoa: Methods and Protocols, Volume 1*, Methods in Molecular Biology, vol. 3013, https://doi.org/10.1007/978-1-0716-5142-1_12,

strand [3] and show high synteny among related species [3, 7–9]. Trypanosomatid genomes are organized into polycistronic transcription units, where multiple genes are transcribed together as a single precursor RNA [3]. This RNA undergoes trans-splicing, during which a spliced leader (SL) sequence is added to the 5′ end, followed by polyadenylation at the 3′ end, resulting in mature mRNAs [3]. Unlike in many other organisms, gene expression in trypanosomatids is primarily regulated post-transcriptionally, as transcription occurs constitutively and lacks specific promoters for individual genes [3]. These parasites, responsible for diseases such as Chagas disease, leishmaniasis, and sleeping sickness, pose significant global health challenges [10].

To better understand trypanosome biology, researchers have employed various omics technologies. Genomics has provided insights into DNA structure and organization, highlighting the presence of polycistronic transcription units and post-transcriptional regulation [11, 12]. Proteomics has enabled the identification and quantification of proteins across different life cycle stages [13, 14], while metabolomics has explored metabolites and metabolic pathways crucial for parasite survival [15, 16]. Among these approaches, transcriptomics has proven fundamental for studying dynamic gene expression in response to environmental stimuli, revealing key mechanisms of differentiation and adaptation [17–19].

Transcriptomic studies have shed light on how trypanosomatids regulate gene expression to adapt to their complex life cycles and host environments [20]. For instance, transcriptomic profiling of *Trypanosoma brucei* has identified differential gene expression patterns that drive its transition between bloodstream and procyclic forms [21]. Similarly, transcriptome analyses of *Leishmania* species suggest that mRNA abundance does not always correlate with protein levels, reinforcing the idea that post-transcriptional mechanisms such as RNA stability and translation efficiency play a major role in determining phenotype [22, 23]. Moreover, single-cell RNA sequencing (scRNA-Seq) has uncovered cellular heterogeneity and stage-specific gene expression in *T. brucei*, providing a deeper understanding of phenotypic variation [24].

Given the strong influence of post-transcriptional regulation in trypanosomatids, transcriptomic data alone may not be sufficient to directly infer phenotypic differences. While transcriptomic studies have successfully identified regulatory networks and adaptive responses, integrating these findings with proteomic analyses and functional validation is essential to establish clear links between gene expression patterns and observable traits [25].

Despite the importance of transcriptomics, there is currently no standardized bioinformatics protocol for analyzing trypanosomatid RNA sequencing data. To address this gap, an optimized method has been developed, integrating multiple bioinformatics tools into a structured workflow to enhance the efficiency and accuracy of RNA sequencing data analysis.

The workflow begins with raw data preprocessing, including sequence cleaning and filtering. Differential expression analysis identifies genes with altered activity under different conditions. Functional annotations assign known functions using databases and comparisons. Gene categorization is based on biological processes and metabolic pathways uses tools like Gene Ontology (GO) and KEGG Orthology (KO) assignment. Results are visually represented through graphs and diagrams.

This method accounts for the complexity of trypanosomatid transcriptomes, including gene redundancy, unusual post-transcriptional regulation, and transcriptome size variability. It provides a framework for researchers to interpret data and apply it to parasite biology studies [26, 27].

The method optimizes and accelerates research on trypanosomatids like *Trypanosoma cruzi* and *Leishmania* spp. By enhancing transcriptomic data analysis and integrating specialized tools like KEGG and TriTrypDB (a trypanosomatids-specific database) [28], it enables precise gene expression profile characterization. This facilitates the identification of essential genes and metabolic pathways, providing insights into parasite survival and adaptation [29].

For Chagas disease, this method can analyze gene expression changes during transitions between parasite forms, improving understanding of immune evasion and oxidative stress resistance. It can also provide insights into drug resistance gene regulation, aiding new therapeutic strategies.

For leishmaniasis, this methodology is useful for analyzing gene expression regulation under adverse conditions, such as temperature shifts and immune pressure. It can identify virulence factors, aiding the design of targeted therapies.

This approach is valuable for studying the transcriptomic plasticity of trypanosomatids and their response to environmental changes. This information is crucial for understanding adaptation mechanisms and developing diagnostic, control, and treatment strategies. Its flexibility and scalability allow application to other parasites and fields like molecular epidemiology, drug resistance monitoring, and targeted therapy development.

2 Materials

This protocol outlines the steps for executing RNA-Seq data processing scripts in a high-performance computing (HPC) environment running a Linux distribution as its base software, along with R scripts for differential expression analysis and the assignment of Gene Ontology (GO) and KEGG Orthology (KO) (Fig. 1). Before the analysis, it is essential that the researcher has performed RNA

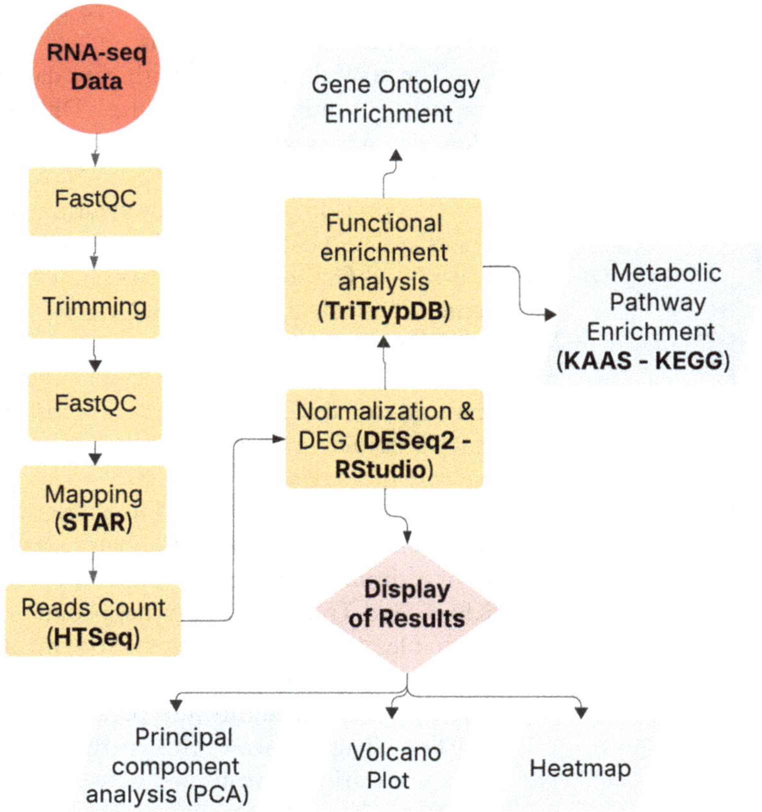

Fig. 1 Workflow for the transcriptomic analysis

extraction and sequencing after conducting the corresponding experiment with their trypanosomatid organism of interest. The RNA-Seq data used in this protocol consists of paired-end reads generated using the Illumina sequencing platform.

Before proceeding with the analysis, ensure that the following software is installed on your Linux terminal:

(a) **SLURM**—Job management system for cluster execution. Run the following command on your Linux terminal to install it:

```
sudo apt update
sudo apt install slurm-wlm
```

For more information visit the website: https://slurm.schedmd.com/documentation.html

(b) **FastQC**—Tool for sequence quality assessment. Run the following command on your Linux terminal to install it:

```
sudo apt install fastqc
```

For more information visit the website:

https://www.bioinformatics.babraham.ac.uk/projects/fastqc/

(c) **Trimmomatic**—Used for sequence trimming and cleaning. Run the following command on your Linux terminal to install it:

wget https://github.com/usadellab/Trimmomatic/releases/download/v0.39/Trimmomatic-0.39.zip

unzip Trimmomatic-0.39.zip (*see* **Note 1**).

Move the folder to /usr/local/bin (or any other path you prefer):

```
sudo mv Trimmomatic-0.39 /usr/local/bin/
```

Run the following command to verify that Trimmomatic is installed correctly:

Trimmomatic version

For more information visit the website:

http://www.usadellab.org/cms/?page=trimmomatic

(d) **Java 11**—Required for running Trimmomatic. Run the following command on your Linux terminal to install it:

```
sudo apt install openjdk-11-jre
```

For more information visit the website:

https://www.oracle.com/java/technologies/javase/jdk11-archive-downloads.html

(e) **STAR/2.7.10b**—Read mapping tool. Run the following command on your Linux terminal to install it:

wget https://github.com/alexdobin/STAR/archive/refs/tags/2.7.10b.tar.gz

```
tar -xvzf 2.7.10b.tar.gz
cd STAR-2.7.10b/source
make STAR
```

Move the folder to /usr/local/bin (or any other path you prefer):

```
sudo cp STAR /usr/local/bin/
```

For more information visit the website:

https://github.com/alexdobin/STAR

(f) **Python/3.8.12**—Used for read counting with HTSeq-count. Run the following command on your Linux terminal to install it:

```
sudo apt install python3.8 python3-pip # For Debian/Ubuntu
```

For more information visit the website:

https://htseq.readthedocs.io/en/master/install.html

(g) **R-4.4.2**—Software for differential expression analysis.

For optimal performance, this software should be installed on a personal computer compatible with Linux, Windows, or macOS, meeting the following recommended requirements: 64-bit multi-core processor (Intel Core i5 or higher), 8 GB of RAM or more, and a minimum of 100 GB of free disk space. Installation is completed by manually downloading the software from its official website: https://cran.r-project.org/bin/windows/base/.

Then install **RStudio.**

(h) **RStudio 2024.12.1 + 563**—Integrated development environment for differential expression analysis and results visualization.

RStudio installation is performed by manually downloading the software from its official website: https://posit.co/download/rstudio-desktop/. It is important to note that RStudio shares the same recommended system requirements as R for optimal performance.

For Linux, dependencies can be installed using:

```
sudo apt update
sudo apt install -y r-base r-base-dev
```

3 Methods

All procedures described below must be executed while ensuring compliance with the minimum computing requirements necessary for the proper operation of each software. The recommended system specifications are as follows:

(a) **Operating System**: Linux (Ubuntu 20.04+ or CentOS 7+).

(b) **Processor**: Multi-core CPU (≥8 cores, Intel Xeon or AMD EPYC recommended).

(c) **Memory (RAM)**: At least 32 GB.

(d) **Storage**: Minimum 1 TB SSD (faster I/O operations are beneficial for RNA-Seq data processing).

(e) **GPU (Optional)**: For accelerated computing in certain workflows.

(f) **Software Environment**: SLURM-based HPC cluster for workload management.

Most steps will be executed using a Bash script (*see* Subheading 3.1) to manage jobs in an HPC cluster with SLURM, a workload management system (*see* **Note 1**). These specifications ensure efficient execution of FastQC, Trimmomatic, STAR, Python-based read counting, and R-based differential expression analysis.

For executing commands in RStudio, it is essential to verify that all required packages are correctly installed and that there are no conflicts with the specific R version in use.

Throughout this document, you will encounter data and information enclosed within parentheses () and emphasized with bold text and quotation marks (““). These placeholders must be replaced with your research-specific data. In the context of bioinformatics pipelines, you will find indications marked with the # symbol, followed by text in *italics*.

3.1 Bash Script Creation

(a) To create a Bash script for execution in a high-performance computing (HPC) cluster, follow these detailed steps:

(b) Create and Open a Script File.

In the terminal, open a text editor like nano or vim to write the script. For example, if using nano, run the following command:

```
nano ("script_name").sh
```

This will open a blank file where you can write your script.

(c) The first line of the script must indicate that it will be executed using Bash, which is defined with the following line:

```
#!/bin/bash
```

This line, known as the “shebang,” ensures that the script runs in the Bash environment.

(d) Since the script will run on a cluster managed by SLURM, #SBATCH directives must be included to define how the job will be executed.

(e) Specify the Cluster Partition:

```
#SBATCH -p ("option")
```

This option defines which partition of the cluster the job will run on. Depending on the system, partitions such as normal, high, or gpu may be available.

(f) Assign a Job Name:

```
#SBATCH --job-name= ("job_name")
```

Assigns a name to the job for easy identification in the job queue.

(g) Define Output and Error Files:

```
#SBATCH --output=%x-%j.star.out
#SBATCH --error=%x-%j.star.err
```

%x represents the job name defined in --job-name.

%j represents the **Job ID**, which is a unique number automatically assigned by the SLURM job scheduler when the job is submitted to the execution queue.

The output file (.out) will store information generated by the program, while the error file (.err) will log any error messages.

(h) Specify Memory Allocation and Execution Time:

```
#SBATCH --mem= ("amount")
```

Defines the amount of memory allocated for job execution. If not specified, the cluster assigns a default amount, which may be insufficient.

```
#SBATCH --time= ("D-HH:MM:SS")
```

Sets the maximum execution time in the format days-hours:minutes:seconds. It is recommended to request only the necessary time, as some clusters may allocate resources based on job duration.

(i) Assign Cluster Resources:

```
#SBATCH --nodes= ("number")
```

Indicates the number of nodes to be used in the cluster.

```
#SBATCH --ntasks-per-node= ("number")
```

Defines the number of tasks to be executed per node.

```
#SBATCH --cpus-per-task= (number)
```

Specifies the number of CPUs each task will use. These options should be adjusted according to the job requirements and cluster resource availability.

(j) Configure Email Notifications:

```
#SBATCH --mail-type= ("event")
#SBATCH --mail-user= ("example@gmail.com")
```

Enables email notifications about the job status. Events such as ALL (all events), BEGIN (job start), END (job completion), and FAIL (job failure) can be configured.

(k) Add Execution Commands:

After defining the #SBATCH directives, include the commands the script will execute. These commands will be detailed in later sections.

(l) Save and Close the File:

If using nano, save the changes by pressing:

CTRL + X

Y (to confirm)

Enter

(m) Run the following command to execute the script in SLURM:

```
sbatch ("script_name").sh
```

Each of the parameters mentioned can be modified or removed according to the researcher's needs. Additionally, Bash scripts can be written in any text editor (*see* **Note 2**) and then transferred to the HPC environment for execution, as demonstrated with various files in Subheading 3.2.

3.2 Uploading Files

(a) After obtaining RNA-Seq sequencing data, it should be stored on a server or computing system capable of running the required software. To facilitate processing, files should be organized into individual folders for each sample.

(b) Creating Folders:

To create a folder for each sample, use the following command in the terminal:

```
mkdir ("Folder_Name")
```

(c) Transferring and Organizing Files:

Each folder should contain the paired-end (FASTQ) sequencing files with the following naming convention:

```
("Sample_01_R1").fastq.gz #Read 1 (Forward)
("Sample_01_R2").fastq.gz #Read 2 (Reverse)
```

To transfer these files from a local computer to the HPC, secure file transfer protocols should be used. Tools such as **scp** (Secure Copy), which is typically included in most operating systems, provide secure methods for remote file management (*see* **Note 3**).

(d) Create a file named **("file_name")**.txt in the directory where the organized folders are located, using commands directly on the HPC (as explained in Subheading 3.1) or by using the text editor of your choice (*see* **Note 2**), keeping in mind that you must then transfer this file to the HPC (*see* **Note 3**). The file should contain a list of the sample folder names, each on a separate line, with no additional whitespace (*see* **Note 4**).

3.3 Quality Evaluation with FastQC

(a) To optimize the execution of the processes, it is necessary to create a Bash script called "**("name_fastqc")**.sh" inside the working directory on the HPC. This can be done directly on the HPC using commands (as explained in Subheading 3.1) or using the text editor of your choice (*see* **Note 2**), keeping in mind that you must then transfer this file to the HPC (*see* **Note 3**).

(b) From this point forward, whenever the creation of a Bash script is mentioned, the same options will apply, it can be created directly on the HPC or using a text editor, with the need to transfer it to the HPC.

(c) The script must contain the following:

```
#!/bin/bash
#SBATCH -p ("normal")
#SBATCH --mem=("5G")
#SBATCH --job-name= #Job name
#SBATCH --output=("%x-%j.fastqc.out")
#SBATCH --time=("1-00:00:00")
#SBATCH --error=("%x-%j.err")
#SBATCH --mail-type=("ALL")
#SBATCH --mail-user= #Electronic mail to receive notifica-
tions

# Load FastQC module
module load fastqc

# Create folder for quality results
mkdir -p ("route/location/folder")/fastqc_results

# Read each sample name from the.txt file and process it
while IFS= read -r line; do

# Quality evaluation before trimming
fastqc -o ("route/location/folder")/fastqc_results ("route/
location/RNAseq_data")/"$line"/"$line"_1.fq.gz ("route/loca-
tion/RNAseq_data")/"$line"/"$line"_2.fq.gz
done < ("route/location/file.txt")/("file_name.txt")
```

(d) Save the script and run it with the command sbatch **("name.sh")**.

3.4 Trimming

(a) Create a Bash script named "(**"name"**).sh" in your HPC directory. The script should contain the following:

```
#!/bin/bash
#SBATCH -p ("normal")
#SBATCH --mem=("5G")
#SBATCH --job-name= #Job name
#SBATCH --output=("%x-%j.trimming.out")
```

```
#SBATCH --time=("1-00:00:00")
#SBATCH --error=("%x-%j.err")
#SBATCH --mail-type=("ALL")
#SBATCH --mail-user= #Electronic mail to receive notifica-
tions

#Load modules
module load trimmomatic
module load java11

#Create folder for the trimmed files

mkdir -p ("route location folder")/trimming

#Read each sample name from the.txt file

cat ("route location file.txt")/("file name.txt") | tr -d '\r'
| while read line

do
java -jar /opt/ohpc/pub/apps/trimmomatic/0.38/bin/trimmo-
matic-0.38.jar PE \
("route location folders with RNA.seq data")/"$line"/
"$line"_1.fq.gz \
("route location folders with RNA.seq data")/"$line"/
"$line"_2.fq.gz \
-threads 10 \
("route location trimming folder")/"$line"_1_paired_trim.fq.
gz \
("route location trimming folder")/"$line"_1_unpaired_trim.
fq.gz \
("route location trimming folder")/"$line"_2_paired_trim.fq.
gz \
("route location trimming folder")/"$line"_2_unpaired_trim.
fq.gz \

ILLUMINACLIP:/opt/ohpc/pub/apps/trimmomatic/0.38/bin/adap-
ters/TruSeq3-PE.fa:2:30:10:2:TRUE \
SLIDINGWINDOW:4:20 MINLEN:100 AVGQUAL:20 TRAILING:20 (see
Note 5)

#Create flag file to indicate that the sample has been
processed

touch ("route location trimming folder")/trimm_med_"$line"
done
```

(b) Save the script and run it with the command sbatch **("name.sh")** (*see* **Note 6**).

3.5 Mapping Against Reference Genome

(a) Once the sequences have been trimmed (trimming), it is recommended to perform a new quality control (QC) assessment to verify data integrity before proceeding to mapping.

(b) Then, create a script for the mapping of samples against the reference genome of the organism of interest, which must have been previously downloaded from the corresponding database (such as TriTrypDB) (*see* **Note 7**). It is important to highlight that, in the case of *Trypanosoma cruzi*, the DTU of study must be considered, since this information is crucial to obtain precise and relevant results in genomic analysis.

(c) Follow the following scheme for the creation of the script to carry out the mapping:

```
#!/bin/bash
#SBATCH -p ("normal")
#SBATCH --job-name= #Job name
#SBATCH --output=("%x-%j.star.out")
#SBATCH --nodes=("2")
#SBATCH --ntasks-per-node=("10")
#SBATCH --cpus-per-task=("1")
#SBATCH --error=("%x-%j.star.err")
#SBATCH --mail-type=("ALL")
#SBATCH --mail-user= #Electronic mail to receive notifica-
tions

#Load modules
module load star/2.7.10b

#Create folder for the mapped files
mkdir -p ("route location folder")/mappings

#Location of folders of interest for mapping
input_dir="("route location trimming folder")"
output_dir="("route location mappings folder")"
genome_dir="("route location reference genome folder")" (see
Note 7)

#Process files with the new naming pattern
for file in ${input_dir}*_1_paired_trim.fq.gz
do
# Extract the sample name from the file
sample_name=$(basename "$file" "_1_paired_trim.fq.gz")
STAR --genomeDir "$genome_dir" \
--readFilesCommand zcat --readFilesIn "${input_dir}${sample_-
name}_1_paired_trim.fq.gz" "${input_dir}${sample_name}
_2_paired_trim.fq.gz" \
```

```
 --outSAMtype BAM SortedByCoordinate --limitBAMsortRAM
16000000000 --outSAMunmapped Within \
 --twopassMode Basic --outFilterMultimapNmax 1 --quantMode
GeneCounts --runThreadN 10 --outFileNamePrefix "${output_dir}
${sample_name}"
done
```

(d) Save the script with the name you want "file_name.sh" and run it with the command sbatch **("file_name.sh")** (*see* **Note 6**).

3.6 Counting the Number of Reads Mapped to each Gene with HTSeq-Count

(a) Create a bash script to count the number of reads mapped to each gene following the following format:

```
#!/bin/bash
#SBATCH -p ("normal")
#SBATCH --job-name= #Job name
#SBATCH --output=("%x-%j.htseq.out")
#SBATCH --nodes=("1")
#SBATCH --ntasks-per-node=("10")
#SBATCH --cpus-per-task=("1")
#SBATCH --error=("%x-%j.HTSEQ.err")
#SBATCH --mail-type=("ALL")
#SBATCH --mail-user= #Electronic mail to receive notifica-
tions

#Load modules
module load python/3.8.12

#Create folder for the resulting files.
mkdir -p ("route location folder")/HTSEQ

 for file in ("route location mappings folder")/*Aligned.
sortedByCoord.out.bam; do
filename=$(basename "$file")
filename_noext="${filename%.*}"
 python -m HTSeq.scripts.count -f bam --stranded=no --mod-
e=intersection-nonempty -r name "$file" \
("route location folder GFF file") (see Note 8) \
> "("route location HTSEQ folder")/${filename_noext}.txt"
done
```

(b) Save the script with the name you want **("name_file")**.sh and run it with the command sbatch **("name_file.sh")** (*see* **Note 6**).

(c) At the end, you will have a set of.txt files with the number of reads assigned to each gene, which you can then analyze to see which genes are over or under expressed.

3.7 Normalization and Differential Expression Analysis Using the DESeq2 Package in RStudio

(a) Once the counting of the number of reads that have been mapped to each gene with HTSeq-count is executed, the.txt files are downloaded to a folder inside a computer (*see* **Note 3**) and analyzed with an operating system compatible with the RStudio software (Windows, macOS, and Linux).

(b) The RStudio software is executed, and the following libraries are installed using the commands:

```
install.packages("DESeq2")
install.packages("ggplot2")
install.packages("pheatmap")
install.packages("EnhancedVolcano")
install.packages("ggrepel")
install.packages("ggfortify")
install.packages("writexl")
install.packages("openxlsx")
```

(c) Once installed, data preparation is carried out by executing the following R code:

```
# Define the work directory
directory2 <- ("Location of our.txt files obtained from HTSeq-count")
setwd(directory2) #Sets the directory as the current work directory

#List the count files (generated by HTSeq-count)
sampleFiles2 <- list.files(directory2, pattern = "\cr.txt$")
# Lists only.txt files

#Assign conditions based on filename pattern matching
sampleCondition <- ifelse(grepl("treated", sampleFiles2, ignore.case = TRUE), "C2", ifelse(grepl("control", sample-Files2, ignore.case = TRUE), "C1", NA)) (see Note 9)

#Check for any NA values that may need manual assignment
if (any(is.na(sampleCondition))) {warning("Some files do not match 'treated' or 'control' patterns. Please check and assign manually.")} (see Note 9)

#Create a sample table
sampleTable2 <- data.frame(sampleName = sampleFiles2,
fileName = sampleFiles2,
condition = sampleCondition)

#Convert the 'condition' column to a factor (categorical variable)
sampleTable2$condition <- factor(sampleTable2$condition) (see Note 10)
```

(d) To conduct the DESeq2 analysis, the subsequent commands will be executed in the following order.

```
#Load the DESeq2 library
library(DESeq2)

#Create the DESeqDataSet object
ddsHTSeq <- DESeqDataSetFromHTSeqCount(sampleTable = sample-
Table2,
directory = directory2,
design = ~ condition)

#Filtering of genes with low expression
smallestGroupSize <- 3 # Size of the smallest group
keep <- rowSums(counts(ddsHTSeq) >= 10) >= smallestGroupSize
#Genes with at least 10 counts in at least 3 samples (see Note
11)
dds <- ddsHTSeq[keep,] #The filter is applied

#Reorder the levels of the condition for data normalization
dds$condition <- factor(dds$condition, levels = c("C1",
"C2"))
dds$condition <- relevel(dds$condition, ref = "C1") #"C1"
will be the reference group

#Execute the differential expression analysis
dds <- DESeq(dds) (see Note 12)
```

(e) The following commands will be executed to extract and explore the results.

```
#Obtain the results of the analysis
res <- results(dds, contrast = c("condition", "C1", "C2"))
#Specifies the comparison

#Order the results by p-value
resOrdered <- res[order(res$pvalue),]

#Summary of the results
summary(res)

#Counting of significant genes (adjusted by multiple tests)
sum(res$padj < 0.05, na.rm = TRUE) #Genes with adjusted
p-value < 0.05
```

```
#Results with an adjusted p-value threshold of 0.05
res05 <- results(dds, alpha = 0.05)
summary(res05)

#Handling of NA values in adjusted p-value (optional)
res05$padj[is.na(res05$padj)] <- 0 #Assigns 0 to NAs (see Note
13)

#Results with an adjusted p-value threshold of 0.01
res01mod <- res05[res05$padj < 0.01,] (see Note 14)
```

(f) Once the execution of the previous commands is finished, the transformation of the data and its respective visualization is carried out, for which the following commands must be executed:

```
#Normalized transformation of the data (for visualization and
other analyses)
ntd <- normTransform(dds) #Variance stabilizing transforma-
tion

#Regularized logarithmic transformation (for visualization
and clustering)
rld <- rlog(dds) #Logarithmic transformation that reduces the
influence of high counts

#MA plot (to visualize changes in expression vs. statistical
significance)
plotMA(res05, ylim = c(-2, 2), colSig = "red") #Significant
genes in red
```

(g) Finally, the data is exported to an.xlsx file for later analysis, using the following commands:

```
#Load the writexl library
library(writexl)

#Convert the ordered results to a data frame and add a column
with the gene IDs
df <- as.data.frame(resOrdered)
df$Gene_ID <- rownames(resOrdered)

#Move the Gene_ID column to the beginning (optional)
df <- df[, c("Gene_ID", setdiff(names(df), "Gene_ID"))]

#Export to Excel
write_xlsx(df, path = ("file_name.xlsx")) #The name of the
file is changed to the researcher's taste.
```

3.8 Visualization of Results in RStudio

(a) For the visualization of the results obtained, different types of graphs can be generated in RStudio, such as principal component analysis (PCA), Heatmaps, and Volcano plots.

(b) First, for the realization of the PCA, the following commands are executed in RStudio:

```
#Perform PCA

pca_res <- prcomp(t(assay(rld)), scale = TRUE)
explained_var <- pca_res$sdev^2/sum(pca_res$sdev^2) * 100

#Preparing Data for ggplot

pca_data <- as.data.frame(pca_res$x)
pca_data$condition <- colData(rld)$condition
pca_data$sample <- colnames(assay(rld))

#Create PCA Plot
ggplot(pca_data, aes(x = PC1, y = PC2, color = condition,
shape = condition)) +
geom_point(size = 4) + #Plots individual samples with size 4
geom_text(aes(label = sample), vjust = -1, hjust = 0.5, size
= 3.5) + #Adds sample labels
stat_ellipse(type = "norm", linetype = 2) + #Draws normal
distribution ellipses for each group
theme_minimal() + #Uses a clean, minimalistic theme
labs(title = "PCA of Gene Expression Data",
x = paste0("PC1 (", round(explained_var[1], 1), "%)"),
#Labels PC1 with variance explained
y = paste0("PC2 (", round(explained_var[2], 1), "%)")) +
#Labels PC2 with variance explained
scale_color_manual(values = c("C1" = "#E69F00", "C2" =
"#56B4E9")) + #Color palette
scale_shape_manual(values = c("C1" = 16, "C2" = 17)) + #Uses
distinct point shapes for differentiation
theme
(legend.title = element_blank()) #Removes legend title for
clarity
```

(c) Second, for the realization of the Volcano Plot, the following commands are executed in RStudio:

```
#Load Volcano Plot data
volcano <- read.csv(("MG.csv"))
row.names(volcano) <- volcano[, 1]
new_colnames <- c("gene", "baseMean", "log2FoldChange",
"lfcSE", "stat", "pvalue", "padj")
colnames(volcano) <- new_colnames
```

```
#Define Color Palette
keyvals <- ifelse(
volcano$log2FoldChange < -2.0, '#1E90FF', #Dodger Blue for
"low" ifelse(volcano$log2FoldChange > 2.0, '#FFA500', #Orange
for "high"
'#808080')) #Gray for "mid"
keyvals[is.na(keyvals)] <- '#808080' #Gray for NAs
names(keyvals)[keyvals == '#FFA500'] <- 'high'
names(keyvals)[keyvals == '#808080'] <- 'mid'
names(keyvals)[keyvals == '#1E90FF'] <- 'low'

#Create Volcano Plot
EnhancedVolcano(volcano, x = "log2FoldChange", y = "padj",
lab = rownames(volcano), colCustom = keyvals)
```

(d) Finally, for the realization of the Heatmap, the following commands are executed in RStudio:

```
#Select Genes
selected_genes <- rownames(res01mod) #Make sure res01mod is
defined
rld_mat <- assay(rld)[selected_genes,]

#Scale and Clean Data
rld_mat_scaled <- t(scale(t(rld_mat)))
rld_mat_scaled[is.na(rld_mat_scaled)] <- 0
rld_mat_scaled[!is.finite(rld_mat_scaled)] <- 0
colnames(rld_mat_scaled) <- c(""Name replica treated 1", "
Name replica treated 2", " Name replica treated 3 ", " Name
replica control 1", " Name replica control 2", " Name replica
treated 3"") (see Note 15)

#Annotations and Color Palette
annotation_col <- data.frame(Condition = colData(dds)$condi-
tion)
rownames(annotation_col) <- colnames(rld_mat_scaled)
ann_colors <- list(
Condition = c(C1 = "lightyellow", C2 = "lightblue")
)
heatmap_colors <- colorRampPalette(c("blue", "white", "red"))
(50)

#Create Heatmap
pheatmap(rld_mat_scaled,
cluster_rows = TRUE,
cluster_cols = TRUE,
show_rownames = FALSE,
show_colnames = TRUE,
```

```
annotation_col = annotation_col,
annotation_colors = ann_colors,
main = ("Heatmap Differentially Expressed Genes"),
color = heatmap_colors)
```

3.9 Functional Enrichment Analysis

(a) Open the.xlsx file generated in Subheading 3.6 using Microsoft Excel or any other compatible spreadsheet software.

(b) Apply a filter to the "Log2FoldChange" column.

(c) Create two new spreadsheets:

In the first sheet, extract and copy all genes that have a Log2FoldChange value ≥2.

In the second sheet, extract and copy all genes that have a Log2FoldChange value ≤ −2.

(d) For each of the spreadsheets created in the previous step, copy the gene IDs that are in the first column.

(e) Go to the TriTrypDB database website: https://tritrypdb.org/tritrypdb/app/.

(f) In the main menu, select "Searches", then "Genes", then "Annotation, curation and identifiers" and, finally, "List of IDs".

(g) In the "List of IDs" interface, paste the list of gene IDs that you copied previously.

(h) Press the "Get Answer" button.

(i) TriTrypDB will display a page with the annotation information corresponding to each of the genes entered.

(j) Within the results page, select the "Analyze Results" option.

(k) Use the "Gene Ontology Enrichment" option to perform the gene ontology analysis with the uploaded genes.

(l) Select the corresponding organism, the desired type of ontology (Biological Process, Cellular Component or Molecular Function) and set the P-value cutoff to 0.05. Click "Submit" to start the analysis (Fig. 2).

(m) TriTrypDB will display a table with the GO terms associated with your genes, the number of genes per term and the Fold Enrichment value.

(n) Download the GO terms data file by clicking the "download" button, located in the upper left corner of the data table. The file will be downloaded in.tsv format.

(o) Open the downloaded.tsv file using a spreadsheet editor. Microsoft Excel is recommended for better compatibility (Fig. 3).

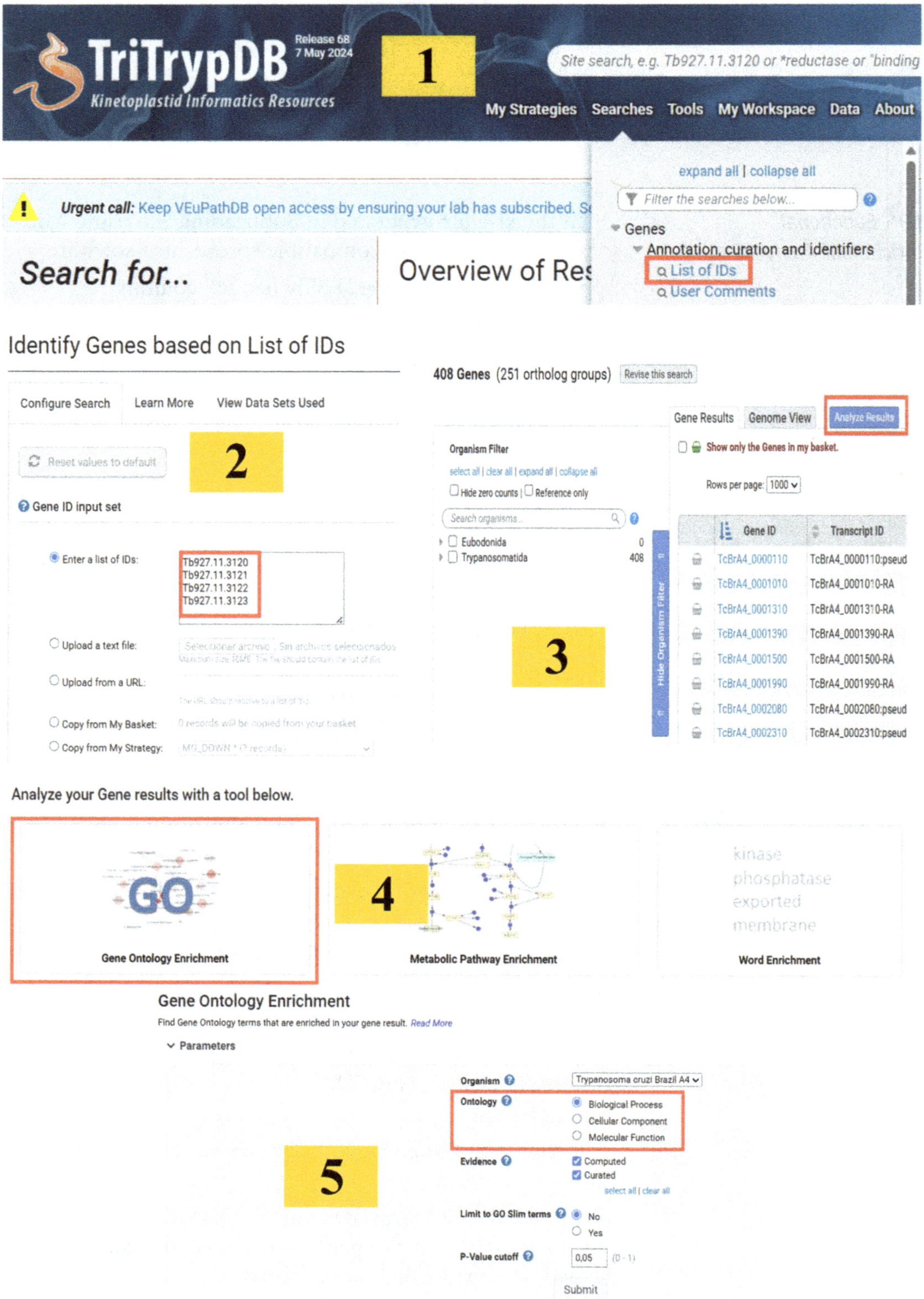

Fig. 2 Pathway to functional enrichment analysis in TriTrypDB. This series of figures illustrates the workflow within TriTrypDB, guiding users from the initial pathway to gene ID listing and interface, through the visualization of gene annotation results, to the application of post-annotation analysis tools. The sequence culminates in the options interface for Gene Ontology enrichment analysis, showcasing the platform's ability to facilitate comprehensive gene function exploration

(p) Delete the following columns: "ID", "Bgd count", "Result gene list", "Pct of bgd", "Fold enrichment", "Odds ratio", "P-value", "Benjamini", and "Bonferroni". These columns are not required for further analysis.

(q) Change the name of the "Name" column to "Term" and the "Result count" column to "Count". This will facilitate data interpretation.

(r) Add a new column named "Condition". In this column, specify whether the genes are "Up-regulated" or "Down-regulated", as appropriate for each row (*see* **Note 16**).

(s) Save the modified file in.CSV format and assign a descriptive name (e.g., (**"name.csv"**)).

(t) Start RStudio and run the following code:

```
#Reading the CSV File
("CC") <- read.csv(("name.csv"), sep = ",") (see Notes 17 and
18)

#Define Color Palette
base_color <- c("orange", "#66CD00")

#Creating the Dotplot with ggplot2
("DotplotCCvs") <- ggplot(("CC"), aes(x = Condition, y = Term,
color = Condition)) +
geom_count(aes(size = Count)) +
theme
(panel.background = element_rect(fill = NA, color = NA),
plot.background = element_rect(fill = NA, color = NA),
axis.line = element_line(color = "black")) +
scale_color_manual(values = base_color) +
theme
(axis.text.x = element_text(angle = 90, vjust = 1, hjust =
1, size = 7, colour = "black"),
axis.text.y = element_text(size = 10),
strip.text.y = element_text(angle = 0, vjust = 0, hjust =
0, size = 10, colour = "black"),
strip.text.x = element_text(size = 10)) (see Note 19)

#Show the Dotplot
print(("dotplotCCvs"))
```

(u) Repeat the steps from the creation of the.CSV file for each type of ontology (Cellular Component, Biological Process and Molecular Function).

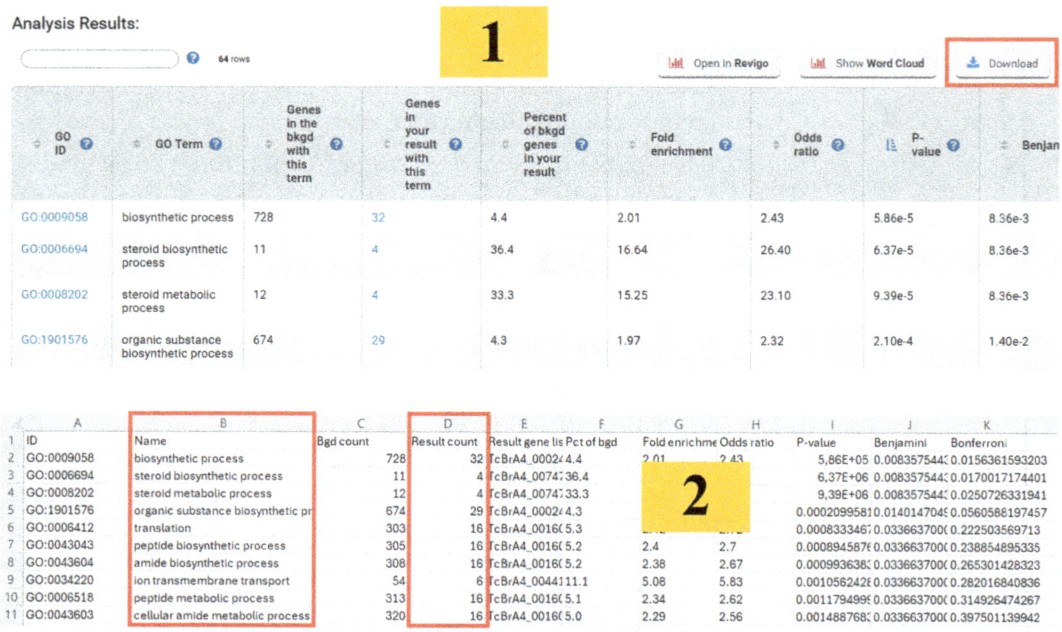

Fig. 3 Viewing and downloading results of gene ontology enrichment analysis in TriTrypDB. This figure illustrates the process of visualizing and exporting gene ontology enrichment results in TriTrypDB. **Panel 1** displays the interface showing the enriched GO terms, while **Panel 2** highlights the key data columns relevant for downstream analysis

3.10 Analysis of Metabolic Pathways with KAAS—KEGG

(a) In the TriTrypDB gene annotation results interface, click the "Download" option.

(b) In the next interface, choose "FASTA—Sequence retrieval, configurable" as the download format.

(c) Select "Text File" as the download type and click "Get Sequences" to obtain the.fasta file with the nucleotide sequences of your genes (Fig. 4).

(d) Open the KAAS-KEGG website https://www.genome.jp/kegg/kaas/

(e) Select the "Partial Genome—KAAS job request (SBH method)" option.

(f) Upload the.fasta file downloaded from TriTrypDB.

(g) Enter your email address to receive notifications about the status of the analysis.

(h) In the "Organism list" section, select the organism of interest.

(i) Click "Compute" to start the analysis (Fig. 5).

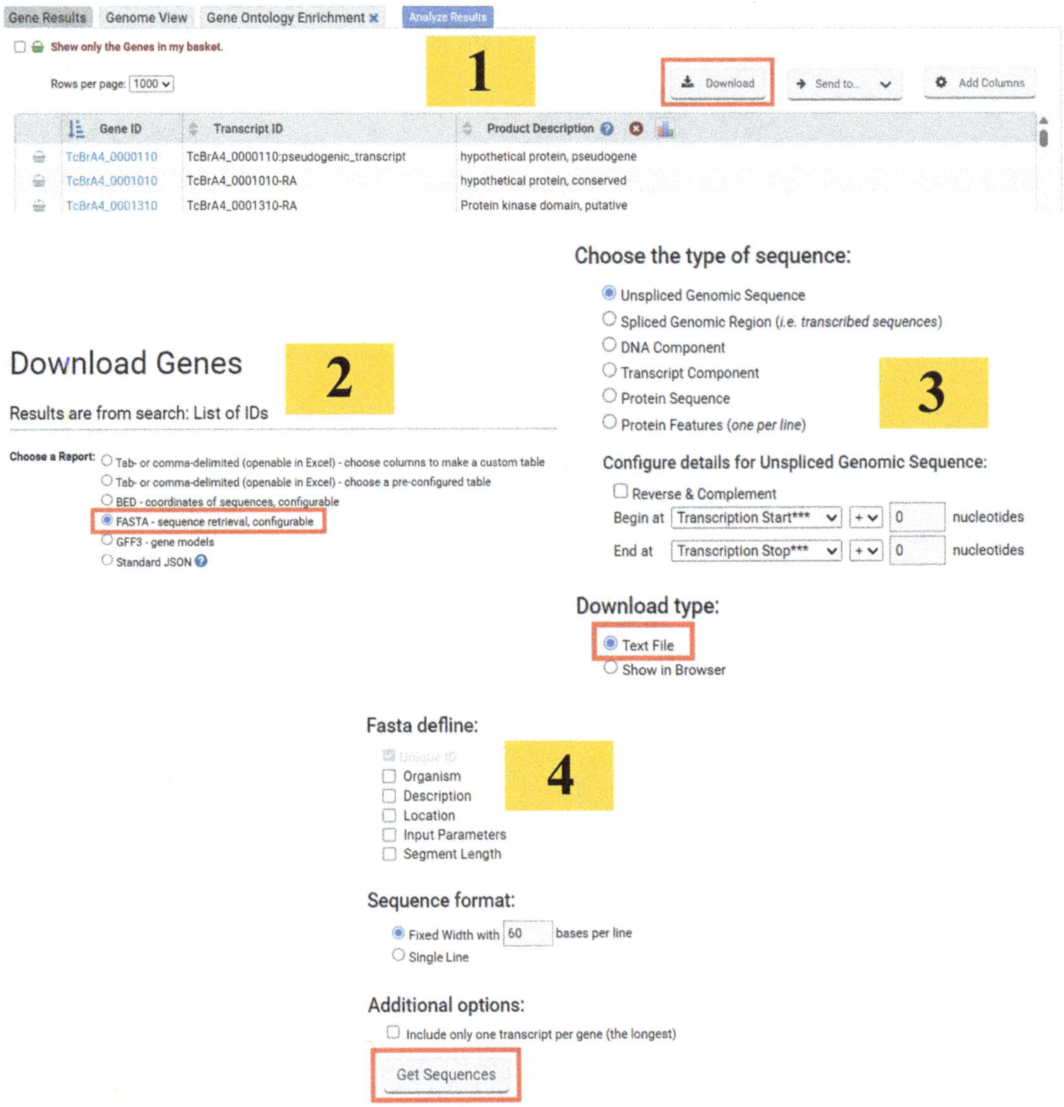

Fig. 4 Nucleotide sequence download process in TriTrypDB. Interface for nucleotide sequence downloads (**Panel 1**). This tool offers a selection of various file formats, ensuring compatibility with a wide range of bioinformatics applications (**Panels 2 and 3**). Once the format preferences are configured, the system allows for the efficient download of the desired sequences (**Panel 4**)

(j) Check your email and click on the link to confirm the submission of the job to KAAS-KEGG.

(k) Once the analysis is finished, you will receive an email with a link to the results.

(l) Open the link provided to explore the metabolic pathways identified and obtain detailed information.

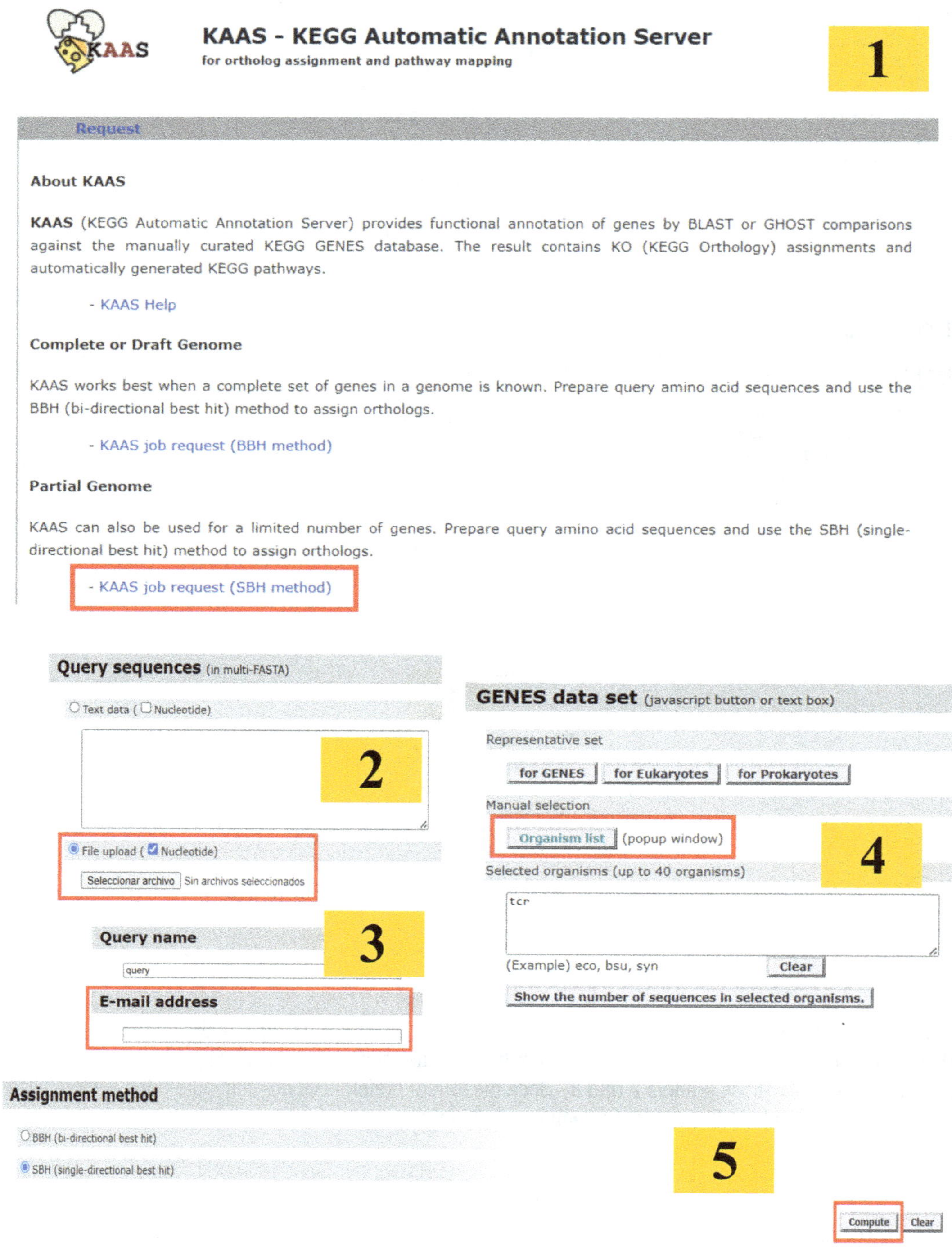

Fig. 5 KAAS-KEGG analysis workflow: from sequence input to results. The KAAS-KEGG analysis initiates from the KAAS-KEGG home page, where users access the sequence annotation tool (**Panel 1**). The process begins with the interface for loading sequences in KAAS-KEGG, enabling users to input their data (**Panel 2**). Subsequently, the system presents the add email interface, allowing users to receive notifications upon completion of the analysis (**Panel 3**). The final interface for loading sequences and selecting the organism of interest in KAAS-KEGG consolidates the input parameters, ensuring accurate annotation (**Panel 4**). Finally, the final step for analysis with KAAS-KEGG executes the annotation, generating KEGG pathway and module assignments based on the provided sequences and organism selection (**Panel 5**)

4 Notes

1. If you do not have unzip, install it with:

   sudo apt install unzip -y

2. To create Bash scripts, Python files, or.txt files, you can use any available text editor instead of creating them in the HPC terminal. However, we especially recommend the Atom text editor due to its great versatility. You can download it from the following link: https://atom-editor.cc/.

3. The following commands illustrate typical file transfers from a local system to an HPC using SCP (Secure Copy):

Transferring a local file:

scp (“**/path/to/local/file**”) user@remote_host:(“**/path/to/destination/**”)

Transferring a local directory (recursive):

scp -r (“**/path/to/local/directory**”) user@remote_host:(“**/path/to/destination/**”)

The -r flag enables **recursive** copying, including all subdirectories and files.

4. The (“**file_name**”).txt file that will be created with the names of the folders where the files are stored must not contain symbols or punctuation marks that may be confusing. An example of how the.txt file should look like is the following:

```
DA_BENZ_1 #Folder name 1
DA_BENZ_2 #Folder name 2
DA_BENZ_3 #Folder name 3
DA_CTR_1 #Folder name 4
DA_CTR_2 #Folder name 5
DA_CTR_3 #Folder name 6
```

5. Change the parameters according to your data. Those placed in the script refer to the following:

   ILLUMINACLIP:/opt/ohpc/pub/apps/trimmomatic/0.38/bin/adapters/TruSeq3-PE.fa:2:30:10:2:TRUE: This is the command to remove the Illumina adapters. If other adapters were used for your samples, change them for what you used.

   ILLUMINACLIP: Indicates that an adapter clip will be made.

   /opt/ohpc/pub/apps/trimmomatic/0.38/bin/adapters/TruSeq3-PE.fa: Path to the file that contains the sequence of the TruSeq3 adapters for paired-end reads (PE—Paired-

End). This file is crucial, since it contains the sequences of the adapters that were added during the preparation of the sequencing library.

2:30:10:2:TRUE: These are the specific parameters for the adapter clip:

2: Maximum number of mismatches allowed in the adapter sequence when searching for matches.

30: Seed word (length of the adapter sequence that is initially searched).

10: Score threshold for adapter alignment.

2: Maximum number of adapters that can be found in a read.

TRUE: Indicates that the adapters should be removed at the end of the reads.

SLIDINGWINDOW:4:20: This parameter activates the trimming by sliding window.

SLIDINGWINDOW: Indicates that the sliding window method will be used.

4: Window size (4 bases).

20: Minimum average Phred quality required within the window. If the average quality within the window is less than 20, the window slides and the next section of the read are evaluated. The bases with low quality are trimmed.

MINLEN:100: This parameter establishes the minimum length required for a read after trimming. If a read, after removing the adapters and applying the sliding window, has a length less than 100 bases, it is discarded.

AVGQUAL:20: This parameter establishes the minimum average Phred quality required for a complete read after trimming. If the average quality of the entire read is less than 20, it is discarded.

TRAILING:20: This parameter removes the bases with a Phred quality less than 20 at the end of the read (the "tail"). This step is done after the adapter clip and the trimming by sliding window.

6. You can monitor the progress of the job using the commands squeue, sacct, or smap. Check the output files (Trimmed.out) and errors (%x-%j.err) for any problems.
7. Step-by-step guide to downloading a reference genome from TriTrypDB.

(a) **Prerequisites**

Before starting, ensure you have:

Access to a Linux server with an internet connection.

wget installed for downloading files.

gunzip installed for extracting files if necessary.

(b) **Access the TriTrypDB Website**

Open a web browser and go to the official TriTrypDB site: https://tritrypdb.org

Use the search function in the database to locate the desired reference genome. For this guide, search for the Trypanosoma cruzi Brasil A4 reference genome.

(c) **Obtain the Download Link**

Within the organism's page, navigate to the "Genomes" section.

Look for the FASTA files corresponding to the reference genome.

Copy the URL of the files you want to download (usually. fasta.gz or.tar.gz files).

(d) **Download the Files to the Server**

Use wget to download the files from the server terminal. For example:

wget https://tritrypdb.org/common/downloads/release-68/TcruziBrazilA4/fasta/data/TriTrypDB-68_TcruziBrazilA4_Genome.fasta

(e) **Extract the File**

Use the following command to extract it:

```
gunzip TriTrypDB-68_TcruziBrazilA4_Genome.fasta.gz
```

(f) **Verify File Integrity**

After downloading and extracting, verify that the file is complete:

```
ls -lh TriTrypDB-68_TcruziBrazilA4_Genome.fasta
```

To check the first lines of the file:

```
head TriTrypDB-68_TcruziBrazilA4_Genome.fasta
```

(g) **Prepare the File for Mapping**

If you are using tools such as bwa, bowtie2, or hisat2, generate genome indexes:

For bwa:

```
bwa index TriTrypDB-68_TcruziBrazilA4_Genome.fasta
```

For bowtie2:

```
bowtie2-build TriTrypDB-68_TcruziBrazilA4_Genome.fasta Tcru-
ziBrazilA4.
```

8. Location of the GFF file, which contains the annotation of genes of the genome of our organism of interest. It is usually downloaded at the same time when we download our reference genome from the repository of the TriTrypDB database.
9. Ensure that sample conditions (Treatment (C2) vs Control (C1)) are correctly assigned based on filename patterns to prevent misclassification. This method minimizes errors due to incorrect manual ordering and allows flexibility for varying numbers of control and treatment samples. If filenames do not include "treated" or "control," manual assignment is required.

   If manual assignment is needed, use the following command:

```
#Define sample conditions manually
sampleCondition <- c("C1", "C1", "C1", "C2", "C2", "C2")
```

   Ensure that the conditions are listed in the same order as the corresponding files in the directory. Incorrect assignments can lead to errors in downstream analyses. Adjust the command by adding or removing "C1" or "C2" as needed to match the number of control and treatment samples.
10. At the end of the execution of the commands, the routes to the files are defined, the count files of HTSeq-count are listed (assuming they are.txt files), and a table is created that relates the names of the files with the biological conditions of each sample. It is crucial that the order for sample Condition corresponds to the order of the listed files.
11. Filtering genes with low counts improves the accuracy of differential expression analysis. A threshold of 10 counts is used to remove genes prone to technical noise. Retaining only genes with at least 10 counts in a minimum of three samples, corresponding to the smallest group, provides adequate read depth for statistical significance.

12. At the end of the execution of the commands, the DESeqDataSet object is created, which contains the counts, the information of the samples and the experimental design. Genes with low expression are filtered to reduce noise. The reference group is defined for the comparisons. Finally, the analysis is executed with DESeq().
13. NA values in adjusted *p*-value columns, resulting from low gene counts or inadequate statistical data, hinder downstream analysis, particularly when identifying significant genes. Assigning 0 to these NAs prevents their removal, classifying them as non-significant and maintaining result consistency.
14. At the end of the execution of the commands, the results of the specified comparison are extracted. The results are ordered, and the number of significant genes is calculated. NA values are handled in the adjusted p-values (if they exist).
15. We can rename our data so that they are displayed in a more aesthetic way in the Heatmap, placing a shorter name that we can identify which sample corresponds to.
16. For the creation of the.csv file that will allow us to do the ontology analysis we must follow the following example format:

```
Term,Count,Condition
Pathogenesis,100,Up-regulated
Cellular and Subcellular Movement,53,Up-regulated
Microtubule-Based Processes,21,Up-regulated
Cell Projection Organization and Assembly,30,Up-regulated
Metabolic Processes,228,Down-regulated
Biosynthetic Processes,218,Down-regulated
Nitrogen and Organonitrogen Metabolism,59,Down-regulated
Gene Expression and Translation,35,Down-regulated
```

17. The name of the file that is created "CC" (Cellular Component), can be changed to "BP" (Biological Process) or "MF" (Molecular Function), according to the group of ontological terms that you want to graph in R. However, you can also put the name you want.
18. Make sure that the separator in your CSV file is ",". If it is different, adjust the sep argument in the read.csv() function.
19. Explanation of the commands:

    ggplot(CC, aes(x = Condition, y = Term, color = Condition)): Initializes the graph with the data of CC and defines the variables for the x and y axes, as well as the color of the points.

    geom_count(aes(size = Count)): Creates points whose size is proportional to the gene count (Count).

`theme (...)`: Customizes the appearance of the graph, including the background, axes, and text.

`scale_color_manual(values = color_base)`: Applies the defined color palette.

`axis.text.x = element_text(angle = 90,...)`: Adjusts the orientation and size of the x-axis text to improve legibility.

References

1. Wang Z, Gerstein M, Snyder M (2009) RNA-Seq: a revolutionary tool for transcriptomics. Nat Rev Genet 10(1):57–63
2. Wolf JB (2013) Principles of transcriptome analysis and gene expression quantification: an RNA-seq tutorial. Mol Ecol Resour 13(4): 559–572
3. Maslov DA, Opperdoes FR, Kostygov AY et al (2019) Recent advances in trypanosomatid research: genome organization, expression, metabolism, taxonomy and evolution. Parasitology 146(1):1–27
4. Porcel BM, Denoeud F, Opperdoes F et al (2014) The streamlined genome of Phytomonas spp. relative to human pathogenic kinetoplastids reveals a parasite tailored for plants. PLoS Genet 10(2):e1004007
5. Jackson AP (2015) Genome evolution in trypanosomatid parasites. Parasitology 142(S1): S40–S56
6. Günzl A (2010) The pre-mRNA splicing machinery of trypanosomes: complex or simplified? Eukaryot Cell 9(8):1159–1170
7. Ghedin E, Bringaud F, Peterson J et al (2004) Gene synteny and evolution of genome architecture in trypanosomatids. Mol Biochem Parasitol 134(2):183–191
8. Peacock CS, Seeger K, Harris D et al (2007) Comparative genomic analysis of three Leishmania species that cause diverse human disease. Nat Genet 39(7):839–847
9. Flegontov P, Butenko A, Firsov S et al (2016) Genome of Leptomonas pyrrhocoris: a high-quality reference for monoxenous trypanosomatids and new insights into evolution of Leishmania. Sci Rep 6(1):23704
10. Witschel M, Rottmann M, Kaiser M et al (2012) Agrochemicals against malaria, sleeping sickness, leishmaniasis and Chagas disease. PLoS Negl Trop Dis 6(10):e1805
11. Peña-Callejas G, González J, Jiménez-Cortés JG et al (2022) Enfermedad de Chagas: biología y transmisión de Trypanosoma cruzi. TIP. Revista especializada en ciencias químico-biológicas, 25
12. Cruz-Saavedra L, Schwabl P, Vallejo GA et al (2022) Corrigendum: genome plasticity driven by aneuploidy and loss of heterozygosity in Trypanosoma cruzi. Microb Genom 8(7): 000873
13. Atwood JA III, Weatherly DB, Minning TA et al (2005) The Trypanosoma cruzi proteome. Science 309(5733):473–476
14. de Godoy LMF, Marchini FK, Pavoni DP et al (2012) Quantitative proteomics of Trypanosoma cruzi during metacyclogenesis. Proteomics 12(17):2694–2703
15. Fall F, Mamede L, Schioppa L et al (2022) Trypanosoma brucei: metabolomics for analysis of cellular metabolism and drug discovery. Metabolomics 18(4):20
16. Trochine A, Creek DJ, Faral-Tello P et al (2014) Benznidazole biotransformation and multiple targets in Trypanosoma cruzi revealed by metabolomics. PLoS Negl Trop Dis 8(5): e2844
17. Queiroz R, Benz C, Fellenberg K et al (2009) Transcriptome analysis of differentiating trypanosomes reveals the existence of multiple post-transcriptional regulons. BMC Genom 10:1–19
18. Carvalho de Lima PL, de Sousa Lopes L, Nunes Rosón J et al (2024) Comprehensive analysis of nascent transcriptome reveals diverse transcriptional profiles across the Trypanosoma cruzi genome underlining the regulatory role of genome organization, chromatin status, and cis-acting elements. bioRxiv:2024–04
19. Bartholomeu DC, SMR T, Cruz AK (2021) Genomics and functional genomics in Leishmania and Trypanosoma cruzi: statuses, challenges and perspectives. Mem Inst Oswaldo Cruz 116:e200634

20. Clayton C (2019) Regulation of gene expression in trypanosomatids: living with polycistronic transcription. Open Biol 9(6):190072
21. Siegel TN, Hekstra DR, Wang X et al (2010) Genome-wide analysis of mRNA abundance in two life-cycle stages of Trypanosoma brucei and identification of splicing and polyadenylation sites. Nucleic Acids Res 38(15): 4946–4957
22. Oliveira C, Faoro H, Alves LR et al (2017) RNA-binding proteins and their role in the regulation of gene expression in Trypanosoma cruzi and Saccharomyces cerevisiae. Genet Mol Biol 40(01):22–30
23. De Gaudenzi JG, Noé G, Campo VA et al (2011) Gene expression regulation in trypanosomatids. Essays Biochem 51:31–46
24. Vigneron A, O'Neill MB, Weiss BL et al (2020) Single-cell RNA sequencing of Trypanosoma brucei from tsetse salivary glands unveils metacyclogenesis and identifies potential transmission blocking antigens. Proc Natl Acad Sci 117(5):2613–2621
25. Clayton C, Shapira M (2007) Post-transcriptional regulation of gene expression in trypanosomes and leishmanias. Mol Biochem Parasitol 156(2):93–101
26. Kelly S, Kramer S, Schwede A et al (2012) Genome organization is a major component of gene expression control in response to stress and during the cell division cycle in trypanosomes. Open Biol 2(4):120033
27. Inchausti L, Bilbao L, Campo VA et al (2025) Single-cell RNA-seq reveals trans-sialidase-like superfamily gene expression heterogeneity in Trypanosoma cruzi populations. bioRxiv:2025-01
28. Amos B, Aurrecoechea C, Barba M et al (2022) VEuPathDB: the eukaryotic pathogen, vector and host bioinformatics resource center. Nucleic Acids Res 50(D1):D898–D911
29. Kanehisa M, Furumichi M, Tanabe M et al (2017) KEGG: new perspectives on genomes, pathways, diseases and drugs. Nucleic Acids Res 45(D1):D353–D361. https://doi.org/10.1093/nar/gkw1092

Chapter 13

Small RNA Profiling in Trypanosomes

Florencia Díaz-Viraqué, Gonzalo Greif, María Rosa García-Silva, and Carlos Robello

Abstract

Trypanosomatids express a diverse repertoire of small non-coding RNAs despite lacking the canonical RNA interference machinery in most species. These small RNAs—including fragments derived from tRNAs, rRNAs, and other non-coding sources—are thought to play regulatory roles in parasite biology and host–parasite interactions. In this chapter, we provide a detailed protocol for the isolation, library preparation, sequencing, and computational analysis of small RNAs in trypanosomes. The method is optimized to preserve short RNA species while minimizing artifacts such as adaptor dimers and concatamerization. It includes quality control and size selection steps to ensure the reproducibility and high quality of the libraries, allowing the study of the transcriptome of small RNAs in these parasites.

Key words Small RNAs, Non-coding RNAs, Protist parasites, Trypanosomes, Next-generation sequencing

1 Introduction

Small RNAs are non-coding RNA molecules that play crucial roles in the post-transcriptional regulation of gene expression across diverse eukaryotic organisms. They are involved in various biological processes, including development, stress response, and genome stability, by modulating mRNA stability and translation [1].

In trypanosomatids—a family within the order Kinetoplastea whose members are exclusively parasitic—the canonical pathways for small RNA biogenesis and function appear to be either absent or highly divergent. RNA interference (RNAi), a pathway conserved across most eukaryotes, is present in *Trypanosoma brucei* and *Leishmania braziliensis*, but is absent in other *Leishmania* species and *Trypanosoma cruzi* [2].

Unlike many other eukaryotes, these parasites lack key components of the RNA interference (RNAi) machinery, such as Dicer

Paul A. M. Michels et al. (eds.), *Euglenozoa: Methods and Protocols, Volume 1*, Methods in Molecular Biology, vol. 3013, https://doi.org/10.1007/978-1-0716-5142-1_13,

and Argonaute proteins [2, 3]. Instead, *T. cruzi* possesses a Piwi-like protein, TcPIWI, capable of binding small RNAs derived from tRNAs and rRNAs, suggesting alternative pathways for small RNA activity [4, 5].

Despite the absence of this classical pathway, several studies showed that *T. cruzi* expresses a variety of other small non-coding RNAs. García-Silva et al. [4] first demonstrated the presence of small RNAs derived from transfer RNAs (tRNAs) and ribosomal RNAs (rRNAs), suggesting that alternative biogenesis pathways are operational in this parasite. Similarly, Franzen et al. [6] characterized diverse small RNA populations in *T. cruzi* and highlighted their potential roles, noting that these sncRNAs might participate in gene regulation, parasite adaptation, and host interactions. Furthermore, Bayer-Santos et al. [7] expanded on these findings by profiling multiple classes of small RNAs across different developmental stages, providing evidence that non-canonical pathways are likely involved. The studies by Franzen and García-Silva collectively indicate that although *T. cruzi* does not employ the classical miRNA pathway, it has evolved a distinct repertoire of small RNAs that could fulfill regulatory functions. These may include tRNA fragments, rRNA-derived small RNAs, and other novel small RNA species, which could contribute to the regulation of gene expression, parasite differentiation, or immune evasion strategies. Understanding the biogenesis, composition, and roles of small RNAs in trypanosomes remains critical, particularly given the parasite's impact on human health and its unique molecular biology.

In this chapter, we describe the experimental procedures for preparing small RNA libraries, sequencing, and computational tools for data analysis in *T. cruzi*, which can be adapted to the rest of the trypanosomes. We aim to provide a comprehensive approach to characterize the small RNA landscape in this organism.

2 Materials

2.1 Parasites

All procedures involving parasites must be conducted in Class II biosafety cabinets. Synchronized *T. cruzi* epimastigotes are cultured under axenic conditions in liver infusion tryptose (LIT) medium supplemented with 10% (v/v) heat-inactivated fetal bovine serum at 28 °C.

2.2 Solutions, Reagents, and Consumables

1. LIT medium: 5.0 g/L of Liver Infusion Broth, 4.4 g/L of NaCl, 0.4 g/L of KCl, 2.2 g/L of glucose, 5 g/L of tryptose, 11.6 g/L of Na_2HPO_4, 15 g/L of Yeast Extract. Adjust pH to 7.2 and filter-sterilize. Add 25 mg of hemin, 100 U/mL of penicillin, and 0.1 mg/mL of streptomycin.

2. 10× phosphate-buffered saline (PBS): 1.37 M NaCl, 27 mM KCl, 100 mM Na_2HPO_4, and 18 mM KH_2PO_4, pH 7.4 in DEPC water. Autoclave. Prepare a fresh 1× solution in sterile conditions for parasite washes and chill at 4 °C.
3. Kit for preparation of RNA minipreps (e.g., Direct-zol™ RNA MiniPrep kit by Zymo Research, but equivalent kits from other suppliers are also suitable).
4. Qubit RNA HS Assay Kit (*Thermo Fisher Scientific*) or an equivalent RNA quantification kit suitable for low-concentration samples.
5. Reagents provided in the NEBNext Multiplex Small RNA Library Prep set for Illumina kit (*New England Biolabs*).
6. Absolute ethanol.
7. 80% ethanol freshly prepared in nuclease-free water.
8. 1.5 and 2.0 mL DNA low-binding tubes (e.g., Eppendorf LoBind or equivalent).
9. RNase-free disposable pellet pestles.
10. Sterile scalpel.
11. Gel filtration column (e.g., Corning®, Costar®, Spin-X® Centrifuge Tube Filters (Cellulose Acetate Filters) (Sigma Aldrich #CLS8162).
12. TE buffer (10 mM Tris–HCl, 1 mM EDTA, pH 8.0).
13. Linear Acrylamide.
14. 3 M sodium acetate pH 5.5.

2.3 Equipment

1. Class II microbiological safety cabinet.
2. Thermal cycler.
3. Agarose gel electrophoresis equipment or Agilent 2100 Bioanalyzer (*Agilent Technologies*).
4. UV transilluminator.
5. Fluorescence-based quantification equipment (Qubit, DeNovix, etc.).
6. Microcentrifuge.
7. Rotator.

3 Methods

The method described here aims to capture the complete range of small RNAs from trypanosomes using the NEBNext® Multiplex Small RNA Library Prep Set for Illumina®, combined with paired-end sequencing. First, the conversion of small RNAs into indexed libraries for next-generation sequencing on the Illumina platform is detailed, followed by a description of genome-wide profiling analysis.

3.1 Sample Preparation

1. Extract total RNA from *T. cruzi* using a kit that retains RNA fragments ≥17 nucleotides (e.g., Direct-zol™ RNA MiniPrep Kit, Zymo Research), following the manufacturer's instructions. Use approximately 1–5 × 10^7 epimastigotes harvested from exponential-phase axenic cultures (typically 3–4 days of growth), or an equivalent number of cell culture-derived trypomastigotes collected from infected cell supernatants at 5–6 days post-infection.
2. Include the treatment with RNase-free DNase to remove contaminating DNA.
3. Measure RNA concentration using an assay selective for RNA, such as the Qubit RNA HS Assay Kit (*Thermo Fisher Scientific*), and assess quality using an Agilent 2100 Bioanalyzer (*Agilent Technologies*; *see* **Note 1**).
4. High-quality total RNA (RNA Integrity Number, RIN > 7) should be used to continue the protocol. Small RNA fragments should have a 5′ phosphate and 3′ OH to ligate and must be free of ATP (*see* **Note 2**).

3.2 Library Preparation

To generate small RNA libraries directly from total RNA, start from 100 ng to 1 μg of total RNA. We describe the library preparation using the NEBNext Multiplex Small RNA Library Prep set for Illumina kit (*New England Biolabs*). Except when indicated, it is recommended to premix all the reagents and use them immediately. A schematic overview of the main steps is shown in Fig. 1a.

3.2.1 Ligation of 3′ SR Adaptor

1. The first step involves adding the 3′ end adaptor. Start with 100 ng–1 μg of total RNA. Add the following components to a sterile, nuclease-free polymerase chain reaction (PCR) tube.

Input RNA (100 ng–1 μg)	1–6 μL
3′ SR adaptor for Illumina (green cap)	1 μL
Nuclease-free water	As needed to adjust to the final volume
Total	7 μL

2. Incubate the sample in a preheated thermal cycler for 2 min at 70 °C, then transfer it to ice.

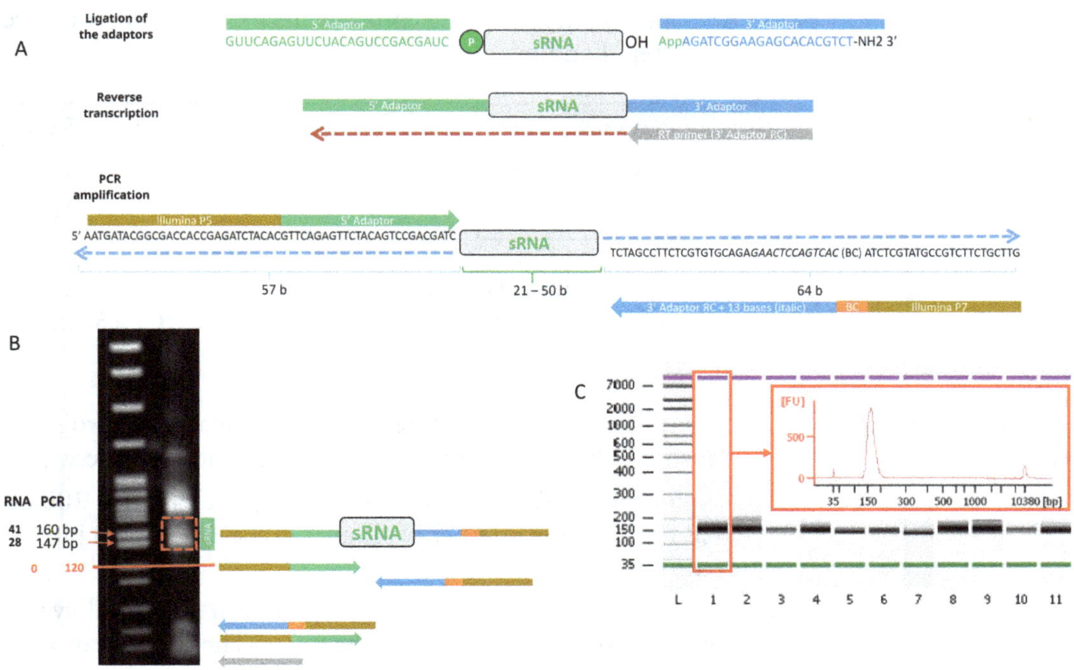

Fig. 1 (**a**) Schematic overview of the main steps of library construction. (**b**) The 140–150 bp bands correspond to adapter-ligated constructs derived from the 21- and 30-nucleotide small RNA, respectively. The smaller bands correspond to adapter dimers (127 bp) or excess primers (~70–80 bp). (**c**) Electropherogram trace of the gel size–selected purified library from *Trypanosoma cruzi* small RNA libraries after size selection

3. Add and mix the following components:

3′ Ligation reaction buffer (green cap)	10 μL
3′ ligation enzyme mix (green cap)	3 μL
Total	20 μL

4. Incubate for 1 H at 25 °C in a Thermal Cycler (*See* **Note 3**)

3.2.2 Hybridization of the Reverse Transcription Primer

During this step, the reverse transcription primer hybridizes with the 3′ end adaptor. This step is critical for preventing adaptor-dimer formation. The reverse transcription primer (SR RT Primer) hybridizes with the 3′ end adaptor (3′ SR Adaptor), not only with the 3′ SR Adaptor ligated to the RNAs in the previous step, but also with the excess of 3′ SR Adaptor (that remains free after the 3′ ligation reaction), and transforms the single-stranded DNA adaptor into a double-stranded DNA (dsDNA) molecule. dsDNAs are not substrates for ligation mediated by T4 RNA Ligase 1 and therefore do not ligate to the 5’ SR Adaptor in the subsequent ligation step.

1. Add and mix the following components to the ligation mixture from **step 4** and mix well.

Nuclease-free water	4.5 μL
SR RT primer for Illumina (pink cap)	1 μL
Total	25.5 μL

2. Place the mix in a thermal cycler with a heated lid set to >85 °C and run the following program: 5 min at 75 °C, 15 min at 37 °C, 15 min at 25 °C, and hold at 4 °C.

3.2.3 Ligation of 5′ SR Adaptor

1. During the 5 min before the end of the thermal cycler program, resuspend the 5′ SR adaptor in 120 μL of nuclease-free water.
2. Aliquot the 5′ SR adaptor into a separate, nuclease-free 200 μL PCR tube, for the number of samples in the experiment plus an excess of 10%.
3. Incubate the adaptor at 70 °C for 2 min in a thermal cycler, then immediately transfer the tube to ice. The denatured adaptor should be kept on ice and used within 30 min. If not used immediately, it should be stored in the freezer for later use (*see* **Note 4**).
4. Add and mix the following components to the ligation mixture from **step 2** of the previous Subheading 3.2.2 and mix well. In this case, do not premix reagents.

5′ SR Adaptor for Illumina (denatured) (yellow cap)	1 μL
5′ ligation reaction buffer (10×) (yellow cap)	1 μL
5′ ligation enzyme mix (yellow cap)	2.5 μL
Total	30 μL

5. Incubate for 1 h at 25 °C in a thermal cycler.

3.2.4 Reverse Transcription (RT)

1. Mix the following components in a sterile, nuclease-free tube. Reagents should be premixed. Use immediately.

Adaptor-ligated RNA from **step 4** of the previous Subheading 3.2.3	30 μL
First Strand synthesis reaction (red cap)	8 μL
Murine RNase inhibitor (red cap)	1 μL
ProtoScript II reverse transcriptase (red cap)	1 μL
Total	40 μL

2. Incubate for 60 min at 50 °C.
3. Immediately proceed to PCR amplification.

Safe stopping point: If you do not plan to proceed immediately to PCR amplification, heat-inactivate the RT reaction at 70 °C for 15 min. The samples can then be safely stored at temperatures ranging from −15 to −25 °C.

3.2.5 PCR Amplification of the Library

1. Add and mix the following components to the RT reaction mix from **step 2** of the previous Subheading 3.2.4 and mix well:

LongAmp Taq 2× Master Mix (blue cap)	50 μL
SR primer for Illumina (blue cap)	2.5 μL
Index (X) primer (*see* **Note 5**) (blue cap)	2.5 μL
Nuclease-free water	5 μL
Total	100 μL

Perform the PCR using the following PCR cycling conditions:

PCR step	T (°C)	Time	Cycles
Initial denaturation	94	30 s	1
Denaturation Annealing Extension	94 62 70	15 s 30 s 15 s	15–25*
Final extension	70	5 min	1
Hold	4	∞	

* The number of PCR cycles may need to be optimized if clear and well-resolved bands are not observed in the electrophoresis gel image. Amplification parameters may vary depending on the amount of input RNA, the tissue type, and the species analyzed. For an input of 100 ng of total RNA, 15 PCR cycles are recommended. Samples with a high proportion of small RNAs may necessitate a reduced number of cycles; 12 cycles are sufficient.

Safe stopping point: If you do not plan to proceed, it is safe to store the library at −20 °C after PCR. Avoid leaving the sample at 4 °C overnight, if possible.

3.2.6 Quality Control Check and Size Selection

The following steps describe the procedure for quality control checks and size selection using a 6% polyacrylamide gel.

1. Mix 25 μL of the PCR product with 5 μL of Gel Loading Dye Blue (6×).
2. Load 5 μL of Quick-Load pBR322 DNA-MspI Digest in one well on a 6% PAGE ten-well gel.

3. Load two wells with 15 μL each of the mixed amplified cDNA construct and loading dye onto a 6% PAGE ten-well gel.
4. Run the gel for 1 h at 120 V, or until the blue dye has reached the bottom of the gel. Do not let the blue dye exit the gel.
5. Remove the gel from the apparatus and stain with SYBR Gold nucleic acid gel stain in a clean container for 2–3 min. Then view the gel on a UV transilluminator. Fig. 1b shows typical results from *T. cruzi* total RNA libraries.
6. Use a sterile scalpel to excise the gel region corresponding to 140–150 bp bands to retrieve small RNAs (dashed red square, Fig. 1b).
7. Place the gel slices from the same sample in one 1.5 mL tube. Crush the gel slices with RNase-free Disposable Pellet Pestles, and then soak in 250 μL of DNA Gel Elution buffer (1×).
8. Rotate end to end for at least 2 h at room temperature.
9. Transfer the eluate and the gel debris to the top of a gel filtration column (for example, Corning®, Costar®, Spin-X® Centrifuge Tube Filters (Cellulose Acetate Filters) (Sigma Aldrich #CLS8162).
10. Centrifuge the filter for 2 min at >14,000× *g*.
11. Recover the eluate and add 1 μL Linear Acrylamide, 25 μL 3 M sodium acetate pH 5.5, and 750 μL of 100% ethanol. Mix by inverting the tube.
12. Precipitate in a dry ice/methanol bath or at −80 °C for at least 30 min.
13. Spin in a microcentrifuge >14,000× *g* for 30 min at 4 °C and discard the supernatant without disturbing the pellet.
14. Wash the pellet by adding 1 mL of freshly prepared 80% ethanol. Mix by inverting and then spin in a microcentrifuge at 14,000× *g* for 30 min at 4 °C. Gently remove the supernatant using a pipette, taking care not to disturb the pellet.
15. Air-dry the pellet at room temperature for up to 10 min and resuspend it in 12 μL of TE buffer.
16. Run 1 μL of the library in the Bioanalyzer (*Agilent Technologies*) using the Agilent High Sensitivity DNA Kit (*Agilent Technologies*). Check to obtain the expected size and purity of the band (Fig. 1c).

3.3 Library Sequencing

1. Quantify the libraries using Qubit™ dsDNA Quantification Assay Kit (Thermo)or equivalent.
2. Determine the concentration at nM using the formula: 1 ng/μ L of library equals a library 3 nM/average size (150 bp)/ 500 bp.
3. Dilute the libraries to 2 nM and run 50 bp single-end reads on an Illumina MiSeq (Illumina) at a final concentration of 10 pM. Up to six libraries can be pooled and sequenced in the same

run. The number of pooled libraries depends on the desired coverage for each sample and can be increased when using high-throughput platforms such as NovaSeq.

3.4 Computational Analysis

1. The analysis of small RNA samples generally begins with the identification and removal of adapter sequences. Adapter-trimmed reads are then aligned to reference databases, such as mature miRNA sequences from miRBase, other RNA species, or the genome. Once small RNAs are identified, quantification and comparative analyses, including the identification of differentially expressed small RNAs, can be performed.
2. The first step of the analysis involves removing low-quality reads (typically defined as reads containing more than 10–20% bases with Phred quality scores below 20, reads shorter than 15–17 nucleotides after adapter trimming, or reads containing ambiguous bases) and reads lacking the 3′ adapter sequence. The presence of this adapter at the 3′ end ensures that the full-length molecule is captured rather than degradation products.
3. The trimmed and high-quality reads are then mapped to the genome of interest and compared with known sequences. Depending on their genomic loci of origin, reads are classified into categories such as miRNAs, tRNA-derived fragments, rRNA-derived fragments, snoRNAs, snRNAs, and other classes of small non-coding RNAs [5, 8, 9].

4 Notes

1. Unlike most eukaryotes, which typically display two prominent ribosomal RNA (rRNA) bands, trypanosomatids exhibit three distinct rRNA bands when analyzed by agarose gel electrophoresis.
2. Prepare all solutions with RNase-free reagents. It is recommended to handle RNA in a clean workspace using pipettes and materials dedicated exclusively to RNA work and RNase-free tubes. Decontaminate surfaces and equipment with RNase/DNase removal solutions. Always wear gloves when handling RNA.
3. When working with methylated RNAs such as Piwi-interacting RNAs (piRNAs), it is recommended to use longer incubation times at lower temperatures (e.g., 18 h at 16 °C) to enhance ligation efficiency. However, note that these conditions may also promote the formation of concatamerization products.
4. Store the remaining resuspended 5′ SR adaptor at −80 °C. Please minimize freeze–thaw cycles; the 5′ SR adaptor could be aliquoted. Denature aliquots before use.

5. The number of PCR cycles may need to be optimized if clear and distinct bands are not observed in the gel image. Amplification conditions may vary depending on RNA input, tissue type, and species; for 100 ng total RNA input, 15 PCR cycles are recommended. If the sample contains a high abundance of small RNA, fewer cycles—typically fewer than 12—may be sufficient.

Acknowledgments

This work was supported by FOCEM—Fondo para la Convergencia Estructural del Mercosur (COF 03/11) (CR); ACIP grant (ACIP 1846-24) from Institut Pasteur Paris; Universidad de la República (Uruguay), CSIC Iniciacion 22320200200121UD (FDV); and Agencia Nacional de Investigación e Innovación (ANII, Uruguay), Proyecto de Investigación Fundamental "Fondo Clemente Estable" FCE_3_2022_1_172653 (FDV). FDV received fellowships from the ANII (POS_NAC_2016_1_129916) and CAP (Universidad de la República, BFPD_2021_1#45569540). FDV GG, MRG, and CR are members of the Sistema Nacional de Investigadores (SNI-ANII, UY) and researchers in the Programa de Desarrollo de Ciencias Básicas (PEDECIBA, UY).

References

1. Morris KV, Mattick JS (2014) The rise of regulatory RNA. Nat Rev Genet 15:423–437. https://doi.org/10.1038/nrg3722
2. Lye LF, Owens K, Shi H et al (2010) Retention and loss of RNA interference pathways in trypanosomatid protozoans. PLoS Pathog 6: e1001161. https://doi.org/10.1371/journal.ppat.1001161
3. Knip M, Constantin ME, Thordal-Christensen H (2014) Trans-kingdom cross-talk: small RNAs on the move. PLoS Genet 10:e1004602. https://doi.org/10.1371/journal.pgen.1004602
4. Garcia-Silva MR, Frugier M, Tosar JP et al (2010) A population of tRNA-derived small RNAs is actively produced in Trypanosoma cruzi and recruited to specific cytoplasmic granules. Mol Biochem Parasitol 171:64–73. https://doi.org/10.1016/j.molbiopara.2010.02.003
5. Garcia-Silva MR, Sanguinetti J, Cabrera-Cabrera F et al (2014) A particular set of small non-coding RNAs is bound to the distinctive Argonaute protein of Trypanosoma cruzi: insights from RNA-interference deficient organisms. Gene 538:379–384. https://doi.org/10.1016/j.gene.2014.01.023
6. Franzén O, Arner E, Ferella M et al (2011) The short non-coding transcriptome of the protozoan parasite Trypanosoma cruzi. PLoS Negl Trop Dis 5:e1283. https://doi.org/10.1371/journal.pntd.0001283
7. Bayer-Santos E, Lima FM, Ruiz JC et al (2014) Characterization of the small RNA content of Trypanosoma cruzi extracellular vesicles. Mol Biochem Parasitol 193:71–74. https://doi.org/10.1016/j.molbiopara.2014.02.004
8. Michaeli S, Doniger T, Gupta SK et al (2012) RNA-seq analysis of small RNPs in Trypanosoma brucei reveals a rich repertoire of non-coding RNAs. Nucleic Acids Res 40(3): 1282–1298. https://doi.org/10.1093/nar/gkr786
9. Fernandez-Calero T, Garcia-Silva R, Pena A et al (2014) Profiling of small RNA cargo of extracellular vesicles shed by Trypanosoma cruzi reveals a specific extracellular signature. Mol Biochem Parasitol 199:19–28. https://doi.org/10.1016/j.molbiopara.2015.03.003

Chapter 14

Desorption Electrospray Ionization–Mass Spectrometry Imaging (DESI-MSI) of Metabolites in Heart Samples Infected with *Trypanosoma cruzi*

Dan Chen, Guilherme M. P. Carrara, Laura-Isobel McCall, and Zhibo Yang

Abstract

Chagas disease is an infectious disease caused by the parasite *Trypanosoma cruzi*, with heart failure being a severe complication of its chronic stage. *T. cruzi* has three lifecycle stages: epimastigote, trypomastigote, and amastigote. Amastigotes are the intracellular form that invade mammalian host cells, replicate in their cytoplasm, and cause tissue damage. Due to the small size of *T. cruzi* amastigotes, fine spatial resolution would be beneficial. Current metabolomics research on Chagas disease primarily focuses on studying metabolites through widely used bulk analysis techniques such as liquid chromatography-tandem mass spectrometry (LC-MS/MS). However, conducting comprehensive molecular studies of Chagas disease is constrained by the inherent limitations, including sample loss and lack of spatial resolution, of these traditional analytical tools. Desorption electrospray ionization mass spectrometry imaging (DESI-MSI) is an advanced analytical method that combines ambient sampling and ionization with mass spectrometry to produce spatially resolved chemical maps of sample surfaces. This chapter describes the general procedures of using DESI-MSI techniques to study metabolic alterations in mouse hearts infected with *T. cruzi*, with high spatial resolution, preserving tissue integrity and offering new insights into potential therapeutic targets.

Key words Chagas disease, *Trypanosoma cruzi*, Mass spectrometry imaging (MSI), Desorption electrospray ionization (DESI), Metabolites, Lipids

1 Introduction

Chagas disease, caused by the protistan parasite *Trypanosoma cruzi* transmitted through triatomine insects, is endemic in the Americas, but has now become a worldwide public health challenge [1, 2]. Chagas disease has two phases: acute phase and chronic phase. The symptoms range from mild fever and swelling in the acute phase to severe complications, such as cardiomyopathy, arrhythmias, and sudden cardiac death, in the chronic phase. Cardiac complications represent the most common severe manifestation of chronic Chagas disease, emphasizing the importance of

Paul A. M. Michels et al. (eds.), *Euglenozoa: Methods and Protocols, Volume 1*, Methods in Molecular Biology, vol. 3013, https://doi.org/10.1007/978-1-0716-5142-1_14,

heart-focused research. The progression of *T. cruzi*–induced cardiomyopathy results in complex alterations in the metabolic profile, making it an ideal candidate for mass spectrometry (MS)-based studies [3]. Several lipids change abundance during infection with *T. cruzi* and in response to treatment. Some phosphatidylethanolamines (PEs) were abundant in patients with symptoms, while 10-hydroxydecanoic acid was reduced, and several sphingolipids returned to normal levels post-treatment [4].

In these studies, untargeted metabolomics using liquid chromatography-tandem mass spectrometry (LC-MS/MS) was a powerful approach to find potential biomarkers and understand aspects of disease pathogenesis. However, one limitation for this type of studies is the lack of spatial resolution. The distribution of many metabolites, which are broadly defined as small molecules (molecular weight <1500 Da), is heterogeneous across different regions of tissues, and the roles of some metabolites can be localized [5, 6]. Bulk analysis of the whole organ is not able to provide this type of spatial information, thus hindering the understanding of disease progression and metabolic pathways [7–9]. "Chemical cartography," which combines molecular networking and three-dimensional (3D) modeling to create chemical maps of tissues, can reveal the spatial and molecular patterns associated with infection. Tissues are systematically sectioned, and the metabolites are extracted from the tissue homogenates for LC-MS/MS analysis. Implementations of chemical cartography in cardiac Chagas disease found region-specific distributions of nucleoside derivatives and eicosanoids linked to the immune responses or tissue damage, and greater metabolic impact of infection at sites of greater disease symptoms, including the heart apex and pericardial regions [5, 10–12].

Unlike LC-MS/MS, which requires homogenization and analyte extraction during sample preparation, and results in sample and spatial resolution loss, mass spectrometry imaging (MSI) preserves the in situ spatial distribution of biomolecules with minimal sample preparation [8, 13]. This approach is invaluable in studying tissue heterogeneity, disease pathology, and drug localization. In the parasitology field, MSI has been applied in the differentiation of healthy and infected tissue, discovery of parasite biomarkers, and investigation of parasite–host interactions [14–16]. When acquiring data during an MSI experiment, the plane of the tissue slice is divided into pixels along the x-axis and y-axis. By moving the tissue on a stage, the analytes in each pixel are ionized and detected by MS. Each pixel contains full mass spectra covering a wide range of biomolecules. The ion image of each *m/z* is then reconstructed using the coordinates of each pixel and the intensity of the *m/z* at each pixel. The ion images enable a direct visualization of the heterogeneity of molecular distributions [17–19].

Depending on their sampling and ionization environment, MSI techniques can be categorized into two major groups: vacuum-based (e.g., MALDI-MSI, SIMS-MSI) and ambient-based (e.g., DESI-MSI, AP-MALDI-MSI, nanoDESI-MSI, single-probe MSI) techniques [7, 20–24]. Due to their higher sensitivity, high spatial resolution, and commercial availability, vacuum-based methods have been widely used for a broad range of studies. Among them, MALDI-MSI, which uses lasers to desorb and ionize matrix-associated analytes under vacuum, has emerged as a powerful tool to study infectious diseases. In a study of cutaneous leishmaniasis using mouse models, MALDI-MSI was applied to distinguish healthy, infected, and drug-treated tissue. Lower peak intensities above *m/z* 1300 were observed in healthy tissues, while infected tissue samples had high-intensity peaks around *m/z* 1050, and drug-treated tissue samples had a high-intensity peak around *m/z* 1180 [15]. Scott et al. revealed progressive depletion of host arachidonic acid (AA)-containing phospholipids during *Francisella* infection in the mouse spleen using MALDI-MSI. Phosphatidylinositol (PI) (18:0/24:0), PI 36:4, phosphatidylethanolamine (PE) 38:4, and phosphatidic acid (PA) 38:4 decreased in the white pulp by 48–60 h post infection. This timing concurs with spleen damage and progression to lethal sepsis [25]. Using LA-ICP-MSI and MALDI-MSI, Cassat et al. unveiled the variance of distribution of protein and metal ions associated with disseminated infection caused by *Staphylococcus aureus*. Calcium and calprotectin were enhanced in abscesses, while iron, zinc, manganese, and copper were decreased in the infected tissue. δ-Hemolysin and CsbD-like protein from *S. aureus* co-localized to the center of the abscess and can be used as markers of bacteria presence [26]. Perry et al. used MALDI-MSI and fluorescence labeling to reveal that lysyl-phosphatidylglycerols (lysyl-PGs) (e.g., lysyl-PG(31:0), lysyl-PG (33:0)) were highly correlated with some but not all *S. aureus* bacteria. Even within the same abscess, some bacteria foci had high levels of lysyl-PG, while some were absent. This was confirmed in the human heart valve [27]. Hulme et al. studied *Salmonella* Typhimurium infection in murine mesenteric lymph nodes using MALDI-MSI. Palmitoylcarnitine was co-localized with areas in which the density of immune cells was disrupted and *S.* Typhimurium present [28]. These studies demonstrate that MSI has great advantages in studying infectious diseases, especially the spatial heterogeneity of metabolite and lipid distribution, and the discovery of biomarkers of infection.

Recent advancement in sampling and ionization techniques has resulted in rapid development of ambient MSI methods. Among them, DESI-MSI, which ionizes molecules using charged solvent droplets to bombard the tissue surface and generate analyte ions under ambient pressure, has become a popular technique due to multiple advantages [29]. (1) Little or no sample preparation.

Unlike MALDI-MSI, which requires complex sample preparation, tissue slices in DESI-MSI studies do not need any coating of matrix molecules, greatly reducing potential artifacts (e.g., interference between matrix molecules and analyte dislocation) and complexity in experiments and data analysis. (2) The ambient environment is more friendly to the volatile molecules [7, 17, 30]. In vacuum-based MSI, samples are in high-vacuum chambers. This high-vacuum environment can cause sample dehydration or evaporation of small molecules like metabolites before ionization. In ambient pressure, volatile molecules can stay longer on the tissue for in situ detection. Historically, DESI was developed for direct sample surface analysis with relatively low spatial resolution (e.g., typically 50–200 μm) [31]. However, with advancement in instrumentation and optimized parameters, the performance of DESI-MSI has improved to reach a fine spatial resolution of ca. 20 μm [32, 33]. A recent study demonstrated the use of DESI-MSI with slow spray solvent delivery and well-optimized temperature to achieve cellular-level imaging of cultured HCT-116 cells [34]. In addition, because of the minor influence of DESI-MSI experiments on cellular morphology, the same tissue slice sample can be used for staining after DESI-MSI measurement [35]. This further expands the application of DESI-MSI to biological samples.

In our studies, DESI-MSI was used to detect changes of metabolite profiles caused by Chagas disease in mouse heart tissue. In this chapter, we describe the sample preparation, DESI-MSI experiments, and data analysis of mouse heart tissue slices infected with *T. cruzi*. This procedure could, however, be adapted to other intracellular pathogens expressing X-gal, for example, *Leishmania* spp. parasites, or to use any other convenient reporter protein such as fluorescent reporters.

2 Materials

2.1 General Material and Equipment

1. Mice, for example, female 4-week-old C3H/HeNCrl (ordered 14 g or larger) (*see* **Note 1**).
2. *T. cruzi* (LacZ), Tulahuen strain [36], Biodefense and Emerging Infections Research Resources Repository (*BEI* Resources) catalog number NR-18959.
3. Scissors and tweezers.
4. Disposable and absorbent papers.
5. 20-mL syringe.
6. 21G needle.
7. Isoflurane vaporizer.
8. Liquid nitrogen.

9. Styrofoam box.
10. Dry ice.
11. Standard tissue slides (typically 75 mm × 25 mm, 1 mm thickness).
12. Epredia M-1 Embedding Matrix.
13. Microtome.
14. Plastic molds.
15. Slide scanner (e.g., PathScan).
16. Glass cutter.
17. −80 °C freezer.
18. Sonicator.
19. Incubator.
20. Square Petri dish (10 cm × 10 cm).
21. M-class UPLC system (Waters Corp, Milford, MA).
22. 500-μL syringe (e.g., Hamilton).
23. Syringe pump (e.g., Harvard Apparatus Standard Infusion PHD Ultra Syringe Pumps).
24. Synapt-G2Si Qtof with DESI-XS stage (Waters Corp, Milford, MA).
25. High-Definition Imaging (HDI) Software (Waters Corp, Milford, MA).
26. MassLynx (Waters Corp, Milford, MA).
27. Waters DataConnect (https://microapps.on-demand.waters.com/).
28. Waters MSI segmentation (https://microapps.on-demand.waters.com/).
29. Python 3 (https://www.python.org/).
30. R (https://www.r-project.org/).

2.2 Solutions

1. Rehydration solution: 2 mM of magnesium chloride hexahydrate with 1× phosphate-buffered saline dissolved in deionized water.
2. X-gal (5-bromo-4-chloro-3-indolyl-beta-D-galactopyranoside) stock solution: 20 mg/mL in dimethylformamide. Fresh preparation every time.
3. X-gal staining solution: 4 mM of potassium hexacyaniferrate (III) and 4 mM of potassium hexacyanoferrate (II) trihydrate dissolved in rehydration solution. Use this solution to dilute X-gal stock solution to 1 mg/mL as the staining solution. Freshly mix before usage.

4. Spray solution: methanol: water (95:5, v/v, all ultra-high performance liquid chromatography (UHPLC) grade) with 0.1% of formic acid and 200 pg/μL of leucine enkephalin.
5. Calibration solution: 0.1 mM sodium hydroxide in 24-mL acetonitrile (UHPLC grade), 5.4 mL UHPLC grade water, 300 μL formic acid (UHPLC grade).

3 Methods

3.1 Preparation of Glutaraldehyde-Fixed, M1-Embedded Tissue Section

This subsection explains how to collect hearts from mice, prepare embedded hearts and details of tissue cryosectioning, tissue slice storage, and tissue slice management.

3.1.1 Harvesting Mouse Hearts

1. Euthanize infected mice according to institutional IACUC-approved protocols (*see* **Note 2**).
2. Perfuse the mouse hearts with 20 mL of 1× PBS.
3. Harvest the hearts and soak them in glutaraldehyde (1:100 dilution in PBS of 70% glutaraldehyde stock) for 5 min.
4. After the fixation with glutaraldehyde, wash the hearts twice with PBS.
5. Place hearts in a cryovial tube and maintain in an ice bath (dry ice plus 70% ethanol) until storage at −80 °C.

3.1.2 Heart Embedding and Cryosectioning

1. Embed frozen heart in M1 embedding matrix. Freeze to a solid chunk at −20 °C (*see* **Note 3**).
2. Set the temperature of the microtome to −20 °C.
3. Drop a thick layer of embedding material on the specimen holder and freeze to a solid block at −20 °C.
4. Mount the frozen embedding material block with specimen holder in the cryostat. Tighten the screws to ensure it is secure and tightly held.
5. Set the slice thickness to 18 μm by adjusting the cryostat's control panel.
6. Adjust the height of the embedding material block to an appropriate height relative to the microtome blade.
7. Section the embedding material block, adjusting the angle of the cryostat blade to ensure complete and flat slices.
8. Section the embedding material block to get a flat surface parallel to the blade.
9. Mount the embedded sample block onto the flat surface of the embedding material block on the specimen holder in the cryostat, ensuring it is secure and tightly held. A thin layer of

embedding material can be applied to glue the sample block to the flat surface of the embedding material block. Lower down the height of the specimen holder to an appropriate position relative to the blade. Do not change the angle of the blade.

10. Trim excessive embedding material to obtain flat and complete tissue slices if needed.
11. Section the heart into 18-μm thickness slices with transverse orientation.
12. Collect the tissue slices by carefully transferring them to a regular histological tissue glass slide with a brush. Prevent tearing apart or folding of the tissue.
13. Number each tissue slide.

3.1.3 Tissue Slice Management and Storage

1. Scan each tissue slice using a tissue scanner.
2. Keep a registration record for each tissue slice. The registration record should include tissue information (tissue type, sample ID, species, collection date), cryosectioning details (section date, embedding material, operator, cryostat ID if multiple cryostats are used), sectioning parameters (temperature, thickness, blade type, angle, section orientation), and storage details (slide ID, storage location, storage conditions).
3. Store the slices in a −80 °C freezer before the DESI-MSI experiment.

3.2 DESI-MSI Operation

This section describes the details of DESI-MSI operation, including manual mass calibration using DESI XS stage, stage origin calibration, and data acquisition.

3.2.1 Mass Calibration

Mass calibration can be performed using standard auto-calibration according to Waters Synapt G2-Si QTof user manual (replace the DESI XS stage with a regular ESI housing first) or manually calibrated using DESI XS stage. The following steps describe manual mass calibration using the DESI XS stage.

1. Place a clean glass slide in slot A of the DESI XS stage. Turn on the DESI XS stage and then disconnect the DESI Source Control software. Position the sprayer over the clean glass slide using Omni Spray2-D software and ensure a clean background. Start the flow of calibration solution with 2 μL/min (*see* **Notes 4** and **5**).
2. In the MassLynx software sidebar, click the MS Tune tab to open the MS Tune page, and set the cone temperature to 150 °C. Turn on voltage and API gas (TRP gas is automatically turned on with voltage on. The spray solvent carrier gas is N_2 and the pressure is set as 0.4 bar). Adjust voltage and sprayer

position to get a good calibration signal. Navigate to *Setup* > *System View*, set acquisition mode (Acq Mode) to TDC mode, and save settings.

3. Choose *Positive Sensitivity* mode. Acquire a 1-min runtime file and save with this name format: Date_Mass_Cal_data_Pos_-Sen. Sum across ~30 s of data, and process with automatic peak detection. Resolve potential issues: If peaks are marked by "!", sum across fewer scans or enable *Attenuate*. Save the spectrum when acceptable.
4. Switch to *Positive Resolution* mode. Repeat data acquisition, processing, and saving steps as shown above.
5. Switch to *Negative Sensitivity* and *Negative Resolution* modes. Repeat the above steps.
6. In MS Console, select *IntelliStart* > *Configure* > *Create Calibration* > *Start* > *Next*. In Calibration Profile Editor, select *Add new calibration*. Use saved raw files and spectra to configure each mode: Positive Sensitivity, Positive Resolution, Negative Sensitivity, and Negative Resolution.
7. Validate calibrations for each mode. Accept results if they are within acceptable ppm range (e.g., all calibrant peaks need to be found and <5 ppm). Ensure the calibration file name is updated on the MS Tune page. Uncheck *Attenuate* and reset Acq Mode to ADC. Exit *System View* and save settings.
8. Open the *MS Console* tab to access the Waters MS console page. Click *μBMS* and set the flow rate to 1.2 μL/min. Deliver the spray solvent at 1.2 μL/min using the Waters M-class UPLC system for a stable flow. Purpose: (1) To clean calibration solution residue from the sprayer. (2) To stabilize the flow rate.
9. Click *Go back to Origin* in the OmniSpray software. Close OmniSpray and reconnect the DESI control software.

3.2.2 Stage Origin Calibration

The position of the sprayer can drift when performing the manual mass calibration or during the ionization source housing/stage switching. The relative distance between the sprayer and the origin (reference point) of the DESI XS stage can be changed and needs to be calibrated to ensure sampling of the desired regions.

1. Mark a corner of a clean glass slide with a small spot using a black Sharpie pen. Place the slide in slot A of the DESI XS stage with the marker at the upper left corner.
2. Turn on the DESI XS stage. Disconnect the DESI control software and open OmniSpray 2D software. On the MS Tune page, turn on API gas and voltage. Set the mass analyzer to *Negative Resolution Mode*.

3. Position the sprayer near the black marker spot. Start a 20-min data acquisition. Manually adjust the sprayer height and position using OmniSpray 2D software. Monitor the intensity of *m/z* 666.038 ± 0.05 (black Sharpie ink signal). Record the coordinates of the highest intensity (*see* **Note 6**).
4. Place the instrument in *Standby Mode*. Click *Go Back to Origin* in OmniSpray 2D software and close the software. Enter the recorded coordinates as *DESI Configuration X and Y* in the DESI stage control software. Reconnect the DESI stage control software. *The instrument is now ready for use.*

3.2.3 Sample Loading

1. Take out the sample tissue slides from the −80 °C freezer. Place each slide in a square Petri dish, ensuring one tissue per well. Allow the tissue to thaw and dry, which usually takes a few minutes. While the slide remains flat on the bench, gently wipe away any condensed water from the glass surface using a Kim-Wipe (*see* **Note 7**).
2. Open the shield cover of the DESI XS stage. Place the sample slides onto the designated sample slot. If using glass slides, cut them using a glass cutter to retain only the portion containing the tissue sample.
3. Load tissue slices (e.g., infected and uninfected mouse hearts) into the same sample slot. This setup helps reduce overall experimental time and minimizes batch effects.
4. Take a full-view picture of the sample slides to serve as a reference for the experiment. Ensure the layout of the tissue slices in the image matches their exact arrangement in the sample slot. For cut slides, include the corner of the sample slot in the image for orientation. Use a photo scanner or a phone camera to capture the image. This reference will be used to define the regions for the DESI-MSI scan.

3.2.4 DESI-MSI Data Acquisition Workflow

1. In HDI software, navigate to the *Acquire* tab and click *Create New Plate*. Load the prepared image of tissue slices under *Image File*. Set *Sample Source* to **DESI**, select the current slide location (A or B), and choose the *Plate Type* (typically 76 mm × 26 mm). Click **Next**.
2. Use the crosshair and zoom to define the four corners of the sample slot. Confirm and click **Next** to preview the defined plate, adjust if needed, and click **Finish** and proceed to define MS parameters.
3. Under the *Acquire* tab: Set *Target Instrument* to **Synapt** and *Experiment Type* to **DESI MS**. Set up the following parameters (example values are shown below): Mass range (**100–1200 *m/z***), Trap collision energy (**4 V**), Transfer cell collision energy (**2 V**), Polarity (**Positive Ionization**), and Analyzer (**Resolution Mode**).

4. Go to the *Pattern* tab under *Acquire*: Select the slide slot of the created plate (A or B). Use the drawing tool to draw a pattern over desired regions for DESI scanning. Set the **X/Y Pixel Size** (e.g., 50 μm) and set **Rate** (stage moving speed) (e.g., 100 μm/s). Review the estimated scan time (actual time is longer than estimated time because it includes stage travel time).
5. Save the plate definition as a **.pdm** file named with sample info and experiment date. Click the MassLynx logo on the toolbar to export the sample list to MassLynx as a.txt worksheet with the same name.
6. In MS Tune, turn on **Spray Voltage** and **API Gas**. Make sure to get stable and clean background spectra before scanning the actual tissue samples. In MassLynx, go to *File* > *Import Worksheet* and load the.txt file. Select the analyte row and click **Run** to start acquisition (*see* **Notes 8–12**).
7. Post-Acquisition: Turn off **Spray Voltage** (API gas auto-off) in MS Tune. Stop UPLC flow in the MS Console.

3.3 X-Gal Staining, Adapted from [37]

1. Using 1 mL pipette tips, carefully add rehydration solution to a square Petri dish containing tissue slices (same slices used in the DESI-MSI experiment) until they are fully immersed.
2. Rehydrate the tissue slices for 5 min at room temperature.
3. Aspirate the rehydration solution with minimal disturbance to the tissue slices.
4. Add the same volume of X-gal staining solution to each well of the Petri dish, ensuring the rehydrated tissue slices are fully immersed.
5. Incubate the tissue slices with X-gal solution at 37 °C for 16–18 h.
6. Aspirate the staining solution with minimal disturbance to the tissue slices.
7. Add cold 1× PBS to rinse and wash the stained tissue slices thoroughly.
8. Aspirate the staining solution carefully with minimal disturbance to the tissue slices.
9. Repeat washing and aspiration three times.
10. Capture a complete image of the stained tissue slices using a tissue scanner.
11. Locate intracellular parasites using a microscope with 4× magnification.
12. Save high-resolution images of the infected cells and record their locations.

3.4 Data Preprocessing and Image Construction

1. Load the raw data folder into HDI in the Process module, Build Sample List tab.
2. Set output folder and output file name.
3. Specify experiment type as MS.
4. Define the number of most intense peaks (usually top 1000), depending on the computer memory available (the minimum RAM recommended by the HDI software is 32 GB).
5. Set ***m/z* window** to 0.02 and **Mass Resolution** to 20,000 (for Synapt G2-Si).
6. Enable Lock Mass. Set leucine enkephalin $[M + H]^+$ *m/z* 556.2771 as internal lock mass.
7. Click **Process** to generate a.txt output file.
8. Load the.txt file into Waters MSI segmentation.
9. Perform tissue segmentation, mask the background, and apply normalization based on the total ion current (TIC) for each pixel (*see* **Note 13**).
10. Save the processed file with these adjustments.
11. Load the normalized.txt file into HDI.
12. Click the desired *m/z* to generate the ion image.

3.5 Feature Selection Based on ROI Data from Infected and Uninfected Mouse Hearts

3.5.1 Region of Interest (ROI) Data Extraction

1. Import the.txt file into the HDI Analysis module. Navigate to the Data tab, select the loaded dataset, and click on the *m/z* with maximum intensity to generate the ion image. Import X-gal-stained images of infected and uninfected heart tissue slices and overlay them with the ion image.
2. Locate an infected cell in the infected heart tissue slice. Using the **DrawROI Brush** in the toolbar, choose **DrawROI Rectangle**, and set the ROI size to 1 × 1 pixel^2 (one spectrum per pixel; adjust as needed). Mark the center of the infected cell with a single pixel and click the check icon to confirm. If the infected cell spans more than one pixel, mark additional pixels individually, confirming each selection (typically, 4–6 pixels are exported per infected spot).
3. Under the ROI tab, select all pixels corresponding to the infected cell (Spot 1). Click **Export MVA** to export the spectra of the selected pixels to the desired output folder. Repeat the process for all infected cells and export spectra for corresponding spots in uninfected tissue slices using the same approach.
4. Starting from the infected cell, identify neighboring pixels in successive layers:
   - **Layer 1 (L1):** The 2 × 2 pixel blocks adjacent to the center in four directions.

- **Layer 2 (L2):** The 2 × 2 pixel blocks adjacent to L1 in four directions.
- **Layer 3 (L3):** The 2 × 2 pixel blocks adjacent to L2 in four directions.
- **Layer 4 (L4):** The 2 × 2 pixel blocks adjacent to L3 in four directions.

Export the spectra of each pixel in layers L1–L4 for all infected cells. Repeat for uninfected tissue slices by extracting pixels from the corresponding locations, and export them layer by layer.

5. Group the exported data as follows:
   - Infected cell pixels: **P_center** group.
   - Layers adjacent to infected cells: **PL1**, **PL2**, **PL3**, **PL4** groups.
   - Uninfected layers: **CTL1**, **CTL2**, **CTL3**, **CTL4** groups.

   Merge data from all infected pixels and uninfected pixels to create a comprehensive ROI dataset. Include pixel IDs, sample group information (e.g., P_center, PL1, CTL1), *m/z* values, and corresponding normalized intensities for all *m/z*.

3.5.2 Feature Selection

1. Perform Wilcoxon test on the ROI dataset to calculate the *p*-value of each *m/z* (ions) in each layer between infected and uninfected hearts.
2. Perform Random Forest analysis to select the features of importance. Use a balanced sampling approach, where the number of samples for each class is set to the size of the smallest group.
3. Filter the features (*m/z*) satisfying the following three criteria at the same time (Fig. 1):
   - P_center versus CT_center: $p \leq 0.05$, and MeanDescreaseAccuracy >1.3. This is to select features with significant changes at parasite colonization spots in infected hearts, compared to corresponding regions of uninfected hearts.
   - P_center versus PLx (x = 1–4 for layer 1 to layer 4 around parasites): $p \leq 0.05$. This is to select features that are spatially different in the infected hearts.
   - CT_center versus CTLx (x = 1–4 for layer 1 to layer 4 around the corresponding spots in uninfected hearts): $p > 0.05$. This is to select features that are not spatially different in the uninfected hearts.
4. Generate boxplots of normalized ion intensity distribution for center and each layer to compare the relative abundances among layers (Fig. 2).

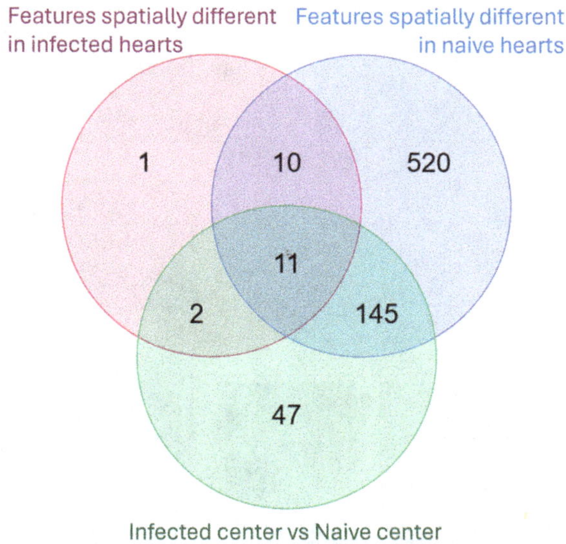

Fig. 1 Example output data: ions spatially different only in infected hearts

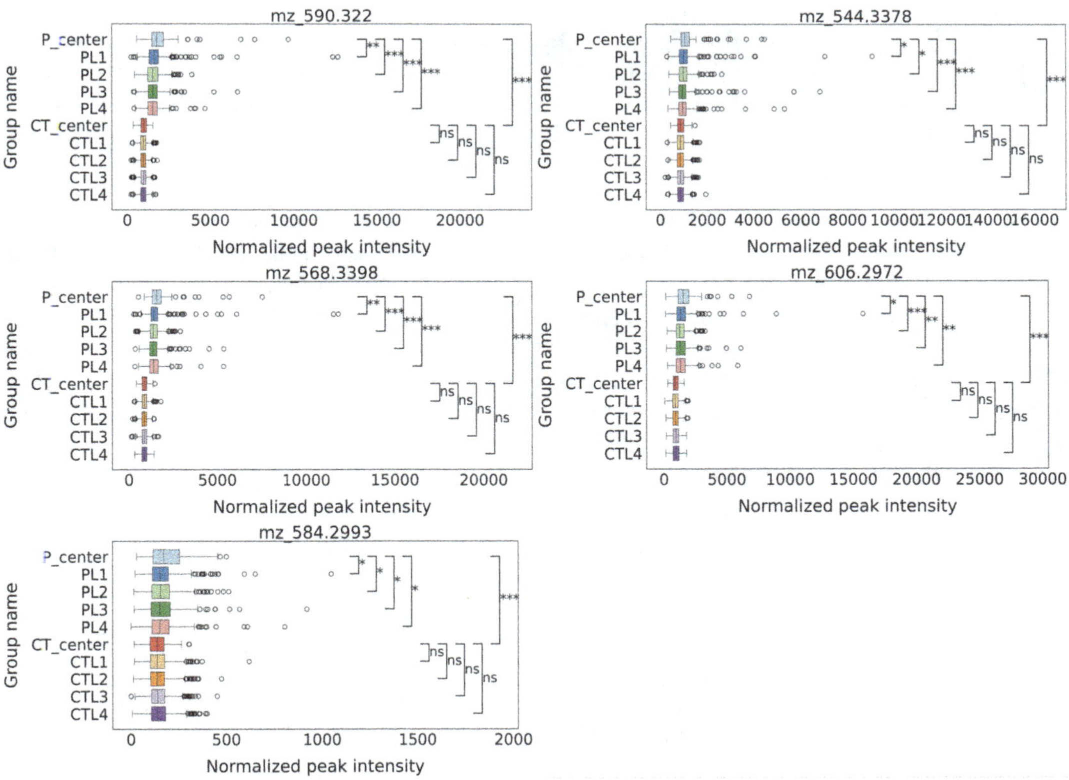

Fig. 2 Comparison of normalized intensity across the parasite spots and corresponding regions in naive hearts. ($p \leq 0.0001$: ***, $p \leq 0.01$:**, $p \leq 0.05$:*, $p > 0.05$: ns)

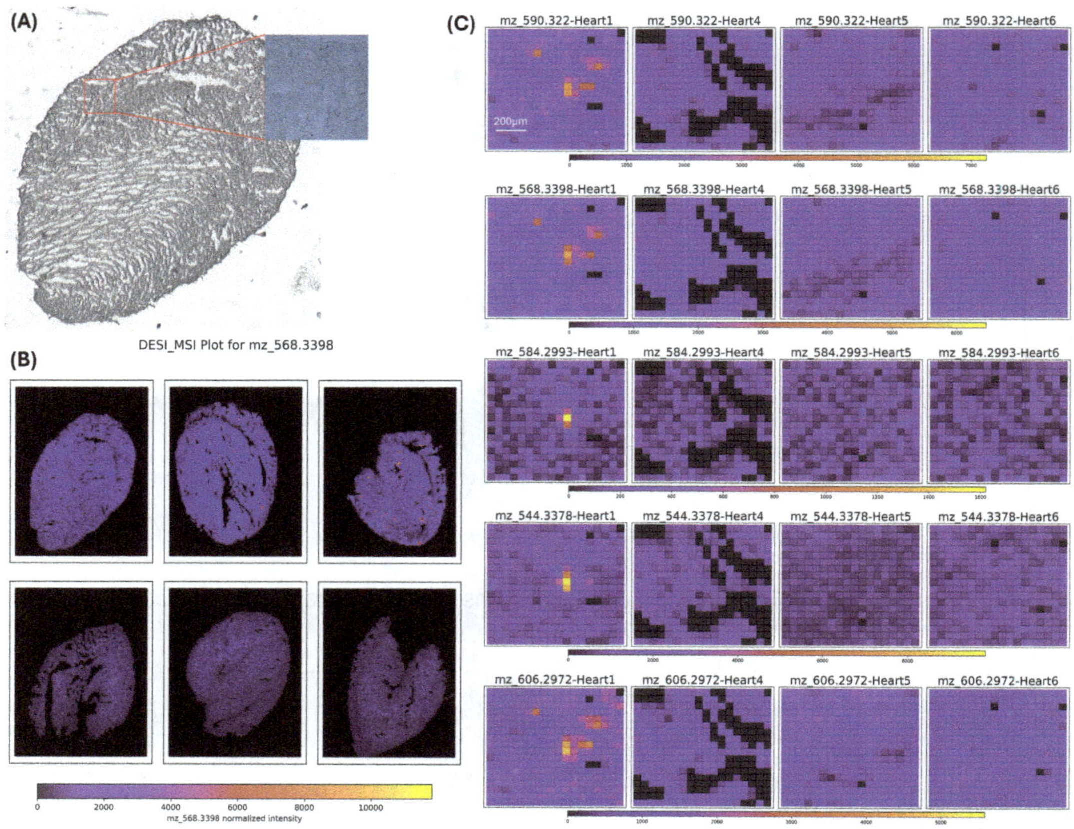

Fig. 3 Example output data: DESI-MSI reveals specific localized metabolic impacts of *Trypanosoma cruzi* infection. (**a**) Microscopic images of stained slices after the DESI-MSI experiment and zoomed-in region. (**b**) DESI-MSI ion images of *m/z* 568.3398: infected hearts (top row) versus uninfected hearts (bottom row). (**c**) DESI-MSI ion images of the parasite-infected spots (column 1) and corresponding spots in uninfected hearts (columns 2–4). From top to bottom: *m/z* 590.322 (tentative annotation, lysophosphatidylcholine (LPC) 22:6 $[M + Na]^+$ or lysophosphatidylethanolamine (LPE) 22:1;0 $[M + K]^+$), *m/z* 568.3398 (tentative annotation, LPC 20:3 $[M + Na]^+$ or LPC 22:6 $[M + H]^+$), *m/z* 584.2993 (tentative annotation, PE 24:6;0 $[M + H]^+$), *m/z* 544.3378 (tentative annotation, LPC 20:4 $[M + H]^+$ or LPC 18:1 $[M + Na]^+$), and *m/z* 606.2972 (tentative annotation, LPC 22:6 $[M + K]^+$)

5. Generate DESI-MSI ion images for selected features:
   - Whole heart plots: Load the dataset in HDI and choose TIC normalization. Click the *m/z* of a selected feature and export the ion image as a png file (*see* **Notes 14** and **15**).
   - Region plots: Select the regions with parasites and surrounding layers, and then create a heatmap using the coordinates and TIC normalized intensity of each *m/z* of the pixels in the region (Figs. 3 and 4) (*see* **Note 15**).

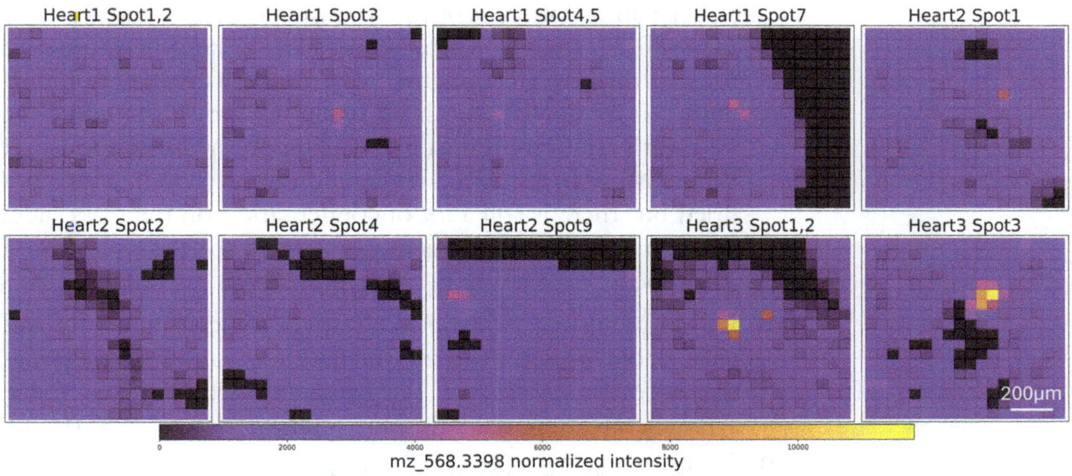

Fig. 4 DESI-MSI plot of *m/z* 568.3398 (tentative annotation, LPC 20:3 $[M + Na]^+$ or LPC 22:6 $[M + H]^+$) in different parasitized regions of infected hearts

3.6 Annotation

Annotate ions by matching the accurate *m/z* with (1) LC-MS/MS bulk analysis of tissue extraction [5, 10, 11, 38]; (2) LipidMaps database (https://www.lipidmaps.org/) by exact mass matching within 0.05 *m/z* and keeping annotations within 15 ppm.

4 Notes

1. Use 125 LacZ Tulahuen parasites for infecting C3H/HeNCrl mice, to enable survival into the chronic stage. Previous tests (not published) using female BALB/c mice infected with 250 LacZ Tulahuen parasites led to a mortality rate of 100% in less than 3 weeks.
2. Sample collection was performed under IACUC protocol number IACUC-23-083 at San Diego State University.
3. It is very important to carefully choose an orientation to embed the organ. The expected orientation of the tissue (sagittal or transverse) should be parallel to the blade, so when embedding the organ, your desired orientation (sagittal or transverse) should be upward.
4. Turn on the syringe pump with the calibration solution. Start with a higher flow rate (e.g., 4 μL/min) to flush away the air bubbles in the tubing first then decrease the flow rate to 2 μL/min.
5. The DESI Source Control software cannot work with Omni Spray2-D software at the same time.
6. Spray voltage 1.0 kV is a good start. Use no more than 1.5 kV for DESI XS.

7. Water condensation happens when taking tissue slides out of the −80 °C freezer. The slices should be kept horizontal to prevent the flow of condensation from destroying the original lipid distribution of the sliced tissue. Covering the lid of the Petri dish can prevent dust from falling on the tissue surface. If the moisture in the sliced tissue and the surrounding area has dried but there is still a lot of condensation on the glass slide, you can carefully wipe tissue-free sections of the slide with a Kimwipe while keeping the slide horizontal.
8. UPLC pump can provide stable spray solution flow rate and yield stable ion signal during the data acquisition. The alternative is to use an external syringe pump with a large-volume syringe containing enough spray solution.
9. Sonicating and degassing the spray solution before installation is helpful in removing air dissolved in the solvent and further improving the stability of solvent spray.
10. Stable flowrate and solvent spray are critical for obtaining good data quality. Keep the ion signal stable and avoid artifacts due to unstable spray.
11. Rate (stage moving speed): usually set as twice the x pixel size per second and will result in 2 spectra per second. For Synapt G2-Si, approximately 0.485 s/scan = 2 spectra/s.
12. The spray voltage and UPLC flow do not stop automatically after data acquisition. Check and turn off the spray voltage and API gas, and stop the UPLC flow to avoid waste of resources.
13. Analytes are chemically diverse across the tissue slices and the long acquisition time of MSI experiment can introduce potential variability. TIC normalization is a widely used normalization method in MSI to compensate for global intensity fluctuation.
14. Optional (whole heart plots): Segment the tissue from the background using Waters MSI Segmentation and get the pixel ID belonging to the tissue and surrounding area. Mask the intensity of all *m/z* as zero for pixels not belonging to tissue for better visualization. Create a heatmap using the coordinates and the normalized intensity of each *m/z* for each pixel belonging to the tissue.
15. The code used in this work can be found at https://github.com/dandandan001/DESI-MSI_data_process

Acknowledgments

This project was made possible in part by grant number 2023-331941 from the Chan Zuckerberg Initiative DAF, an advised fund of Silicon Valley Community Foundation.

Authors wish to thank Dr. Steven Foster in the Mass Spectrometry Core Facility of the University of Oklahoma, Dr. Wei Rao from Waters UK for helping to manually calibrate the Synapt G2-Si using DESI XS stage, and Drs. Anthony Midey and Hernando Olivos from Waters US for helping with the instrumental setup.

References

1. Alonso-Padilla J, Cortés-Serra N, Pinazo MJ et al (2019) Strategies to enhance access to diagnosis and treatment for Chagas disease patients in Latin America. Expert Rev Anti-Infect Ther 17(3):145–157. https://doi.org/10.1080/14787210.2019.1577731
2. Schijman AG, Alonso-Padilla J, Britto C, Herrera Bernal CP (2024) Retrospect, advances and challenges in Chagas disease diagnosis: a comprehensive review. Lancet Reg Health Am 36:100821. https://doi.org/10.1016/j.lana.2024.100821
3. Kwakye-Nuako G, Middleton CE, McCall L-I (2024) Small molecule mediators of host-T. cruzi-environment interactions in Chagas disease. PLoS Pathog 20:e1012012. https://doi.org/10.1371/journal.ppat.1012012
4. Gabaldón-Figueira JC, Ros-Lucas A, Martínez-Peinado N et al (2024) Changes in lipid abundance are associated with disease progression and treatment response in chronic Trypanosoma cruzi infection. Parasit Vectors 17:459. https://doi.org/10.1186/s13071-024-06548-3
5. McCall L-I, Morton JT, Bernatchez JA et al (2017) Mass spectrometry-based chemical cartography of a cardiac parasitic infection. Anal Chem 89:10414–10421. https://doi.org/10.1021/acs.analchem.7b02423
6. Laro J, Xue B, Zheng J et al (2025) Severe acute respiratory syndrome coronavirus 2 infection unevenly impacts metabolism in the coronal periphery of the lungs. iScience 28:111727. https://doi.org/10.1016/j.isci.2024.111727
7. He MJ, Pu W, Wang X et al (2022) Comparing DESI-MSI and MALDI-MSI mediated spatial metabolomics and their applications in cancer studies. Front Oncol 12:891018. https://doi.org/10.3389/fonc.2022.891018
8. Wheeler K, Gosmanov C, Sandoval MJ, Yang Z, McCall L-I (2024) Frontiers in mass spectrometry-based spatial metabolomics: current applications and challenges in the context of biomedical research. TrAC Trends Anal Chem 175:117713. https://doi.org/10.1016/j.trac.2024.117713
9. Liu G-X, Li Z-L, Lin S-Y et al (2023) Mapping metabolite change in the mouse brain after esketamine injection by ambient mass spectrometry imaging and metabolomics. Front Psychiatry 14:1109344. https://doi.org/10.3389/fpsyt.2023.1109344
10. Liu Z, Ulrich vonBargen R, Kendricks AL et al (2023) Localized cardiac small molecule trajectories and persistent chemical sequelae in experimental Chagas disease. Nat Commun 14:6769. https://doi.org/10.1038/s41467-023-42247-w
11. Hoffman K, Liu Z, Hossain E et al (2021) Alterations to the cardiac metabolome induced by chronic T. Cruzi infection relate to the degree of cardiac pathology. ACS Infect Dis 7:1638–1649. https://doi.org/10.1021/acsinfecdis.0c00816
12. Dean DA, Gautham G, Siqueira-Neto JL et al (2021) Spatial metabolomics identifies localized chemical changes in heart tissue during chronic cardiac Chagas Disease. PLoS Negl Trop Dis 15:e0009819. https://doi.org/10.1371/journal.pntd.0009819
13. Lan Y, Zou Z, Yang Z (2024) Single Cell mass spectrometry: towards quantification of small molecules in individual cells. Trends Analyt Chem 174:117657. https://doi.org/10.1016/j.trac.2024.117657
14. Yu B, Zhan R, Hu Y, Lv Z (2024) Mass spectrometry imaging: an emerging technology in medical parasitology. Anal Chem 96(20):8011–8020. https://doi.org/10.1021/acs.analchem.3c05341
15. Barajas-Solano C, Muñoz B, Chicano-Gálvez E et al (2022) Discriminator for cutaneous Leishmaniasis using MALDI-MSI in a murine model. J Am Soc Mass Spectrom 33(6):

952–960. https://doi.org/10.1021/jasms.2c00015

16. Kadesch P, Hollubarsch T, Gerbig S et al (2020) Intracellular parasites Toxoplasma gondii and Besnoitia besnoiti, unveiled in single host cells using AP-SMALDI MS imaging. J Am Soc Mass Spectrom 31:1815–1824. https://doi.org/10.1021/jasms.0c00043
17. Soudah T, Zoabi A, Margulis K (2023) Desorption electrospray ionization mass spectrometry imaging in discovery and development of novel therapies. Mass Spectrom Rev 42:751–778. https://doi.org/10.1002/mas.21736
18. Spruill ML, Maletic-Savatic M, Martin H et al (2022) Spatial analysis of drug absorption, distribution, metabolism, and toxicology using mass spectrometry imaging. Biochem Pharmacol 201:115080. https://doi.org/10.1016/j.bcp.2022.115080
19. Jain SK (2017) The promise of molecular imaging in the study and treatment of infectious diseases. Mol Imaging Biol 19:341–347. https://doi.org/10.1007/s11307-017-1055-0
20. Ščupáková K, Balluff B, Tressler C et al (2020) Cellular resolution in clinical MALDI mass spectrometry imaging: the latest advancements and current challenges. Clin Chem Lab Med 58:914–929. https://doi.org/10.1515/cclm-2019-0858
21. Fletcher JS, Vickerman JC (2010) A new SIMS paradigm for 2D and 3D molecular imaging of bio-systems. Anal Bioanal Chem 396:85–104. https://doi.org/10.1007/s00216-009-2986-3
22. Zhang H, Lu KH, Ebbini M et al (2024) Mass spectrometry imaging for spatially resolved multi-omics molecular mapping. NPJ Imaging 2:20. https://doi.org/10.1038/s44303-024-00025-3
23. Rao W, Pan N, Yang Z (2016) Applications of the single-probe: mass spectrometry imaging and single cell analysis under ambient conditions. J Vis Exp. https://doi.org/10.3791/53911
24. Rao W, Pan N, Yang Z (2015) High resolution tissue imaging using the single-probe mass spectrometry under ambient conditions. J Am Soc Mass Spectrom 26:986–993. https://doi.org/10.1007/s13361-015-1091-4
25. Scott AJ, Post JM, Lerner R et al (2017) Host-based lipid inflammation drives pathogenesis in infection. Proc Natl Acad Sci USA 114:12596–12601. https://doi.org/10.1073/pnas.1712887114
26. Cassat JE, Moore JL, Wilson KJ et al (2018) Integrated molecular imaging reveals tissue heterogeneity driving host-pathogen interactions. Sci Transl Med 10. https://doi.org/10.1126/scitranslmed.aan6361
27. Perry WJ, Grunenwald CM, Van de Plas R et al (2022) Visualizing Staphylococcus aureus pathogenic membrane modification within the host infection environment by multimodal imaging mass spectrometry. Cell Chem Biol 29:1209–1217.e4. https://doi.org/10.1016/j.chembiol.2022.05.004
28. Hulme HE, Meikle LM, Wessel H et al (2017) Mass spectrometry imaging identifies palmitoylcarnitine as an immunological mediator during salmonella Typhimurium infection. Sci Rep 7:1–13. https://doi.org/10.1038/s41598-017-03100-5
29. Takáts Z, Wiseman JM, Gologan B, Cooks RG (2004) Mass spectrometry sampling under ambient conditions with desorption electrospray ionization. Science 306:471–473. https://doi.org/10.1126/science.1104404
30. Slijkhuis N, Towers M, Claude E, van Soest G (2025) MALDI versus DESI mass spectrometry imaging of lipids in atherosclerotic plaque. Rapid Commun Mass Spectrom 39:e9927. https://doi.org/10.1002/rcm.9927
31. Qi K, Wu L, Liu C, Pan Y (2021) Recent advances of ambient mass spectrometry imaging and its applications in lipid and metabolite analysis. Metabolites 11:780. https://doi.org/10.3390/metabo11110780
32. Tillner J, Wu V, Jones EA et al (2017) Faster, more reproducible DESI-MS for biological tissue imaging. J Am Soc Mass Spectrom 28:2090–2098. https://doi.org/10.1007/s13361-017-1714-z
33. Wu V, Tillner J, Jones E et al (2022) High resolution ambient MS imaging of biological samples by desorption electro-flow focussing ionization. Anal Chem. https://doi.org/10.1021/acs.analchem.2c00345
34. Colwell N, Chen D, Yang Z (2024) Achieving single-cell resolution via desorption electrospray ionization mass spectrometry imaging (DESI-MSI). ChemRxiv
35. Eberlin LS, Ferreira CR, Dill AL et al (2011) Nondestructive, histologically compatible tissue imaging by desorption electrospray ionization mass spectrometry. Chembiochem 12:2129–2132. https://doi.org/10.1002/cbic.201100411
36. Buckner FS, Verlinde CL, La Flamme AC, Van Voorhis WC (1996) Efficient technique for screening drugs for activity against

Trypanosoma cruzi using parasites expressing beta-galactosidase. Antimicrob Agents Chemother 40:2592–2597. https://doi.org/10.1128/AAC.40.11.2592

37. Buckner FS, Wilson AJ, Van Voorhis WC (1999) Detection of live Trypanosoma cruzi in tissues of infected mice by using histochemical stain for beta-galactosidase. Infect Immun 67:403–409. https://doi.org/10.1128/IAI.67.1.403-409.1999
38. Hossain E, Khanam S, Dean DA et al (2020) Mapping of host-parasite-microbiome interactions reveals metabolic determinants of tropism and tolerance in Chagas disease. Sci Adv 6: eaaz2015. https://doi.org/10.1126/sciadv.aaz2015

Chapter 15

Microbiome Profiling in Chagas Disease: Sample Collection, Sequencing, and Analysis

Sergio Castañeda, Juan David Ramírez, and Cristina Poveda

Abstract

Chagas disease, caused by Trypanosoma cruzi, leads to chronic cardiac and gastrointestinal complications. Emerging evidence shows the gut microbiome plays a key role in modulating disease severity, with shifts in microbial composition influencing immune responses and metabolic pathways. Here, we describe a workflow for microbiome analysis in *T. cruzi*–infected mice. Methods included sample collection from feces and gastrointestinal tissues, DNA extraction, sequencing, and quality control. Then, we outline bioinformatic analyses covering taxonomic profiling, diversity assessment, and microbial network construction. Finally, protocols for functional prediction tools are also included to explore microbial capabilities and the identification of signatures associated with disease progression.

Key words *Trypanosoma cruzi*, Chagas disease, Gut microbiome, Sequencing, Microbial diversity, Host–microbiota interaction

1 Introduction

Chagas disease (CD), caused by the protist *Trypanosoma cruzi*, remains a significant public health concern, affecting millions worldwide and leading to chronic cardiac and gastrointestinal complications [1, 2]. While the pathogenesis of CD has traditionally been attributed to direct parasite-induced tissue damage and immune responses, recent research highlights the critical role of the host microbiome, particularly the gut microbiota modulating disease progression and clinical outcomes [3].

Advances in high-throughput sequencing technologies have enabled comprehensive profiling of the microbiome in both human patients and experimental models of CD. Studies employing 16S/18S rRNA gene sequencing, whole genome metagenomics, and mass spectrometry-based metabolomics have revealed that *T. cruzi* infection induces notable perturbations in the composition and function of the gut microbiota [4]. For

Paul A. M. Michels et al. (eds.), *Euglenozoa: Methods and Protocols, Volume 1*, Methods in Molecular Biology, vol. 3013, https://doi.org/10.1007/978-1-0716-5142-1_15,

instance, patients with CD often exhibit a distinct microbial signature, including an enrichment of *Parabacteroides* spp. and reduced abundance of beneficial taxa such as *Akkermansia*, especially among those with cardiac involvement [5]. Experimental models further demonstrate that these microbiome alterations are associated with changes in immune profiles and metabolic pathways, potentially influencing the establishment and persistence of infection [1, 5].

The process of microbiome profiling in the context of CD involves meticulous sample collection, most commonly stool samples, followed by DNA extraction and sequencing to characterize microbial communities. Analytical approaches such as beta diversity metrics, operational taxonomic unit (OTU) classification, and functional pathway analysis are then employed to interpret the data and identify disease-associated microbial shifts. These methodologies have not only improved our understanding of host–parasite–microbiome interactions but also hold promise for identifying novel biomarkers and therapeutic targets to improve patient outcomes.

Despite these advances, significant knowledge gaps remain regarding the causal relationships between microbiome alterations and disease phenotypes, as well as the potential for microbiota-directed interventions in CD management. This chapter provides an overview of current methodologies for microbiome profiling in CD, detailing best practices for sample collection, sequencing strategies, and analytical frameworks, while highlighting key findings and future directions in this rapidly evolving field.

2 Materials

2.1 Sample Collection and Preservation

1. Sterile gloves.
2. Clean, labeled cages.
3. Sterile toothpicks.
4. 2-mL sterile Eppendorf tubes (pre-labeled).

2.2 Endpoint Gastrointestinal Content Collection

1. 70% ethanol spray.
2. Sterile surgical scissors and forceps.
3. Sterile gloves.
4. Sterile 2-mL microcentrifuge tubes (pre-labeled and pre-weighed).
5. Laminar airflow hood (for aseptic sample collection).

2.3 DNA Extraction

1. Stool DNA Isolation Kit. We use the kit from Norgen Stool DNA Isolation Kit (Cat# 27600), on which the protocol description is based.

2. Microcentrifuge tubes (2.0 and 1.5 mL).
3. 70% Ethanol.
4. Vortex mixer.
5. Bead beater such as Omni bead rupture.
6. Microcentrifuge (capable of 14,000 rpm/~20,000 × *g*).
7. Pipettes and pipette tips (1–1000 μL range).
8. Ice bucket.

2.4 Quality Control of Extracted DNA. Qubit dsDNA High Sensitivity (HS) Assay

1. Qubit dsDNA High Sensitivity (HS) Assay Kit (includes Qubit reagent solution).
2. Qubit fluorometer.
3. Pipettes: P1000 micropipette (for measuring 1 and 199 μL).
4. Pipette tips: sterile, RNase-free tips (appropriate for P1000 and P10 micropipettes).
5. 1.5-mL sterile microcentrifuge tubes.
6. DNA sample.

2.5 Quality Control of Extracted DNA. Nanodrop

1. NanoDrop spectrophotometer.
2. P10 or P20 micropipette (for accurate transfer of small volumes, 1–2 μL).
3. Pipette tips for P10 or P20 micropipette.
4. DNA sample.

2.6 Quality Control of Extracted DNA. Integrity Check: Agarose Gel Electrophoresis

1. TAE buffer—Tris-acetate-EDTA buffer for preparing the gel (1× working solution): 40 mM Tris-acetate, 1 mM EDTA.
2. GelRed (a safer alternative to ethidium bromide).
3. 1% agarose powder for gel preparation.
4. 6× Loading dye.
5. Gel casting tray and comb.
6. Electrophoresis apparatus and power supply (100–120 V).
7. Micropipettes: P10/P20.
8. Pipette tips (sterile, RNase-free).
9. DNA marker or ladder for size reference when assessing the DNA bands.
10. Gel documentation system.

2.7 Quality Control of Extracted DNA. Bioanalyzer

1. Agilent DNA High Sensitivity kit (includes chip, gel-dye concentrate, ladder, marker).
2. Vortex mixer.
3. Microcentrifuge.

4. Agilent Bioanalyzer instrument + software.
5. DNA samples (1–1000 pg/μL recommended).

2.8 Metabarcoding and Metagenomics (Shotgun Metagenomics). Shotgun Metagenomics Library Preparation DNA

We use the kit from NEB, which is based on the protocol description.

1. High-quality DNA (10–500 ng).
2. NEBNext Ultra II FS DNA Library Prep Kit.
3. Nuclease-free water.
4. PCR tubes.
5. Thermal cycler.

2.9 Library Construction

We use the kit from NEB, which is based on the protocol description.

1. Fragmented genomic DNA.
2. NEBNext® Ultra™ II DNA Library Prep Kit for Illumina.
3. Indexing adapters (unique dual or single indexes).
4. AMPure XP beads (or equivalent).
5. Nuclease-free water or 0.1× TE: 1 mM Tris–HCl, pH 8.0, 0.1 mM EDTA.
6. Qubit fluorometer (dsDNA HS Assay).
7. Thermal cycler.
8. Magnetic rack.
9. PCR reagents (supplied with kit or separate high-fidelity polymerase).

2.10 Sequencing on Illumina Platform for Metabarcoding Gene Sequencing

1. Final purified and indexed libraries (e.g., 16S V3–V4, ITS1, COI).
2. Qubit dsDNA HS Kit.
3. Bioanalyzer.
4. 10-mM Tris–HCl (pH 8.5).
5. Illumina sequencing reagent kit (e.g., MiSeq v2 or v3, or NextSeq kit).
6. PhiX control (Illumina).
7. Illumina instrument (e.g., MiSeq, iSeq, NextSeq).

2.11 Software

1. *FASTQC*: Program used to evaluate the quality of FASTQ reads. Download link: https://www.bioinformatics.babraham.ac.uk/projects/download.html#fastqc.
2. *MULTIQC*: Download link: https://github.com/MultiQC/MultiQC

3. *DADA2*: Download link: https://github.com/benjjneb/dada2
4. *Phyloseq*: Package in R. Download link: https://joey711.github.io/phyloseq/index.html
5. *R:* R is a free statistical computing and graphics software environment. Download link: https://cran.r-project.org/mirrors.html.
6. *Cytoscape*: Program for visualizing complex networks. Download link: https://cytoscape.org/download.html
7. *Igraph*: R package of the igraph network analysis library. Download link: https://r.igraph.org/
8. *Trimmomatic*: The program used to evaluate and remove low-quality reads and adapter sequences. Download link: http://www.usadellab.org/cms/?page=trimmomatic.
9. *MetaPhlAn4*: Computational tool for profiling the composition of microbial communities for metagenomics shotgun sequencing data. Download link: https://huttenhower.sph.harvard.edu/metaphlan/
10. *StrainPhlAn*: Metagenomic strain-level population genomics tool. Download link: http://segatalab.cibio.unitn.it/tools/strainphlan/
11. *HUMAnN 4.0*: Pipeline for efficiently and accurately profiling the community's presence/absence and abundance of microbial pathways. Download link: https://github.com/biobakery/humann
12. *PhyloPhlAn:* Integrated pipeline for large-scale phylogenetic profiling of genomes and metagenomes. Download link: https://github.com/biobakery/phylophlan
13. *Bowtie 2*: Fast and sensitive read alignment. Download link: https://github.com/BenLangmead/bowtie2
14. *metaSPAdes:* Metagenomic assembler. Download link: https://github.com/ablab/spades
15. *MEGAHIT:* An ultra-fast single-node solution for large and complex metagenomics assembly via succinct de Bruijn graph. Download link: https://github.com/voutcn/megahit
16. *QUAST:* Quality Assessment Tool for Genome Assemblies. Download link: https://github.com/ablab/quast
17. *MetaBAT 2:* An adaptive binning algorithm for robust and efficient genome reconstruction from metagenome assemblies. Download link: https://bitbucket.org/berkeleylab/metabat/src/master/

18. *MaxBin2:* An automatic tool for binning metagenomics sequences. Download link: https://sourceforge.net/projects/maxbin2/
19. *CONCOCT:* A program for unsupervised binning of metagenomic contigs by using nucleotide composition, coverage data in multiple samples, and linkage data from paired-end reads. Download link: https://github.com/BinPro/CONCOCT
20. *DAS Tool:* For genome-resolved metagenomics.
    Download link: https://github.com/cmks/DAS_Tool
21. *GTDB-Tk:* Software toolkit for assigning objective taxonomic classifications to bacterial and archaeal genomes based on the Genome Database Taxonomy (GTDB). Download link: https://github.com/Ecogenomics/GTDBTk
22. *Prokka:* Software tool to rapidly annotate bacterial, archaeal, and viral genomes. Download link: https://github.com/tseemann/prokka
23. *Roary:* Rapid large-scale prokaryote pan genome analysis. Download link: https://github.com/sanger-pathogens/Roary
24. *Abricate:* Mass screening of contigs for antimicrobial resistance or virulence genes. It comes bundled with multiple databases: NCBI, CARD, ARG-ANNOT, Resfinder, MEGARES, EcOH, PlasmidFinder, Ecoli_VF, and VFDB. Download link: https://github.com/tseemann/abricate

3 Methods

3.1 Sample Collection and Preservation

This protocol outlines the aseptic collection of fecal samples from mice to ensure high-quality microbiota characterization via 16S ribosomal RNA (rRNA) gene amplicon sequencing. **Fresh stool collection:** Ensure all materials are prepared in a sterile environment to prevent microbial contamination and label each cage with the corresponding mouse number before beginning fecal collection [6, 7] (*see* **Note 1**).

1. Place each mouse into its designated clean, labeled cage.
2. Allow the mouse to remain in the cage for up to 5 min or until it excretes four to eight fresh fecal pellets. Once collection is complete, promptly return the mouse to its home cage.
3. Label each cage with the appropriate mouse number before beginning fecal collection.
4. Using a sterile toothpick, carefully pick up the fecal pellets.

5. Transfer the pellets directly into a pre-labeled 2-mL Eppendorf tube.
6. Ensure minimal handling to prevent contamination.

3.2 Endpoint Gastrointestinal Content Collection

Given that the microbial species vary along different gastrointestinal (GI) tract sections, endpoint collection of GI contents allows for a comprehensive evaluation of microbiome composition [8].

1. Euthanize mice following Institutional Animal Care and Use Committee (IACUC)-approved protocols. Work inside a laminar airflow hood to minimize environmental contamination.
2. Thoroughly spray the abdomen and thorax with 70% ethanol to minimize contamination.
3. Make a 5- to 10-mm horizontal incision in the abdomen and peel back the skin to expose the peritoneum.
4. Create a second incision, then gently pull the skin and peritoneal layers apart to open the abdominal cavity fully.
5. Carefully dissect and remove the entire GI tract using aseptic technique.
6. Isolate the small intestine, cecum, and colon and carefully slit each section open using sterile surgical scissors.
7. Transfer the contents directly into pre-labeled, pre-weighed sterile 2-mL microcentrifuge tubes. Immediately snap freeze the tubes in liquid nitrogen or dry ice to preserve microbial integrity.
8. Store collected samples at −80 °C until DNA extraction. For alternative preservation, Norogen's Stool Nucleic Acid Collection and Transport Tube (Cat# 45630 and 45660) can be used for room temperature storage, effectively preserving samples for over 2 years.

3.3 DNA Extraction

This protocol utilizes the Stool DNA Isolation Kit (Cat# 27600) from Norgen Biotek Corporation to efficiently extract microbial DNA from fecal samples [9]. DNA Extraction Protocol:

1. Add up to 200 mg of fresh stool sample to a provided Bead Tube and add 1 mL of Lysis Buffer L. Vortex briefly to mix (*see* **Note 2**).
2. Add 100 μL of Lysis Additive A and vortex briefly.
3. Homogenize samples using a bead beater such as Omni bead rupture; settings: S = 5.00, T = 0:20, D = 0:10, C = 02.
4. Centrifuge at 14,000 rpm (~20,000 × *g*) for 2 min.
5. Transfer up to 600 μL of clean supernatant into a DNase-free microcentrifuge tube (*see* **Note 3**).

6. Add 100 μL of Binding Buffer I and mix by inverting the tube a few times. Incubate the mixture on ice for 10 min.
7. Centrifuge at 14,000 rpm (~20,000 × *g*) for 2 min to pellet any remaining debris.
8. Carefully transfer up to 500 μL of clean supernatant into a fresh 2 mL DNase-free microcentrifuge tube.
9. Add an equal volume of 70% ethanol (prepared by the user) to the lysate (e.g., 100 μL ethanol per 100 μL lysate). Vortex to mix.
10. Assemble a spin column with a provided collection tube. Transfer 600 μL of the lysate-ethanol mixture onto the column and centrifuge for 1 min at 14,000 rpm (~20,000 × *g*).
11. Discard the flowthrough and reassemble the spin column with the collection tube (*see* **Note 4**).
12. Repeat **step 9** with the remaining lysate mixture.
13. Add 500 μL of Binding Buffer C to the column and centrifuge for 1 min at 10,000 rpm (~10,000 × *g*). Discard the flowthrough and reassemble the column with its collection tube.
14. Add 500 μL of Wash Solution A and centrifuge for 1 min at 10,000 rpm (~10,000 × *g*). Discard the flowthrough and reassemble the column with its collection tube.
15. Repeat the 500 μL Wash Solution A step, centrifuge at 10,000 rpm (~10,000 × *g*) for 2 min to thoroughly dry the resin, and discard the collection tube.
16. Place the column into a fresh 1.7-mL Elution Tube (provided with the kit). Add 100 μL of Elution Buffer B directly onto the column membrane. Incubate at room temperature for 1 min. Centrifuge for 1 min at 10,000 rpm (~10,000 × *g*) to elute DNA.
17. Store at −80 °C to maintain DNA integrity.

3.4 Quality Control of Extracted DNA

To ensure the quality and integrity of the extracted DNA for downstream sequencing, the following quality control steps are recommended. **Qubit dsDNA High Sensitivity (HS) Assay:** offers precise fluorometric DNA detection in small quantities.

1. Prepare the Qubit dsDNA High Sensitivity (HS) Assay reagent according to the manufacturer's instructions.
2. Mix 1 μL of the DNA sample with 199 μL of the Qubit reagent solution in a microcentrifuge tube.
3. Incubate for 2 min at room temperature to allow proper reaction.
4. Measure the fluorescence using the Qubit fluorometer.
5. Record the DNA concentration, which is reported in ng/μL.

3.5 Quality Control of Extracted DNA, NanoDrop [10]

1. Ensure the NanoDrop spectrophotometer is calibrated according to the manufacturer's instructions.
2. Transfer 1–2 μL of the DNA sample onto the measurement pedestal of the NanoDrop.
3. Close the lid and select the DNA (260/280) measurement mode. Press "Measure" to record the absorbance.
4. Review the 260/280 absorbance ratio, aiming for a ratio between 1.8 and 2.0 for pure DNA (*see* **Note 5**).

3.6 Quality Control of Extracted DNA. Integrity Check: Agarose Gel Electrophoresis

1. Prepare a 1% agarose gel in TAE buffer and add gel red or an alternative DNA stain.
2. Load 5–10 μL of the DNA sample, mixed with 1 μL of 6× loading dye, into the gel wells.
3. Run the gel at 100–120 V for approximately 30–45 min, or until the dye front has migrated sufficiently.
4. Visualize the gel under a UV transilluminator (*see* **Note 6**).

3.7 Quality Control of Extracted DNA. Bioanalyzer [11]

1. To prepare gel–dye mix. Spin down the DNA HS dye concentrate. Add 25 μL of dye concentrate to 250 μL of DNA HS gel matrix (1:10). Vortex well and spin down. Store at 4 °C, protect from light. The product will be stable for 6 weeks.
2. Load the chip: Label the DNA HS chip. Pipette 9 μL of gel–dye mix into the well marked "G." Place the chip in the chip priming station. Apply the plunger for 60 s, then release slowly.
3. Add 9-μL gel–dye mix into the remaining wells marked "G."
4. Add 5-μL DNA HS marker into all sample wells (including ladder well).
5. Add 1 μL of DNA ladder to the ladder well (L).
6. Add 1 μL of each DNA sample to the sample wells (S1–S11).
7. Vortex chip for 1 min at 2400 rpm using the Agilent IKA vortexer.
8. Insert the chip into the Agilent Bioanalyzer. Launch the Agilent 2100 Expert Software, choose the DNA High Sensitivity assay and run.

3.8 Metabarcoding and Metagenomics (Shotgun Metagenomics) Sequencing Protocol [12]. Library Preparation Gene Metabarcoding

The 16S/18S rRNA gene sequencing approach provides a cost-effective method for the taxonomic profiling of microbiota communities in CD samples. The library preparation process involves several critical steps:

1. PCR Amplification of Hypervariable Regions: PCR amplification of the hypervariable regions of the 16S rRNA gene allows for taxonomic profiling of complex microbial communities. In this study, we selected the V3–V4 region, which is widely

Table 1
Comparison of some 16S rRNA hypervariable regions

Region	Primer pair (forward–reverse)	Amplicon size (bp)	Pros	Limitations	References
V1–V3	27F–534R	~5800	Provides high resolution for Actinobacteria and Firmicutes, making it ideal for studying skin and oral microbiota.	May underrepresent Bacteroidetes and exhibits a bias toward Gram-positive bacteria	[12–14, 16]
V3–V4	341F–805R	~550	High taxonomic resolution with excellent performance on Illumina 2 × 300 bp sequencing	Potential primer bias against certain Archaea and low-abundance taxa	[12–14, 16]

validated for microbial diversity assessments and offers a good compromise between resolution and amplicon length. The primers 341F and 805R were used for amplification, generating a ~550 bp fragment compatible with Illumina paired-end sequencing. The choice of the V3–V4 region ensures broad bacterial coverage while minimizing biases associated with extreme GC content or low primer specificity. However, researchers may consider targeting alternative regions such as V4–V5 or V1–V3 based on the specific goals of their study or the expected composition of the microbial community [12–15]. A comparison of primer selection is shown on Table 1.

3.9 Shotgun Metagenomics Library Preparation

1. Prepare the Reaction Mix

   26-μL Nuclease-free water.

   7-μL NEBNext Ultra II FS Reaction Buffer.

   2-μL NEBNext Ultra II FS Enzyme Mix.

   25-μL genomic DNA (adjust volume based on concentration)
2. Incubate for Fragmentation: Incubate at 37 °C for 10–20 min *(shorter time = longer fragments; longer = shorter fragments)*.
3. Incubate at 65 °C for 30 min to inactivate the enzyme.
4. Proceed to end repair, adapter ligation, or cleanup as per your library prep kit (*see* **Note 7**).

Table 2
PCR setting

Step	Temperature	Time	Cycles
Initial denaturation	98 °C	30 s	1
Denaturation	98 °C	10 s	6–12[a]
Annealing/extension	60 °C	75 s	
Final extension	72 °C	5 min	1
Hold	4 °C	∞	

[a]Six cycles if starting with ≥100 ng DNA, 8–12 cycles for low-input libraries (~10–50 ng). Too many cycles can cause overamplification and PCR duplicates

3.10 Library Construction

NEBNext® Ultra™ II DNA Library Prep Kit for Illumina [17].

1. End Repair and A-tailing. Mix DNA with end repair and A-tailing enzymes and buffer. Incubate:

   20–30 min at 20 °C (end repair).

   30 min at 65 °C (A-tailing).

2. Add adapters (typically at 10:1 molar excess) and ligase mix. Incubate at 20 °C for 15–30 min.
3. Cleanup with AMPure XP beads, use a 1.0 × −1.8× bead:DNA ratio to size-select your library (e.g., 1.2× gives ~300–500 bp fragments).
4. Wash twice with 80% ethanol, then elute in 20- to 30-μL Tris-HCl or elution buffer.
5. Library Amplification (PCR). Mix:

   15-μL Adapter-ligated DNA.

   25-μL PCR master mix.

   5 μL each Indexed primers (i5 + i7).

6. Run PCR, PCR conditions are shown in Table 2.
7. Cleanup with AMPure XP beads (usually 1.0×). Elute in 20–30 μL of elution buffer.
8. Library Quantification and Quality Control:

   Concentration Check: Use Qubit dsDNA HS Assay: Expect ~10–50 ng/μL.

   Fragment Size Distribution: Agilent Bioanalyzer (High Sensitivity DNA chip) (*see* **Note 8**).

   Purity Check: NanoDrop: Check A260/280 (~1.8) and A260/230 ratios (~2.0–2.2).

Table 3
Run setting

Setting	Value
Read 1	250–300 cycles
Index 1 (i7)	8 cycles
Index 2 (i5)	8 cycles
Read 2	250–300 cycles

3.11 Sequencing on Illumina Platform for Metabarcoding Gene Sequencing

1. Quantify and Normalize Libraries: Quantify each library using Qubit dsDNA High Sensitivity assay. Dilute each library to 4 nM with 10 mM Tris–HCl (pH 8.5).
2. Pool equal volumes of each 4 nM library to create a multiplexed library pool.
3. Check Pool Quality (Optional but recommended): Run 1 μL of pooled library on Bioanalyzer or TapeStation to check for size (~300–600 bp) and absence of primer/adapter dimers.
4. Denature the pooled library. Mix 5 μL of 4 nM pooled library with 5 μL of 0.2 N NaOH. Incubate for 5 min at room temperature.
5. Dilute the library to the final loading concentration (*see* **Note 9**).
6. Mix denatured PhiX with library at 5–20% of total volume: 5% spike-in for general run, 15–20% recommended for low-diversity libraries (e.g., amplicons). Example: 570 μL pooled library +30 μL PhiX (5% spike-in) = 600 μL for loading.
7. To load your library on the Illumina MiSeq using the v3 600-cycle kit, add 600 μL of the prepared library mix (*see* **Note 10**).
8. Run the sequencing with the appropriate read setup: for the MiSeq v3 use 2 × 300 bp for full-length 16S V3–V4 or 2 × 250 bp for most ITS or COI amplicons, depending on your target region. Running settings are shown in Table 3.

3.12 Sequencing Platform Selection, Run Configuration, and Quality Control

When selecting a sequencing platform for your study, it is crucial to consider the scale of your project, desired read length, and sequencing depth. Illumina sequencing platforms such as MiSeq and NovaSeq offer different capabilities suited for small-scale or large-scale projects, respectively. Below is a breakdown of platform selection, run configuration, and quality control procedures that will guide you through the setup of your sequencing run.

Platform: To select a sequencing platform for your research, it is crucial to consider factors such as the scale of the study, read length, and output requirements. Two popular Illumina platforms, MiSeq and NovaSeq, are commonly used in different types of sequencing projects. Each platform is optimized for specific applications, offering varying read lengths and output capabilities to best suit the needs of both small-scale and large-scale studies. Selecting the right sequencing platform is essential for achieving the best results for your study's needs. **MiSeq:** The platform is ideal for small-scale studies, offering a read length of 2 × 300 bp and a read output of 25 million reads per run. It is well-suited for targeted sequencing applications, such as amplicon sequencing. **NovaSeq:** The platform is designed for large-scale projects, providing higher throughput and greater sequencing capacity, making it perfect for high-volume sequencing needs.

1. Run Configuration: Once the platform is selected, it is important to configure the sequencing run for optimal performance. Key parameters include read length, loading concentration, and sequencing depth (Table 4).
2. Quality Control: Maintaining quality control throughout the sequencing process ensures that your data is reliable and usable for downstream analysis. Real-time monitoring and post-run quality control are essential steps in assessing data quality (Table 5).

Table 4
MiSeq and NovaSeq run parameters for 16S rRNA amplicon sequencing

Parameter	MiSeq setting	NovaSeq setting	Considerations
Read length	2 × 300 bp	2 × 250 bp	MiSeq uses 2 × 300 bp for optimal taxonomic resolution.
Loading concentration	8–12 pM with 10–20% PhiX spike-in	8–12 pM with 10–20% PhiX spike-in	Standard concentration for proper cluster generation.
Sequencing depth	50,000–100,000 reads/sample	100,000–200,000 reads/sample (low-biomass)	Adjust according to sample type; higher depth for low-biomass.

Table 5
Key quality control measures for Illumina Sequencing [18–20]

QC step	Tool/method	Purpose
Real-time monitoring	Cluster Density, Q30 Scores	Track during sequencing to ensure good quality (Q30 > 80% is ideal).
Post-run QC	bcl2fastq (Illumina)	Demultiplex samples after sequencing.
Initial quality check	FastQC	Check for sequencing issues like adapter contamination or low-quality bases.

3.13 Bioinformatics Workflow for Metabarcoding in CD

Metabarcoding is a high-throughput sequencing approach that targets conserved genomic regions (e.g., 16S rRNA for bacteria, 18S rRNA for eukaryotes) to characterize microbial communities with taxonomic resolution [19, 21]. In CD research, this method enables the identification of pathogen co-infections, gut microbiota changes, and host-associated microbial shifts linked to disease progression [3, 22–25]. By amplifying and sequencing hypervariable regions, metabarcoding provides a cost-effective alternative to shotgun metagenomics for large-scale ecological studies, though it is limited to pre-selected marker genes. The bioinformatics workflow involves rigorous preprocessing to mitigate PCR and sequencing errors, precise taxonomic classification using curated reference databases, and advanced ecological analyses to uncover clinically relevant microbiome patterns [26]. This section details optimized pipelines for Chagas-specific applications, emphasizing marker selection, computational tools, and validation strategies to ensure robust biological interpretations.

1. Data Preprocessing: To ensure high-quality, artifact-free sequencing data by removing technical noise (adapters, low-quality reads, host contamination) while preserving biological signals relevant to CD.
2. Quality Assessment: *See* Table 6.
3. Quality Filtering: DADA2 is a robust and widely adopted quality filtering and denoising algorithm for high-throughput amplicon sequencing data [29]. Unlike traditional clustering methods, DADA2 applies a model-based error correction approach to distinguish true biological sequences from sequencing artifacts, allowing for the resolution of Amplicon Sequence Variants (ASVs) at single-nucleotide resolution [30, 31]. This high level of precision is especially valuable in CD studies, where subtle microbial shifts may hold clinical significance. DADA2 performs essential preprocessing steps, including quality filtering, PhiX reads removal, and trimming

Table 6
Quality assessment tools for metabarcoding data

Tool	Purpose	Advantages	Limitations	References
FastQC	Visualizes raw sequencing data quality: per-base quality, GC content, sequence duplication.	Identifies systematic errors (e.g., adapter contamination, quality drop at 3′ ends).	Does not perform trimming or filtering.	[27]
MultiQC	Aggregate quality reports from multiple samples.	Enables batch QC visualization; ideal for large-scale studies.	Relies on the upstream tools' outputs.	[28]

low-quality base calls [29]. For degraded clinical samples typical in Chagas cohorts, relaxed truncation parameters (e.g., `truncLen`) can help preserve informative reads. Command example:

```
plotQualityProfile(fnFs[1:2])
plotQualityProfile(fnRs[1:2])
out <- filterAndTrim(fnFs, filtFs, fnRs, filtRs, truncLen = c
(260, 230), trimLeft = c(5, 5), maxN = 0, maxEE = c(2, 2),
truncQ = 11, rm.phix = TRUE, compress = TRUE, multithread =
TRUE)
```

4. Taxonomic Classification with DADA2: To achieve high-resolution taxonomic assignment of Amplicon Sequence Variants (ASVs) while minimizing false positives, critical for detecting subtle microbiome changes in CD progression [29]. The DADA2 workflow addresses key challenges in microbiome analysis, especially in low-biomass samples such as those encountered in CD research. It allows for the precise identification of ASVs. It enhances our ability to detect low-abundance pathogens and strain-level variation, which is vital for understanding pathogen dynamics in co-infection scenarios [29].
5. Error Model Learning: Error model learning in the DADA2 workflow plays a crucial role in microbiome analysis by characterizing the specific error profiles associated with Illumina sequencing technologies [29]. This step is essential for distinguishing true biological variants from sequencing errors, which is particularly critical in studies involving low-abundance samples, such as those encountered in CD research. By accurately estimating the error rates, DADA2 can correct errors in the sequencing data, ensuring that only genuine biological variations are retained. The learning process involves estimating these error rates from a subset of the data, typically recommended to be around 1×10^8 bases, and iterating until the error model reaches convergence. This method minimizes the risk of false positives and ensures that the downstream taxonomic classification is based on high-quality data, thus enhancing the reliability of the results in identifying subtle microbiome shifts and pathogen dynamics [29]. Command example:

```
## Learn forward and reverse error rates
errF <- learnErrors(filtFs0, nbases = 1e+05, multithread =
TRUE) # minimum nbases = 1e+08
errR <- learnErrors(filtRs0, nbases = 1e+05, multithread =
TRUE) # minimum nbases = 1e+08
plotErrors(errF, nominalQ=TRUE)
plotErrors(errR, nominalQ=TRUE)
```

6. Dereplication and Sample Inference: Dereplication and sample inference partitions reads into Amplicon Sequence Variants (ASVs) at a single-nucleotide resolution [29]. This approach offers a significant advantage over traditional OTU clustering, as it does not rely on arbitrary thresholds (e.g., the common 97% identity cutoff for OTUs) [29, 31, 32]. By avoiding these arbitrary thresholds, DADA2 retains even rare but biologically relevant variants, ensuring that all true biological diversity is captured. This is particularly valuable in studies where detecting low-abundance pathogens is essential. In the context of CD, optimizing the method to pool data (pool = TRUE) enhances the detection of low-abundance pathogens like *T. cruzi* [29, 30, 33]. Additionally, the parameter omegaA = 1e-40 applies strict chimera removal, a feature especially important for clinical samples where sequencing artifacts like chimeras can skew results [29]. Command example:

```
## Dereplicate reads
derepRs <- derepFastq(filtRs0, verbose = TRUE)
derepFs <- derepFastq(filtFs0, verbose = TRUE)

## Perform denoising and merge paired-end reads
dadaFs <- dada(derepFs, err = errF, multithread = TRUE)
dadaRs <- dada(derepRs, err = errR, multithread = TRUE)

## Merge the denoised reads
mergers <- mergePairs(dadaFs, derepFs, dadaRs, derepRs, ver-
bose = TRUE)
```

7. Taxonomic Assignment. Several reference databases are used for taxonomic classification, each tailored to specific types of microbial communities [30, 34–38]. For example, the *SILVA 138* database is designed for both eukaryotic and prokaryotic organisms and is ideal for 18S and 16S combined studies [35, 36]. For bacterial profiling, the *Greengenes2 2022* database, focused exclusively on 16S rRNA sequences, is commonly used [37]. In the context of CD research, particularly when investigating co-infections with *T. cruzi*, the *PR2 v4.14* database is recommended for eukaryotic 18S sequences, offering better resolution for protists [39]. The classification method utilized by DADA2 is based on the Naive Bayes classifier, which provides high accuracy in assigning sequences to their correct taxonomic groups. The `minBoot = 80` parameter ensures that only classifications with a bootstrap confidence of 80% or higher are retained, further enhancing the reliability of the results [29] (*see* Table 7). For studies focusing on *Trypanosoma* infections, a custom 18S reference database with *T. cruzi*

Table 7
Reference database

Database	Coverage	Recommended for	Link
SILVA [35, 36]	Eukaryotes + Prokaryotes	18S/16S combined studies	https://www.arb-silva.de/
Greengenes [37, 38]	16S only	Bacterial profiling	https://ftp.microbio.me/greengenes_release/
PR2 [39]	Eukaryotic 18S	Trypanosoma co-infections	https://pr2-database.org/

reference strains can be used to improve the specificity of pathogen detection in mixed microbiome samples. Command example:

```
## Assign taxonomy
## Assign taxonomy using the RDP training set 18 (replace with
your reference file)
# Load RDP training set
rdp_file <- paste(dirname(rstudioapi::getActiveDocumentCon-
text()$path), '/database/rdp_train_set_18.fa.gz', sep = '')
ASV_seq_rdp_set18 <- assignTaxonomy(seqtab_all_no_chimeras,
rdp_file, multithread = TRUE)
```

Advantages for CD Research: The DADA2 workflow offers several advantages for CD research, where the detection of low-abundance pathogens and accurate taxonomic resolution is crucial. First, DADA2 can detect strain-level variation within pathogen populations, which is essential for understanding the diversity of *T. cruzi* strains and their impact on disease progression. Additionally, the ability to resolve co-infection dynamics, such as those between *T. cruzi* and gut commensals, is made possible by the high-resolution nature of ASVs. Last, DADA2 outperforms traditional OTU-based methods, especially in low-biomass samples, where it has demonstrated a higher area under the curve (AUC) value of 0.92 compared to 0.78 for OTUs. This makes it a superior choice for studying complex microbiome dynamics in diseases like Chagas [29–31].

8. Diversity Analysis with Phyloseq: Diversity analysis is essential in microbiome studies to understand the microbial community structure and its potential role in disease progression. Phyloseq, a powerful R package, offers a unified framework for calculating various diversity indices, allowing researchers to quantify both the richness and evenness of microbial communities [40]. These indices are vital in characterizing ecological patterns associated with diseases such as CD. Different diversity

Table 8
Diversity indices calculation [43–46]

Index	Formula	Interpretation	Clinical relevance in Chagas disease
Shannon (H′)	$-\Sigma\ (p_i \ln p_i)$	Measures richness and evenness	Decreased in the chronic phase (microbiota disruption)
Simpson (1-D)	$1-\Sigma(p_i^2)$	Dominance-weighted diversity	Higher in acute infections
Chao1	$S\_obs + (F_1^2/2F_2)$	Estimates total richness	Correlates with treatment response
Faith's PD	Sum of branch lengths	Phylogenetic diversity	Predicts disease progression

indices provide insights into the microbial community's complexity and how it changes in response to infection and treatment. For example, the Shannon index measures both richness and evenness, which is often decreased during the chronic phase of CD, indicating microbiota disruption. The Simpson index, a dominance-weighted diversity metric, typically shows higher values during acute infections, reflecting a less diverse microbial community dominated by certain pathogens. The Chao1 index, which estimates total microbial richness, can be correlated with the response to treatments, and Faith's phylogenetic diversity (PD) quantifies the phylogenetic diversity of the microbiota, which has been shown to predict disease progression in CD [20, 26, 32, 41, 42]. The diversity indices calculation is shown in Table 8.

3.14 Ecological Metrics Implementation

This section outlines the approach for analyzing ecological metrics using diversity indices and network analysis to assess microbial community shifts in CD [47]. The goal is to identify and characterize microbial changes that correlate with disease progression and treatment response. The analysis incorporates multiple diversity metrics to understand both the richness and composition of microbial communities, as well as their relationships with clinical outcomes.

Alpha diversity: measures the diversity within a single sample, reflecting microbial richness and evenness [40, 43–46].

1. Import the microbiome count data (e.g., ASVs or OTUs) into R using the Phyloseq package. Command example:

```
# data nececsary to phyloseq object
otu_mat<- read_excel ("/seqtabNoC.xlsx")
tax_mat<- read_excel("/taxTab.xlsx")
samples_df <- read_excel("/data_sequencing_samples.xlsx")
```

2. Perform rarefaction to subsample the data across various sequencing depths. Command example:

```
PS_rare <- rarefy_even_depth(PS, sample.size = 10000)
```

3. Plot the rarefaction curve with 95% confidence intervals (CIs) to visualize diversity saturation in samples.
4. Calculate the Shannon diversity index or Simpson index to assess microbial richness and evenness. The calculation can be performed using the `estimate_richness()` function in Phyloseq. Command example:

```
alphadiv<- estimate_richness(PS)
```

5. Use the Kruskal–Wallis test to determine whether there are significant differences in alpha diversity between groups (e.g., different stages of CD acute vs. chronic).
6. Apply false discovery rate (FDR) correction to adjust for multiple comparisons using the `p.adjust()` function in R. **Expected outcome:** A plot of rarefaction curves with 95% CIs will show the level of sequencing depth and the diversity saturation for each sample. Statistical output will identify whether differences in alpha diversity exist between the groups tested, with adjusted *p*-values accounting for multiple comparisons.

Beta Diversity: quantifies the differences in microbial composition between samples. Bray–Curtis dissimilarity measures compositional differences, while Weighted UniFrac incorporates phylogenetic relationships to provide a phylogeny-aware distance metric. Ordination of beta diversity is performed using principal coordinates analysis (PCoA), followed by statistical testing with PERMANOVA to assess the effect size (R^2) [41, 47–50].

1. Import microbiome count data into R using the Phyloseq package.
2. Compute Bray–Curtis dissimilarity using the `distance()` function in Phyloseq with method set to "bray." Alternatively, calculate Weighted UniFrac using the distance() function with method set to "unifrac" and weights set to TRUE. Command example:

```
distbray = phyloseq::distance(PS, method="bray")
```

3. Perform principal coordinates analysis (PCoA) using the `ordinate()` function with method set to "PCoA." Visualize the ordination results in a PCoA plot to assess clustering patterns based on microbial composition. **Expected Outcome:** PCoA plot will display the clustering of samples based on

microbial composition, with samples from different groups potentially clustering separately. Command example:

```
##Perform Prinicipal Coordinates Analysis (PCoA) on the phy-
loseq object
ordination_bray=ordinate(PS_rare,method="PCoA",distance=-
distbray)
```

4. Use the `adonis()` function to perform PERMANOVA to test for significant differences in microbial composition between groups (e.g., different disease stages) (*see* **Note 11**). **Expected Outcome:** PERMANOVA will return an R^2 value to quantify the proportion of variance explained by the groupings, with statistical significance indicating differences in beta diversity across groups. Command example:

```
brayasv <- adonis2(distbray~Group, data = PS, permutations=
999)
```

Microbial Co-occurrence Network Analysis: To investigate microbial interactions and identify potential keystone taxa associated with CD by constructing correlation-based networks using SparCC (Sparse Correlations for Compositional data) [51, 52].

1. Filter low-abundance taxa to reduce noise.
2. Normalize data using centered log-ratio (CLR) transformation or input raw compositional data directly into SparCC.
3. Run the SparCC algorithm to compute pairwise correlations between microbial taxa.
4. Perform 100–500 bootstrap replicates to estimate *p*-values for the correlations.
5. Retain only correlations with an absolute value of $|r| > 0.3$. Apply FDR correction to control for multiple testing.
6. Keep only those correlations with FDR-adjusted *p*-values <0.05.
7. Construct a co-occurrence network where nodes represent microbial taxa. And edges: represent significant correlations (positive or negative).
8. Calculate node centrality metrics such as:
   (a) Degree Centrality—number of direct connections.
   (b) Betweenness Centrality—frequency of a node acting as a bridge.
   (c) Closeness Centrality—how close a node is to all others in the network.
   (d) Identify keystone taxa: nodes with high centrality values that may influence microbial stability or disease dynamics.

9. Visualize networks using tools like Cytoscape or igraph in R. Color-code edges by correlation direction and nodes by taxonomic classification or abundance [45, 52–54]. **Expected Outcome:** A visual and quantitative map of microbe–microbe interactions in CD samples. Identifying keystone taxa potentially involved in disease progression, co-infections, or treatment response. Highly central nodes with strong connections may serve as biomarkers or therapeutic targets in microbiome modulation strategies.

3.15 Clinical Integration

Microbial community metrics are stratified by disease stage (acute or chronic) to identify disease-specific microbial shifts. Longitudinal modeling is applied to assess treatment effects and monitor microbial changes over time, providing insights into how the microbiome responds to therapy [25, 54, 55].

3.15.1 Differential Taxa Abundance Analysis

The primary goal of differential taxa abundance analysis is to uncover microbial taxa that exhibit statistically significant changes in relative abundance across different clinical or experimental conditions relevant to CD. This approach enables deeper understanding of host-microbiome interactions and contributes to biomarker identification, mechanistic exploration, and therapeutic development [56–58].

3.15.2 Biomarker Discovery

Differential abundance analysis is employed to identify microbial features that are enriched or depleted in relation to disease stage and clinical form. In CD, this includes comparisons between acute and chronic phases, cardiac versus digestive manifestations, and treatment responders versus non-responders. These taxa may serve as biomarkers for diagnosis or disease monitoring [58].

1. Stratify microbiome samples by clinical categories such as:
   (a) Disease stage: acute versus chronic.
   (b) Clinical form: cardiac versus digestive.
   (c) Treatment outcome: responders versus non-responders.
2. Normalize taxonomic abundance data using compositional-aware methods such as CLR (centered log-ratio) or DESeq2's internal normalization to ensure accurate comparisons.
3. Perform differential abundance testing using tools like DESeq2 (for count-based data), LEfSe (Linear discriminant analysis Effect Size), ALDEx2 or ANCOM-BC (for compositional microbiome data). Use appropriate covariate correction to adjust for confounding factors like antibiotic use, sex, age, and geography [56–58]. Command example:

```
#differential abundance analysis
##Show significantly different taxa at ASV level between
```

```
groups
#By group
lefse_model <- run_lefse(
PS,
taxa_rank = "all",
wilcoxon_cutoff = 0.01,
group = "Group",
kw_cutoff = 0.01,
multigrp_strat = TRUE,
lda_cutoff = 2
```

4. Apply a false discovery rate (FDR) correction (e.g., Benjamini–Hochberg) to control for multiple hypothesis testing. Only taxa with FDR < 0.05 and effect size (e.g., $|\log2FC| > 1$) are considered candidate biomarkers. Top candidate taxa are validated using:
   (a) Quantitative polymerase chain reaction (qPCR) for absolute abundance confirmation.
   (b) Metatranscriptomics to verify transcriptional activity.
   (c) Longitudinal data to track predictive value over time.

3.16 Mechanistic Insights

To elucidate functional relationships between microbial community shifts and host pathophysiology in CD by linking differential taxa abundance to biological readouts such as parasite load, systemic inflammation, and disease kinetics [59, 60].

1. Collect metadata including:
   (a) *T. cruzi* burden (e.g., qPCR Ct values).
   (b) Inflammatory cytokine levels (e.g., IL-6, TNF-α, IFN-γ via ELISA or multiplex assays).
   (c) Clinical progression metrics (e.g., ECG abnormalities, fibrosis scores, gastrointestinal motility).
2. Perform multivariate regression or correlation models (e.g., MaAsLin2, linear mixed-effects models) to associate differentially abundant taxa with continuous host parameters.
   (a) Evaluate effect size and direction (e.g., taxa positively correlated with parasite burden) to identify microbial patterns potentially contributing to persistent infection, host immunomodulation, and/or tissue-specific pathology.
   (b) Repeat association testing within stratified subsets (e.g., acute vs. chronic patients) to identify stage-specific microbial mechanisms.
   (c) Use correlation heatmaps, co-abundance networks, and partial redundancy analysis (pRDA) to map taxa–function relationships. Highlight keystone taxa correlated with inflammatory or anti-inflammatory profiles.

3.17 Therapeutic Targeting

To identify microbial taxa that can be modulated or leveraged as clinical tools for diagnosis, therapy, or monitoring treatment outcomes in CD. The focus is on taxa whose abundance patterns are robustly associated with disease stage, immune status, or response to antiparasitic treatment [1, 25, 59].

1. Screen differentially abundant taxa across clinical contrasts (e.g., acute vs. chronic, cardiac vs. digestive, treated vs. untreated) using tools like DESeq2 or ANCOM-BC.
2. Evaluate early-stage enrichment using ROC curve analysis (e.g., AUC > 0.85). Prioritize taxa reproducible across cohorts or timepoints.
3. Focus on commensals negatively associated with parasite load or inflammation. Cross-reference with literature for known immune-modulatory properties (e.g., *Lactobacillus johnsonii*).
4. Apply machine learning classifiers (e.g., random forest, SVM) to baseline microbiota data to predict responder versus non-responder status.
5. Validate with cross-validation and external datasets.
6. Integrate with metagenomics/metatranscriptomics to infer metabolic pathways (e.g., SCFA biosynthesis, tryptophan metabolism).

3.18 Comprehensive Bioinformatics Workflow for Shotgun Metagenomics in CD Research

Shotgun metagenomics is a high-resolution sequencing approach that provides an unbiased characterization of microbial communities by randomly fragmenting and sequencing all DNA present in a sample [61]. Unlike targeted metabarcoding, this method enables simultaneous taxonomic profiling at strain-level resolution, functional pathway analysis, and genome reconstruction through assembly of short reads into metagenome-assembled genomes (MAGs) [62]. This workflow supports high-resolution taxonomic and functional profiling and is optimized for low-biomass samples such as blood or tissue, where host contamination and sequencing depth are critical constraints. Data Preprocessing: Rigorous Quality Control.

3.18.1 Trimmomatic: Advanced Read Processing

Remove adapters and low-quality bases from raw reads, ensuring optimal data retention and integrity, especially for degraded or low-input Chagas samples [1, 63].

1. Launch Trimmomatic and use this command to trim adapters and low-quality regions. `Adapter` trimming, removes leftover sequencing adapters. `Sliding window`, trims if the quality drops below Q20 in a 4-base window. `Leading/Trailing`, trims poor-quality bases from the ends. `Minlen`, tosses reads

shorter than 50 bp. **Output files:** `_trimmed_paired.fastq.gz`, high-quality reads for analysis. Command example:

```
trimmomatic PE -threads 4 -phred33 \
sample_R1.fastq.gz sample_R2.fastq.gz \
sample_R1_trimmed_paired.fastq.gz sample_R1_trimmed_unpaired.
fastq.gz \
sample_R2_trimmed_paired.fastq.gz sample_R2_trimmed_unpaired.
fastq.gz \
ILLUMINACLIP:adapters.fa:2:30:10 \
SLIDINGWINDOW:4:20 \
LEADING:20 TRAILING:20 \
MINLEN:50
```

2. For quality check, run FastQC or MultiQC on the trimmed reads. You need: 90% + of bases with quality ≥ Q30. 90% + of bases with quality ≥ Q30 [27, 28, 64].
   (a) At least 85% of reads retained (for stool).
   (b) At least 60% retained (for blood/tissue).

3.18.2 Host DNA Removal (If Applicable)

Removing host DNA contamination is a crucial preprocessing step [1, 61]. Bowtie2 is commonly used to align raw sequencing reads against host reference genomes (e.g., *Homo sapiens*, *Mus musculus*) and filter out contaminant sequences [65]. This step not only improves the efficiency of downstream microbial profiling but also enhances the accuracy of taxonomic classification by eliminating non-target background reads. In Chagas-specific contexts, using both human and triatomine host genome references can ensure comprehensive contaminant removal, particularly in mixed biological samples such as blood, tissue, or gut homogenates [1]. Command example:

```
# Remove host (e.g., human or mouse) reads
bowtie2 -x host_genome_index -1 clean_R1.fastq.gz -2 clean_R2.
fastq.gz \
--very-sensitive -S host_mapped.sam --un-conc-gz filtered_-
reads.gz
```

3.19 Taxonomic Profiling: High-Resolution Community Analysis

3.19.1 MetaPhlAn4: Ultra-Sensitive Taxonomic Detection

MetaPhlAn4 identifies microbial species down to 0.01% abundance. It is perfect for understanding the microbial community composition in your samples. Prepare your sample (cleaned fastq files) [66–69]. Run MetaPhlAn4. Command example:

```
#metaphlan /path/to/sample_R1.fastq.gz,/path/to/sample_R2.
fastq.gz \
--input_type fastq \
```

```
--bowtie2db /path/to/metaphlan_db \
--output_file /path/to/output/profiled_metagenome.txt \
--bowtie2out /path/to/output/bowtie2_output.bz2 \
--nproc 8
```

3.19.2 StrainPhlAn: Strain-Level Resolution

StrainPhlAn enables high-resolution taxonomic analysis by characterizing strain-level diversity within microbial species [68, 70]. This tool is particularly valuable for distinguishing fine-scale genetic variations across samples, such as those observed in *T. cruzi*. By leveraging species-specific marker genes, StrainPhlAn facilitates comparative analysis and phylogenetic reconstruction at the strain level, offering insights into population structure, transmission patterns, and host-specific adaptations. **Requirements:** At least 5× coverage of marker regions per sample and a minimum 2% allele frequency to call variants. Use well-annotated reference genomes for *T. cruzi* or other organisms of interest.

1. Map the output from MetaPhlAn4. Command example:

```
mkdir -p sams
mkdir -p bowtie2
mkdir -p profiles
for f in fastq/SRS*
do
 echo "Running MetaPhlAn on ${f}"
 bn=$(basename ${f})
  metaphlan ${f} --input_type fastq -s sams/${bn}.sam.bz2
-bowtie2out bowtie2/${bn}.bowtie2.bz2 –
profiles/${bn}_profiled.tsv
done
```

2. Run StrainPhlAn to extract strain-level markers. Command example:

```
mkdir -p consensus_markers
sample2markers.py -i sams/*.sam.bz2 -o consensus_markers -n 8
mkdir -p output
strainphlan -s consensus_markers/*.json -m db_markers/*.fna -r
reference_genomes/*.fna.bz2 -o output -n 8 -c * --mutation_-
rates
```

3. A **phylogenetic tree** (`strainphlan_out/RAxML_bestTree.tree`) will be generated, showing **strain-level relationships** across your samples. Using FigTree, iTOL, or even ETE Toolkit, you can visualize the phylogenetic tree [71].

3.20 Functional Prediction of Microbial Communities Using HUMAnN 4.0

HUMAnN 4.0 (The HMP Unified Metabolic Analysis Network) enables comprehensive functional profiling of metagenomic samples by integrating taxonomic composition with predicted metabolic capabilities [66]. Building upon taxonomic profiles generated by MetaPhlAn4, HUMAnN reconstructs gene family abundances and metabolic pathways, offering insights into microbial contributions to host physiology, disease processes, and environmental adaptation. **Output Files:** gene_families.tsv - Gene family abundance (e.g., UniRef90), pathabundance.tsv - MetaCyc pathway abundance; pathcoverage.tsv - Pathway completeness estimates; and stratified outputs, taxon-specific functional contributions. The HUMAnN key parameters are shown in Table 9.

3.21 Metagenome-Assembled Genomes (MAGs)

Metagenome-Assembled Genomes (MAGs) offer a powerful approach to reconstruct individual microbial genomes directly from complex metagenomic datasets [72]. This is particularly useful when investigating poorly characterized organisms such as *T. cruzi* or identifying novel taxa involved in host–pathogen or host–microbiome interactions [1]. The process involves two main stages: assembly, where raw sequencing reads are converted into contigs, and binning, where these contigs are grouped into putative genomes [72, 73].

1. Assembly: Convert quality-filtered reads into contigs that represent longer continuous DNA sequences [74]. Assemble high-quality contigs from filtered reads using a metagenome-aware assembler [74–77]. Command example:

```
# Option 1: metaSPAdes
metaspades.py -1 filtered_reads.1.gz -2 filtered_reads.2.gz \
-o spades_output -t 16

# Option 2: MEGAHIT
megahit -1 filtered_reads.1.gz -2 filtered_reads.2.gz \
-o megahit_output --presets meta-sensitive
```

2. Evaluate the assembly quality before binning. Metrics to Watch: N50 ≥ 10 kb, longest contig ≥100 kb, and total assembly size consistent with expected microbiome content [78]. Command example:

```
quast.py spades_output/contigs.fasta -o quast_report/
```

3. Map reads back to contigs to calculate coverage depth [65]. Command example:

```
# Index contigs
bowtie2-build contigs.fasta contigs_index

# Map reads
```

Table 9
HUMAnN 4.0. Key Parameters

Parameter	Description
`--threads`	Number of cores to use
`--nucleotide-database`	Path to ChocoPhlAn (nucleotide-level mapping)
`--protein-database`	Path to UniRef90 (protein alignment using DIAMOND)
`--taxonomic-profile`	MetaPhlAn-generated profile to guide alignment
`--search-mode`	`uniref90` (default), or use custom for *T. cruzi*–specific proteins

```
bowtie2 -x contigs_index -1 filtered_reads.1.gz -2 filtered_-
reads.2.gz \
-S aligned.sam -p 16

# Convert and sort
samtools view -bS aligned.sam | samtools sort -o aligned.
sorted.bam

# Generate depth file
jgi_summarize_bam_contig_depths --outputDepth depth.txt
aligned.sorted.bam
```

4. Binning: Cluster contigs into bins representing putative genomes [72, 79–82]. Command example:

```
# MetaBAT2
metabat2 -i contigs.fasta -a depth.txt -o bins_metabat/bin -t
16

# MaxBin2
run_MaxBin.pl -contig contigs.fasta -abund depth.txt -out
bins_maxbin/bin -thread 16

# CONCOCT (if desired)
cut_up_fasta.py contigs.fasta -c 10000 -o 0 --merge_last -b
contigs_10K.bed > contigs_cut.fa

concoct --composition_file contigs_cut.fa --coverage_file
depth.txt -b concoct_output/
```

5. Integrate multiple binning outputs to get consensus high-quality bins [79]. Command example:

```
DAS_Tool -imetabat2.scaffolds2bin.tsv,maxbin2.scaffolds2bin.
tsv \
-l metabat2,maxbin2 -c contigs.fasta --write_bins 1 -o da-
s_output
```

6. Determine completeness and contamination of each genome bin. Look for: >90% completeness, <5% contamination for high-quality bins, and for medium-quality bins: ≥50% completeness, <10% contamination [83]. Command example:

```
checkm lineage_wf das_output_DASTool_bins/ checkm_out/ -t 16
```

7. Assign taxonomy to MAGs using a standardized, updated database [84]. Command example:

```
gtdbtk classify_wf --genome_dir das_output_DASTool_bins/ \
--out_dir gtdbtk_out/ --cpus 16
```

8. Functional Annotation MAGs [85]. Command example:

```
# Prokka
prokka --outdir annotation_out/ --prefix sample_bin1 \
--cpus 8 das_output_DASTool_bins/sample_bin1.fa
```

3.22 Downstream Analysis of MAGs

Following the functional annotation of MAGs using tools like Prokka, a wide array of downstream analyses can be performed to gain further insights into the ecological roles, evolutionary relationships, metabolic capabilities, and pathogenic potential of the microbial community. This section summarizes key analytical approaches and tools used in this phase of metagenomic studies.

1. Phylogenomic Analysis: Phylogenetic reconstruction is essential for inferring evolutionary relationships among MAGs and placing them within a broader taxonomic context. Developed by SegataLab, PhyloPhlAn enables high-resolution phylogenetic placement of MAGs using thousands of universal marker genes [66, 86]. Command example:

```
phylophlan -i ./mags/ -d phylophlan --diversity high \
--fast --nproc 16 -o ./phylophlan_output/
```

2. Pangenome Analysis: Pangenome analysis helps identify core and accessory genes across a group of related MAGs, providing insight into microbial adaptation and evolution. Panaroo and Roay have been used in this context [87, 88]. Command example:

```
# Assuming all MAGs annotated with Prokka
roary -e --mafft -p 8 -f roary_output *.gff
```

3. Detection of Virulence and Resistance Genes: Identifying virulence factors and antibiotic resistance genes (ARGs) allows for assessment of pathogenic potential and public health relevance [89–95]. Command example:

```
abricate --db vfdb prokka_output.gff > vfdb_report.tsv
```

4 Notes

1. Collect stool samples from mice consistent with the study design. Two to three pellets typically yield 100–200 mg. Mouse fecal collection must be completed as quickly as possible in a sterile environment to avoid microbial contamination.
2. If using Norgen's Stool Nucleic Acid Collection and Transport Tube, transfer 400 μL of preserved sample into a Bead Tube and add 600 μL of Lysis Buffer L. Vortex briefly to mix.
3. A thin layer may form on top of the supernatant. Carefully collect the clean supernatant while avoiding excessive contamination with this layer.
4. Ensure all the lysate has passed through the column. If not, centrifuge for another 1 min at 14,000 rpm (~20,000 × *g*).
5. A260/280 = 1.8–2.0: Indicates minimal contamination and high DNA purity; <1.8: May indicate protein contamination, and >2.0: Could indicate contamination by RNA or other organic compounds.
6. Intact DNA: Clear band of high-molecular-weight DNA. Degraded DNA: Smearing or fragmented bands indicate DNA degradation.
7. Run 1 μL on Agilent Bioanalyzer (Subheading 3.7), expect a smear centered at your target size, for example, 300–500 bp.
8. Expected size: ~300–600 bp depending on fragmentation.
9. For MiSeq runs, the recommended final concentration is typically 8–12 pM. To achieve this, mix the appropriate volume of the 20 pM library with HT1 buffer, adjusting as needed to reach the desired concentration.
10. Be sure to check your specific instrument's loading volume, as it may vary slightly.
11. The PERMANOVA test will provide the effect size (R^2) to assess how well the groups are separated.

References

1. Castañeda S, Muñoz M, Hotez PJ, Bottazzi ME, Paniz-Mondolfi AE, Jones KM, Mejia R, Poveda C, Ramírez JD (2023) Microbiome alterations driven by Trypanosoma cruzi infection in two disjunctive murine models. Microbiol Spectr. https://doi.org/10.1128/SPECTRUM.00199-23/SUPPL_FILE/SPECTRUM.00199-23-S0003.XLSX
2. Duarte-Silva E, Morais LH, Clarke G, Savino W, Peixoto C (2020) Targeting the gut microbiota in Chagas disease: what do we know so far? Front Microbiol 11:3083
3. de Souza-Basqueira M, Ribeiro RM, de Oliveira LC, Moreira CHV, Martins RCR, Franco DC, Amado PPP, Mayer MPA, Sabino EC (2020) Gut Dysbiosis in Chagas disease. A possible link to the pathogenesis. Front Cell Infect Microbiol. https://doi.org/10.3389/fcimb.2020.00402
4. Herreros-Cabello A, Bosch-Nicolau P, Pérez-Molina JA et al (2024) Identification of Chagas disease biomarkers using untargeted metabolomics. Sci Rep 14:1–11

5. Pérez-Molinan JA, Crespillo-Andujar C, Trigo E et al (2023) Chagas disease is related to structural changes of the gut microbiota in adults with chronic infection (TRIPOBIOME Study). PLoS Negl Trop Dis. https://doi.org/10.1371/JOURNAL.PNTD.0011490
6. Hu J, He K, Yang Y et al (2024) Amino acid formula induces microbiota dysbiosis and depressive-like behavior in mice. Cell Rep 43: 113817
7. Di Gesù CM, Matz LM, Fultz R, Bolding IJ, Buffington SA (2023) Monospecies probiotic preparation and administration with downstream analysis of sex-specific effects on gut microbiome composition in mice. STAR Protoc 4:102386
8. Ahrends T, Weiner M, Mucida D (2022) Isolation of myenteric and submucosal plexus from mouse gastrointestinal tract and subsequent flow cytometry and immunofluorescence. STAR Protoc 3:101157
9. Belik J, Shifrin Y, Arning E, Bottiglieri T, Pan J, Daigneault MC, Allen-Vercoe E (2017) Intestinal microbiota as a tetrahydrobiopterin exogenous source in hph-1 mice. Sci Rep 7:1–9
10. Scientific T (2000) NanoDrop 2000 / 2000c spectrophotometer. Thermo Sci inc 97
11. Technologies A (2017) Agilent high sensitivity DNA kit quick start guide. Agil Technol, pp 1–4
12. Zhang H, Wang X, Chen A et al (2024) Comparison of the full-length sequence and sub-regions of 16S rRNA gene for skin microbiome profiling. mSystems. https://doi.org/10.1128/MSYSTEMS.00399-24/SUPPL_FILE/MSYSTEMS.00399-24-S0007.XLSX
13. López-Aladid R, Fernández-Barat L, Alcaraz-Serrano V, Bueno-Freire L, Vázquez N, Pastor-Ibáñez R, Palomeque A, Oscanoa P, Torres A (2023) Determining the most accurate 16S rRNA hypervariable region for taxonomic identification from respiratory samples. Sci Rep 13:1–10
14. Paul B (2022) Concatenated 16S rRNA sequence analysis improves bacterial taxonomy. F1000Res. https://doi.org/10.12688/f1000research.128320.1
15. D'Argenio V, D'Argenio V (2018) Human microbiome acquisition and bioinformatic challenges in metagenomic studies. Int J Mol Sci 19:1–12
16. Yang B, Wang Y, Qian PY (2016) Sensitivity and correlation of hypervariable regions in 16S rRNA genes in phylogenetic analysis. BMC Bioinform 17:1–8
17. New England Biolabs (2020) NEBNext Ultra II DNA Library Prep Kit for Illumina. 3.2:5
18. Lappan R, Classon C, Kumar S, Singh OP, de Almeida RV, Chakravarty J, Kumari P, Kansal S, Sundar S, Blackwell JM (2019) Meta-taxonomic analysis of prokaryotic and eukaryotic gut flora in stool samples from visceral leishmaniasis cases and endemic controls in Bihar State India. PLoS Negl Trop Dis 13: e0007444
19. Stensvold CR (2024) Metabarcoding in gut protozoology. Trends Parasitol. https://doi.org/10.1016/J.PT.2024.10.015
20. Wylezich C, Höper D (2021) Meta-ribosomalomics: RNA sequencing is an unbiased method for parasite detection of different sample types. Front Microbiol 12:1616
21. Zinger L, Bonin A, Alsos IG et al (2019) DNA metabarcoding—need for robust experimental designs to draw sound ecological conclusions. Mol Ecol 28:1857–1862
22. Teotônio IMSN, Dias N, Hagström-Bex L et al (2019) Intestinal microbiota—a modulator of the Trypanosoma cruzi-vector-host triad. Microb Pathog 137:103711
23. Robello C, Maldonado DP, Hevia A et al (2019) The fecal, oral, and skin microbiota of children with Chagas disease treated with benznidazole. PLoS One. https://doi.org/10.1371/journal.pone.0212593
24. McCall L-II, Tripathi A, Vargas F, Knight R, Dorrestein PC, Siqueira-Neto JL (2018) Experimental Chagas disease-induced perturbations of the fecal microbiome and metabolome. PLoS Negl Trop Dis 12:1–15
25. Hossain E, Khanam S, Dean DA et al (2020) Mapping of host-parasite-microbiome interactions reveals metabolic determinants of tropism and tolerance in Chagas disease. Sci Adv 6: eaaz2015
26. Huttenhower C, Morgan XC (2014) Meta'omic analytic techniques for studying the intestinal microbiome. Gastroenterology. https://doi.org/10.1053/j.gastro.2014.01.049
27. Babraham Bioinformatics - FastQC A Quality Control tool for High Throughput Sequence Data. https://www.bioinformatics.babraham.ac.uk/projects/fastqc/. Accessed 16 May 2022
28. Ewels P, Magnusson M, Lundin S, Käller M (2016) MultiQC: summarize analysis results for multiple tools and samples in a single report. Bioinformatics 32:3047–3048
29. Callahan BJ, McMurdie PJ, Rosen MJ, Han AW, Johnson AJA, Holmes SP (2016) DADA2: high-resolution sample inference from Illumina amplicon data. Nat Methods 137(13):581–583

30. Odom AR, Faits T, Castro-Nallar E, Crandall KA, Johnson WE (2023) Metagenomic profiling pipelines improve taxonomic classification for 16S amplicon sequencing data. Sci Rep 13(1):–12
31. Allali I, Arnold JW, Roach J et al (2017) A comparison of sequencing platforms and bioinformatics pipelines for compositional analysis of the gut microbiome. BMC Microbiol 17:1–16
32. Hill TCJ, Walsh KA, Harris JA, Moffett BF (2003) Using ecological diversity measures with bacterial communities. FEMS Microbiol Ecol 43:1–11
33. Castañeda S, Poveda C, Suarez-Reyes C, Wu Y, Haugen N, Patiño LH, Weatherhead JE, Ramírez JD (2025) Microbiota dynamics during Ascaris suum larval migration: implications for host microbial communities in a murine model. Microb Pathog 198:107122
34. Terrón-Camero LC, Gordillo-González F, Salas-Espejo E, Andrés-León E (2022) Comparison of metagenomics and metatranscriptomics tools: a guide to making the right choice. Genes (Basel) 13:2280
35. Quast C, Pruesse E, Yilmaz P, Gerken J, Schweer T, Yarza P, Peplies J, Glöckner FO (2013) The SILVA ribosomal RNA gene database project: improved data processing and web-based tools. Nucleic Acids Res 41:590–596
36. Pruesse E, Quast C, Knittel K, Fuchs BM, Ludwig W, Peplies J, Glöckner FO (2007) SILVA: a comprehensive online resource for quality checked and aligned ribosomal RNA sequence data compatible with ARB. Nucleic Acids Res 35:7188–7196
37. McDonald D, Jiang Y, Balaban M et al (2023) Greengenes2 unifies microbial data in a single reference tree. Nat Biotechnol 14:1–4
38. McDonald D, Price MN, Goodrich J, Nawrocki EP, Desantis TZ, Probst A, Andersen GL, Knight R, Hugenholtz P (2012) An improved Greengenes taxonomy with explicit ranks for ecological and evolutionary analyses of bacteria and archaea. ISME J 6:610–618
39. Guillou L, Bachar D, Audic S et al (2013) The Protist Ribosomal Reference database (PR2): a catalog of unicellular eukaryote Small Sub-Unit rRNA sequences with curated taxonomy. Nucleic Acids Res 41:D597–D604
40. PJ MM, Holmes S (2013) Phyloseq: an R package for reproducible interactive analysis and graphics of microbiome census data. PLoS One. https://doi.org/10.1371/journal.pone.0061217
41. Su X (2021) Elucidating the beta-diversity of the microbiome: from global alignment to local alignment. mSystems. https://doi.org/10.1128/MSYSTEMS.00363-21
42. D'Argenio V, Casaburi G, Precone V, Salvatore F (2014) Comparative metagenomic analysis of human gut microbiome composition using two different bioinformatic pipelines. Biomed Res Int. https://doi.org/10.1155/2014/325340
43. Roswell M, Dushoff J, Winfree R (2021) A conceptual guide to measuring species diversity. Oikos 130:321–338
44. Xia Y, Sun J (2023) Alpha diversity. In: Bioinformatic Stat Anal Microbiome Data, pp 289–333
45. Willis AD, Martin BD (2022) Estimating diversity in networked ecological communities. Biostatistics 23:207–222
46. Dolan JR, Forster D, Dunthorn M et al (2016) A metagenomic alpha-diversity index for microbial functional biodiversity. FEMS Microbiol Ecol:1–11
47. Kers JG, Saccenti E (2022) The power of microbiome studies: some considerations on which alpha and beta metrics to use and how to report results. Front Microbiol 12:796025
48. Xia Y, Sun J (2023) Beta diversity metrics and ordination. In: Bioinformatic Stat Anal Microbiome Data, pp 335–395
49. Tang ZZ, Chen G, Alekseyenko AV (2016) PERMANOVA-S: association test for microbial community composition that accommodates confounders and multiple distances. Bioinformatics 32:2618–2625
50. Swenson NG (2011) Phylogenetic beta diversity metrics, trait evolution and inferring the functional beta diversity of communities. PLoS One 6:e21264
51. Jensen IT, Janss L, Radutoiu S, Waagepetersen R (2024) Compositionally aware estimation of cross-correlations for microbiome data. PLoS One 19:e0305032
52. Ma J, Yue K, Shojaie A (2021) Networks for compositional data, pp. 311–336
53. Layeghifard M, Hwang DM, Guttman DS (2018) Constructing and analyzing microbiome networks in R. Methods Mol Biol 1849:243–266
54. Peeters J, Thas O, Shkedy Z et al (2021) Exploring the microbiome analysis and visualization landscape. Front Bioinform 1:774631
55. Sadowsky MJ, Staley C, Heiner C, Hall R, Kelly CR, Brandt L, Khoruts A (2017) Analysis of gut microbiota - an ever changing landscape. Gut Microbes 8:268–275

56. Lin H, Das PS (2020) Analysis of microbial compositions: a review of normalization and differential abundance analysis. npj Biofilms Microbiomes 6:1–13
57. Yang L, Chen J (2022) A comprehensive evaluation of microbial differential abundance analysis methods: current status and potential solutions. Microbiome 10:1–23
58. Segata N, Izard J, Waldron L, Gevers D, Miropolsky L, Garrett WS, Huttenhower C (2011) Metagenomic biomarker discovery and explanation. Genome Biol 12:R60
59. Poveda C, Bottazzi ME, Jones KM (2021) Mining the metabolome for new and innovative Chagas disease treatments. Trends Pharmacol Sci 42:1–3
60. Lewis MD, Francisco AF, Taylor MC, Jayawardhana S, Kelly JM (2016) Host and parasite genetics shape a link between Trypanosoma cruzi infection dynamics and chronic cardiomyopathy. Cell Microbiol 18:1429
61. Quince C, Walker AW, Simpson JT, Loman NJ, Segata N (2017) Shotgun metagenomics, from sampling to analysis. Nat Biotechnol 35:833–844
62. Wajid B, Anwar F, Wajid I, Nisar H, Meraj S, Zafar A, Al-Shawaqfeh MK, Ekti AR, Khatoon A, Suchodolski JS (2022) Music of metagenomics—a review of its applications, analysis pipeline, and associated tools. Funct Integr Genomics 22:3–26
63. Bolger AM, Lohse M, Usadel B (2014) Trimmomatic: a flexible trimmer for Illumina sequence data. Bioinformatics. https://doi.org/10.1093/BIOINFORMATICS/BTU170
64. Wingett SW, Andrews S (2018) FastQ screen: a tool for multi-genome mapping and quality control. F1000Res 7:1338
65. Langmead B, Salzberg SL (2012) Fast gapped-read alignment with Bowtie 2. Nat Methods 9:357
66. Beghini F, McIver LJ, Blanco-Míguez A et al (2021) Integrating taxonomic, functional, and strain-level profiling of diverse microbial communities with biobakery 3. elife. https://doi.org/10.7554/ELIFE.65088
67. Truong DT, Franzosa EA, Tickle TL, Scholz M, Weingart G, Pasolli E, Tett A, Huttenhower C, Segata N (2015) MetaPhlAn2 for enhanced metagenomic taxonomic profiling. Nat Methods 12:902–903
68. Blanco-Míguez A, Beghini F, Cumbo F et al (2023) Extending and improving metagenomic taxonomic profiling with uncharacterized species using MetaPhlAn 4. Nat Biotechnol 41:1633–1644
69. Manghi P, Blanco-Míguez A, Manara S et al (2023) MetaPhlAn 4 profiling of unknown species-level genome bins improves the characterization of diet-associated microbiome changes in mice. Cell Rep. https://doi.org/10.1016/J.CELREP.2023.112464/ATTACHMENT/99E53810-BB35-467B-91C9-1DED3C46B2AB/MMC9.PDF
70. Truong DT, Tett A, Pasolli E, Huttenhower C, Segata N (2017) Microbial strain-level population structure and genetic diversity from metagenomes. Genome Res 27:gr.216242.116
71. Letunic I, Bork P (2021) Interactive Tree Of Life (iTOL) v5: an online tool for phylogenetic tree display and annotation. Nucleic Acids Res 49:W293–W296
72. Yang C, Chowdhury D, Zhang Z, Cheung WK, Lu A, Bian Z, Zhang L (2021) A review of computational tools for generating metagenome-assembled genomes from metagenomic sequencing data. Comput Struct Biotechnol J 19:6301–6314
73. Eisenhofer R, Odriozola I, Alberdi a (2023) impact of microbial genome completeness on metagenomic functional inference. ISME Commun 3:1–5
74. Vollmers J, Wiegand S, Kaster AK (2017) Comparing and evaluating metagenome assembly tools from a microbiologist's perspective - not only size matters! PLoS One 12:e0169662
75. Li D, Liu CM, Luo R, Sadakane K, Lam TW (2015) MEGAHIT: an ultra-fast single-node solution for large and complex metagenomics assembly via succinct de Bruijn graph. Bioinformatics 31:1674–1676
76. Wooley JC, Godzik A, Friedberg I (2010) A primer on metagenomics. PLoS Comput Biol. https://doi.org/10.1371/journal.pcbi.1000667
77. Nurk S, Meleshko D, Korobeynikov A, Pevzner P (2016) metaSPAdes: a new versatile de novo metagenomics assembler
78. Mikheenko A, Saveliev V, Gurevich A (2016) MetaQUAST: evaluation of metagenome assemblies. Bioinformatics 32:1088–1090
79. Sieber CMK, Probst AJ, Sharrar A, Thomas BC, Hess M, Tringe SG, Banfield JF (2018) Recovery of genomes from metagenomes via a dereplication, aggregation and scoring strategy. Nat Microbiol 3:836–843
80. Kang DD, Froula J, Egan R, Wang Z (2015) MetaBAT, an efficient tool for accurately reconstructing single genomes from complex microbial communities. PeerJ 2015:1–15
81. Lamurias A, Sereika M, Albertsen M, Hose K, Nielsen TD (2022) Metagenomic binning with

assembly graph embeddings. Bioinformatics 38:4481–4487

82. Wu Y-W, Tang Y-H, Tringe SG, Simmons BA, Singer SW (2014) MaxBin: an automated binning method to recover individual genomes from metagenomes using. Microbiome 2: 4904–4909
83. Parks DH, Imelfort M, Skennerton CT, Hugenholtz P, Tyson GW (2015) CheckM: assessing the quality of microbial genomes recovered from isolates, single cells, and metagenomes. Genome Res 25:1043
84. Chaumeil PA, Mussig AJ, Hugenholtz P, Parks DH (2020) GTDB-Tk: a toolkit to classify genomes with the Genome Taxonomy Database. Bioinformatics 36:1925–1927
85. Seemann T (2014) No Title. Bioinformatics 30:2068–2069
86. Segata N, Börnigen D, Morgan XC, Huttenhower C (2013) PhyloPhlAn is a new method for improved phylogenetic and taxonomic placement of microbes. Nat Commun 4:1–11
87. Tonkin-Hill G, MacAlasdair N, Ruis C et al (2020) Producing polished prokaryotic pangenomes with the Panaroo pipeline. Genome Biol 21:1–21
88. Page AJ, Cummins CA, Hunt M, Wong VK, Reuter S, Holden MTG, Fookes M, Falush D, Keane JA, Parkhill J (2015) Roary: rapid large-scale prokaryote pan genome analysis. Bioinformatics 31:3691–3693
89. Bikandi J, Atxaerandio-Landa A, Laorden L, Martinez-Ballesteros I, Martinez-Malaxetxebarria I, Arrieta-Gisasola A, Garaizar J (2022) A practical bioinformatics workflow for routine analysis of bacterial WGS data. Microorganisms. https://doi.org/10.3390/microorganisms10122364
90. Feldgarden M, Brover V, Haft DH et al (2019) Validating the AMRFINder tool and resistance gene database by using antimicrobial resistance genotype-phenotype correlations in a collection of isolates. Antimicrob Agents Chemother. https://doi.org/10.1128/AAC.00483-19
91. Jia B, Raphenya AR, Alcock B et al (2017) CARD 2017: expansion and model-centric curation of the comprehensive antibiotic resistance database. Nucleic Acids Res 45:D566–D573
92. Zankari E, Hasman H, Cosentino S, Vestergaard M, Rasmussen S, Lund O, Aarestrup FM, Larsen MV (2012) Identification of acquired antimicrobial resistance genes. J Antimicrob Chemother 67:2640–2644
93. Gupta SK, Padmanabhan BR, Diene SM, Lopez-Rojas R, Kempf M, Landraud L, Rolain JM (2014) ARG-annot, a new bioinformatic tool to discover antibiotic resistance genes in bacterial genomes. Antimicrob Agents Chemother 58:212–220
94. Doster E, Lakin SM, Dean CJ, Wolfe C, Young JG, Boucher C, Belk KE, Noyes NR, Morley PS (2020) MEGARes 2.0: a database for classification of antimicrobial drug, biocide and metal resistance determinants in metagenomic sequence data. Nucleic Acids Res 48:D561–D569
95. Chen L, Zheng D, Liu B, Yang J, Jin Q (2016) VFDB 2016: hierarchical and refined dataset for big data analysis - 10 years on. Nucleic Acids Res 44:D694–D697

Chapter 16

Metabolomics Analysis of Trypanosomes Using Ion Mobility–Enhanced Mass Spectrometry

Andrew MacKenzie, Fraser Massie, and Tessa Moses

Abstract

Metabolomics, the comprehensive study of low-molecular-weight metabolites, provides a direct snapshot of an organism's physiological state and complements genomics and proteomics. Ion mobility (IM) enhanced metabolomics adds crucial gas-phase separation, resolving isobaric and isomeric metabolites for deeper biological insights. Here we describe a method coupling rapid liquid chromatography to IM-enhanced mass spectrometry, which allows high-throughput metabolomics analysis. This method was successfully applied to trypanosomes to elucidate the mode of action of a small-molecule inhibitor in *Trypanosoma brucei*.

Key words Untargeted metabolomics, Global metabolomics, Liquid chromatography, Drift time, Collision cross-section

1 Introduction

Metabolites are low-molecular-weight (often suggested to be under 1500 Da) intermediates or end products of cellular metabolism. They represent a diverse array of small molecules crucial for life, including sugars, amino acids, lipids, organic acids, and various other compounds. Metabolites form and result from intricate biochemical pathways that operate within biological systems. Untargeted or global metabolomics is the comprehensive study of these small-molecule metabolites in complex samples, including biological systems [1, 2]. As a bioanalytical technology, metabolomics aims to detect, identify, and quantify metabolites encompassed in a system (or sample) and to provide a direct snapshot of an organism's physiological state at a specific point in time (or time of sampling). Metabolites are closer to the observable characteristics (or phenotype) than genes and proteins. Metabolomics, therefore, offers a powerful systems-level understanding of biological function, reflecting the integrated effects of genes, proteins, and

Paul A. M. Michels et al. (eds.), *Euglenozoa: Methods and Protocols, Volume 1*, Methods in Molecular Biology, vol. 3013, https://doi.org/10.1007/978-1-0716-5142-1_16,

the environment [3, 4]. Metabolomics complements genomics and proteomics by helping understand interactions between different metabolic processes and their overall impact on an organism [5, 6]. Metabolomics analysis offers a range of applications by being an invaluable tool for understanding biological systems and addressing real-world challenges [7, 8]. Key applications include (1) enhanced understanding of disease states, disease development and progression, as metabolite profiles are significantly altered in disease; (2) disease diagnostics through biomarker discovery; (3) drug discovery and development, as metabolic profiles can assist drug target identification, drug efficacy and also identify potential drug toxicity; (4) understanding microbial metabolism, particularly in the context of host-microbe interactions; and (5) forensics analysis within the context of drug and toxin identification in biological samples.

Ion mobility (IM)-enhanced metabolomics significantly advances the comprehensive analysis of small molecules in biological systems. By integrating IM with mass spectrometry (MS), IM-MS adds a crucial gas-phase separation dimension, based on ion size, shape, and charge [9]. This enhancement significantly improves the resolution and coverage of complex metabolomics samples, allowing for the disentanglement of isobaric and isomeric metabolites, which are often indistinguishable by conventional liquid chromatography (LC) coupled to MS techniques. IM-enhanced metabolomics, therefore, offers deeper insights into metabolic pathways and enhances the overall understanding of biological processes. Recent advancements in instrumentation offer the orthogonal coupling of IM spectrometry with high-resolution, accurate mass quadrupole time-of-flight (qTOF) MS for the metabolomics analysis of complex biological mixtures. By harvesting the power of both chromatographic and gas-phase separation, we developed a rapid IM-qTOF based metabolomics method for the comprehensive analysis of diverse biological systems [10]. This 3.5-min analytical method significantly reduces analytical time, increases throughput, and has been used to probe metabolomics in trypanosomes [11] and other systems [12–14].

2 Materials

Prepare all solutions using MS-grade solvents and analytical-grade reagents. Prepare and store all reagents at 4 °C. Diligently follow local waste disposal regulations when disposing of solvent and reagent waste.

2.1 Resources

1. High-performance liquid chromatography (HPLC)-grade ammonia solution specific gravity (SG) 0.88 (35%).
2. MS-grade formic acid 98–100%.
3. Ammonium acetate for LC-MS.
4. Ammonium formate for LC-MS.
5. ES-TOF reference mass solution kit (G1969-85001).
6. ESI-L low concentration tuning mix (G1969-85000).
7. Duran glass bottles (do not wash with soap).
8. Microcentrifuge tubes.
9. Micropipettes.
10. Glass measuring cylinders (do not wash with soap).
11. Glass LC vials or plastic LC vials with a glass insert and compatible caps.
12. Gloves.
13. MS-grade chloroform.
14. MS-grade methanol.
15. MS-grade water.
16. MS-grade acetonitrile.
17. MS-grade isopropanol.
18. Vortex.
19. Rocking incubator (in 4 °C cold room).
20. Temperature-controlled centrifuge with microcentrifuge tube rotor (set to 4 °C).
21. Samples for metabolite extraction (a minimum of five biological replicates per group).
22. InfinityLab Poroshell 120 HILIC-Z, 2.1 mm × 50 mm, 2.7 μm column (689775-924).
23. InfinityLab Poroshell 120 HILIC-Z, 3.0 mm, 2.7 μm UHPLC guard column (823750-948).
24. Agilent 1290 Infinity II series UHPLC system coupled to an Agilent 6560 IM-qTOF with a dual Agilent jet stream electrospray ionization (ESI) source, nebulizer magnifier (G1946-80049).
25. Agilent MassHunter software suite.

2.2 Extraction Buffer

Prepare 250 mL of extraction buffer by filling a clean 250-mL glass measuring cylinder with 150-mL MS-grade methanol. Then add 50 mL of MS-grade chloroform, followed by 50 mL of MS-grade water. Carefully transfer the extraction buffer into a clean 250-mL Duran glass bottle for storage at 4 °C for up to 6 months.

2.3 10-mM Ammonium Acetate (pH 9.0) Aqueous Buffer

Prepare 100-mM ammonium acetate stock solution by dissolving 3.85 g ammonium acetate powder in 450-mL MS-grade water. Adjust pH to 9.0 using ammonia (approximately 5 mL required) and make up the volume to 500 mL with MS-grade water. Store stock solution at 4 °C. Prepare 1000 mL of aqueous buffer by filling a clean 1000-mL glass measuring cylinder with 100 mL of 100 mM ammonium acetate stock solution. Make up the volume to 1000 mL with MS-grade water. Carefully transfer the aqueous buffer into a clean 1000-mL Duran glass bottle and adjust pH to 9.0 using ammonia, if required.

2.4 10-mM Ammonium Formate (pH 3.0) Aqueous Buffer

Prepare 100-mM ammonium formate stock solution by dissolving 3.15 g ammonium formate powder in 450-mL MS-grade water. Adjust pH to 3.0 using formic acid (approximately 5 mL required) and make up the volume to 500 mL with MS-grade water. Store stock solution at 4 °C. Prepare 1000 mL of aqueous buffer by filling a clean 1000-mL glass measuring cylinder with 100 mL of 100 mM ammonium formate stock solution. Make up the volume to 1000 mL with MS-grade water. Carefully transfer the aqueous buffer into a clean 1000-mL Duran glass bottle and adjust pH to 3.0 using formic acid, if required.

2.5 Tuning Mix

Prepare 100 mL of tuning mix by filling a clean 100-mL glass measuring cylinder with 10 mL of ESI-L low concentration tuning mix (G1969-85000). Make up volume to 100 mL with MS-grade acetonitrile.

2.6 Reference Mix

Prepare 1000 mL of reference mix by filling a clean 1000-mL glass measuring cylinder with 950 mL of MS-grade acetonitrile. Then add 450 μL of 2.5 mM HP-0921, 500 μL of 100 mM ammonium trifluoroacetate, and 1 mL of 5 mM purine (stock reference solutions are part of the ES-TOF reference mass solution kit G1969-85001). Make up volume to 1000 mL with MS-grade water. Carefully transfer the reference mix into a clean 1000-mL Duran glass bottle and store at 4 °C.

3 Methods

Carry out all procedures at room temperature unless otherwise specified. Automated metabolite extraction, quality control (QC) pool preparation, and sample aliquoting protocols using the Agilent Bravo liquid handling platform are described elsewhere [15].

3.1 Intracellular Metabolite Extraction

1. Harvest 40 × 10^6 cells per sample (*see* **Note 1**) by centrifuging cell cultures at 1250× *g* for 10 min to pellet cells. Remove supernatant, retaining 5 μL of the spent media for extracellular metabolite extraction (*see* Subheading 3.2).
2. Quench the cell pellet in 70% (v/v) ethanol/dry ice bath (*see* **Notes 2** and **3**).
3. Resuspend the cell pellet in 200-μL ice-cold extraction buffer. For particularly tight pellets, pipette the cell pellet and extraction buffer mixture to disperse the pellet thoroughly.
4. Place the samples on a rocker at 70 oscillations/min in a 4 °C cold room for 1 h.
5. Collect the intracellular metabolite extracts by centrifuging the samples at 4 °C for 1 min at 21,130× *g*.
6. Transfer 180 μL of the supernatant to fresh 1.5-mL microcentrifuge tubes and store in an ultralow temperature freezer until metabolomics analysis.

3.2 Extracellular Metabolite Extraction

1. To the 5-μL spent media retained in **step 1** of Subheading 3.1 (*see* **Notes 2** and **3**), add 195-μL extraction buffer.
2. Place the samples on a rocker at 70 oscillations/min in a 4 °C cold room for 1 h.
3. Collect the extracellular metabolite extracts by centrifuging the samples at 4 °C for 1 min at 21,130× *g*.
4. Transfer 180 μL of the supernatant to fresh 1.5-mL microcentrifuge tubes and store in an ultralow temperature freezer until metabolomics analysis.

3.3 Preparation of QC Pool and Aliquoting Samples for Metabolomics Analysis

1. Retrieve metabolite extracts from the ultralow temperature freezer and centrifuge at 4 °C for 5 min at 21,130× *g*.
2. Place metabolite extracts and LC vials on wet ice for the rest of the procedure.
3. Aliquot 50 μL of the metabolite extract into a glass LC vial or a plastic LC vial with a glass insert and cap the vial. Alternatively, LC plates can also be used, but ensure these are compatible with the extraction buffer.
4. To prepare QC pool: Transfer 10 μL of every metabolite extract in the experiment to a 1.5-mL microcentrifuge tube to create an experiment-specific representative metabolite extract. Vortex thoroughly to mix the extracts. Transfer 50 μL of the QC pool into an LC vial and cap the vial.
5. Aliquot 50 μL of the extraction buffer into an LC vial to be used as a solvent blank during the metabolomics analysis.

3.4 Agilent 6560 IM-qTOF Cleaning

Carry out all procedures with the instrument in standby mode when the LC solvent flow and MS gas flows are shut off (*see* **Note 4**).

1. Cleaning the electrospray nebulizer needle: Carefully disconnect the tubes and remove the nebulizer from the spray chamber door. Place the nebulizer in a narrow-necked 50-mL Duran glass bottle containing 40-mL MS-grade isopropanol, such that the needle tip of the nebulizer is submerged in the solvent. Sonicate for 10 min, remove from the solvent, and place on lint-free tissue to air dry. Using a magnifier, ensure that the needle is extended 1–2 mm beyond the tip of the nebulizer. Reinstall the cleaned nebulizer in the spray chamber (*see* **Note 5**).
2. Cleaning the spray shield and capillary cap: Carefully unscrew the spray shield and remove the capillary cap (*see* **Note 6**). Place them in a 100-mL Duran glass bottle containing 80-mL MS-grade isopropanol. Sonicate for 10 min, remove the parts from the solvent, and place on lint-free tissue to air dry. Reinstall the cleaned parts in the spray chamber.
3. Cleaning the spray chamber: Use a solvent-resistant spray bottle filled with MS-grade isopropanol to rinse the inside of the spray chamber. Thoroughly wipe the inner surfaces with lint-free tissue to ensure spray residue from previous experiments is removed. Reinstall the spray chamber.

3.5 Agilent 6560 IM-qTOF Mass Calibration Check

1. Tune the mass spectrometer using the tuning mix and parameter settings in Table 1 (*see* **Note 7**).
2. Before data acquisition in positive or negative ionization mode, check the mass calibration using the tuning mix and parameter settings in Table 1.
3. In positive ionization mode, ensure mass accuracy of *m/z* peaks 118.086, 322.048, 622.029, 922.010, 1221.991, 1521.971, and 1521.971 (*see* **Note 8**).
4. In negative ionization mode, ensure mass accuracy of *m/z* peaks 112.986, 301.998, 601.979, 734.007, 1033.988, 1333.969, and 1633.950 (*see* **Note 8**).

3.6 Agilent 6560 IM-qTOF Drift Tube Calibration Check

1. Before data acquisition in positive and negative ionization mode, check the drift tube calibration using the tuning mix and parameter settings in Table 1.
2. Directly infuse the tuning mix into the mass spectrometer to collect mass and drift spectra in 4-bit multiplexed mode for 1.5 min.
3. Demultiplex the resulting data file using PNNL PreProcessor (*see* **Note 9**).

Table 1
Settings used to tune the Agilent 6560 IM-qTOF (*see* Note 21)

Parameter	Positive mode	Negative mode
Drying gas temperature	225 °C	
Drying gas flow	13 L/min	
Nebulizer pressure	60 psi	
Sheath gas temperature	340 °C	
Sheath gas flow	12 L/min	
Capillary voltage	3000 V	
Nozzle voltage	200 V	
Fragmentor voltage	395 V	
Octopole voltage	750 V	
Acquisition range	50–1700 *m/z*	
MS acquisition rate	0.8 frames/s	
Drift tube: Pressure	3.940 Torr	
Gas type	N_2	
IM trap: Fill time	2500 μs	
IM trap: Release time	250 μs	
IM drift tube: Entrance voltage	1175 V	−1175 V
IM drift tube: Exit voltage	250 V	−250 V
IM drift tube: Temperature	31.5 °C	
IM front funnel: High-pressure funnel delta	150 V	−120 V
IM front funnel: High-pressure funnel RF	120 V	−100 V
IM front funnel: Trap funnel delta	180 V	−180 V
IM front funnel: Trap funnel exit	10 V	−10 V
IM front funnel: Trap funnel RF	180 V	−80 V
IM rear funnel: IM hex entrance	41 V	−41 V
IM rear funnel: IM hex delta	−8 V	8 V
IM rear funnel: IM hex RF	600 V	−600 V
IM rear funnel: Rear funnel entrance	240 V	−240 V
IM rear funnel: Rear funnel exit	43 V	−43 V
IM rear funnel: Rear funnel RF	130 V	−90 V
IM trap: Trap entrance	91 V	−91 V
IM trap: Trap entrance grid delta	10 V	−10 V

(continued)

Table 1
(continued)

Parameter	Positive mode	Negative mode
IM trap: Trap entrance grid low	96 V	−96 V
IM trap: Trap exit	90 V	−90 V
IM trap: Trap exit grid 1 delta	4 V	−5 V
IM trap: Trap exit grid 1 low	88.5 V	−86 V
IM trap: Trap exit grid 2 delta	8.5 V	−10.5 V
IM trap: Trap exit grid 2 low	87 V	−84.9 V

Table 2
Solvent gradient used for chromatographic separation

Time (min)	A (%)	B (%)	Flow rate (mL/min)	Max. pressure limit (bar)
0	7	93	0.800	600
1.80	20	80	0.800	600
2.00	30	70	0.800	600
2.30	30	70	0.800	600
2.35	7	93	0.800	600
3.50	7	93	0.800	600

4. Open the demultiplexed file in the Agilent MassHunter IM-MS Browser and extract drift times for the calibrant ions in the tuning mix. Ensure the residual values for each ion is under 0.1 ms (*see* **Note 10**).

3.7 Agilent 1290 Infinity II Series UHPLC Set up for Data Acquisition

1. Connect the InfinityLab Poroshell 120 HILIC-Z, 3.0 mm, 2.7 μm UHPLC guard column to the InfinityLab Poroshell 120 HILIC-Z, 2.1 mm × 50 mm, 2.7 μm UHPLC column and insert it in the column compartment (*see* **Note 11**). Set column to a constant temperature of 30 °C.
2. Connect solvent line A (aqueous buffer) to 10 mM ammonium formate (pH 3) or 10 mM ammonium acetate (pH 9) for data acquisition in positive or negative mode, respectively. Connect solvent line B (organic buffer) to acetonitrile (*see* **Note 12**). Set flow rate to 0.8 mL/min and solvent gradient to 93% solvent B, before turning on the solvent flow.
3. Chromatographic separation in both positive and negative ionization modes is obtained on 1 μL of the metabolite extract using the solvent gradient in Table 2.

4. Before data acquisition on experimental samples, the UHPLC column is conditioned by injecting the QC pool sample five times. After every five experimental samples, data is acquired on the QC pool sample to monitor instrument performance throughout data acquisition. Solvent blanks are obtained at the end of the experimental samples.
5. To convert drift time data to collision cross section values, data is acquired on the tuning mix during data acquisition. The tuning mix is directly infused into the mass spectrometer for 1.5 min to collect accurate mass and drift time data, at the start of the experiment and after every 50 injections (including samples and QC pools).

3.8 Agilent 6560 IM-qTOF Set up for Data Acquisition

1. Set the Dual AJS ESI source to operate with settings in Table 3.
2. Set the mass spectrometer to collect data in IM-QTOF multiplexing mode with a pulsing sequence length of 4-bit, IM trap fill time of 2500 μs, IM trap release time of 250 μs and acquisition rate of 0.8 frames/s.
3. The data collection mass range is set to 50–1700, with reference mass correction allowed by injecting the reference mix continuously into each sample during data acquisition.

3.9 Assessing Data Quality

1. Check the acquired QC pool raw data files using the Agilent MassHunter Qualitative Analysis 10.0 software. Assess system performance, reproducibility and data integrity over time by analyzing trends in the QC samples.

Table 3
Dual AJS ESI source settings (*see* Note 21)

Parameter	Value
Gas temperature	225 °C
Drying gas	13 L/min
Nebulizer	60 psi
Sheath gas temperature	340 °C
Sheath gas flow	12 L/min
VCap	3000 V
Nozzle voltage	200 V
Fragmentor	395 V
Oct 1 RF Vpp	750 V

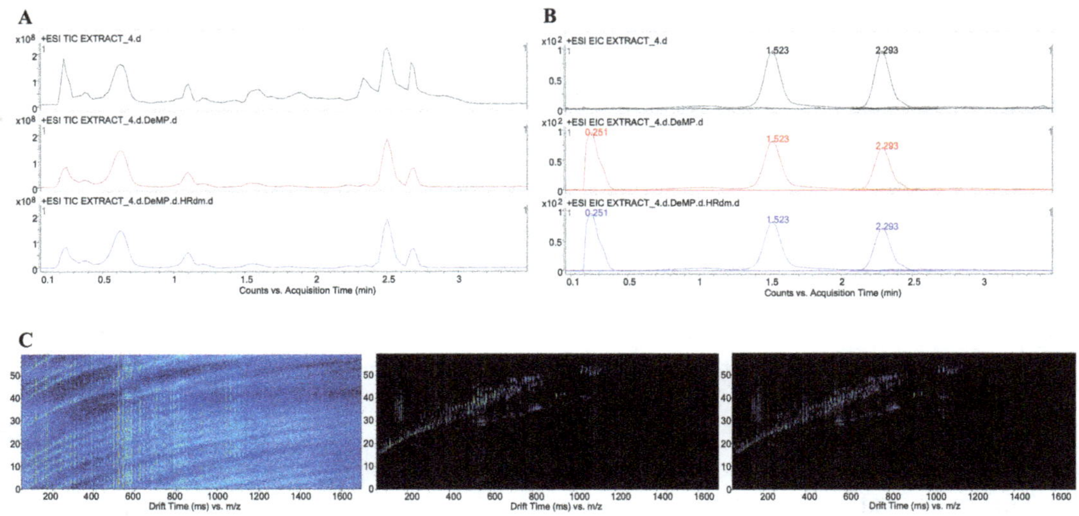

Fig. 1 Chromatograms and mobilograms of a representative trypanosome untargeted metabolomics sample. (**a**) Total ion chromatogram of data acquired in positive ionization mode. The raw (black), demultiplexed (red), and high-resolution demultiplexed (blue) data files are shown here. (**b**) Extracted ion chromatogram (EIC) for target analytes (1) palmitoleic acid (16:1) $[M + Na]^+$ *m/z* 227.2167, 0.251 min; (2) betaine $[M + H]^+$ *m/z* 118.0863, 1.523 min, and (3) L-carnitine $[M + H]^+$ *m/z* 162.1125, 2.293 min. Data shown in raw (black), demultiplexed (red) and high-resolution demultiplexed (blue) data files. Y-axis is set to % of each EIC peak and overlaid to represent multiple EICs on a single plot. (**c**) Mobilograms for retention time frame 1.5 to 1.6 min in the 4-bit multiplexed raw (left), demultiplexed (centre), and high-resolution demultiplexed (right) files

2. Check chromatographic peak shape for symmetry, sharpness, and absence of fronting and tailing (Fig. 1a), which can indicate column issues or poor separation.
3. Note the signal intensities of target analytes (Fig. 1b), to identify any significant drops or fluctuations that might suggest instrument problems or sample degradation.
4. Monitor retention time stability for analytes across the run to detect any drift (Fig. 1b), which can indicate chromatographic inconsistencies or system instability.
5. Assess the baseline for noise levels (Fig. 1a), drift, and presence of spurious peaks. A clean background ascertains accurate peak integration during data processing.

3.10 Processing Raw Data Files

1. The raw data files (.d format) acquired by the Agilent MassHunter Data Acquisition 10.0 software contain 4-bit multiplexed data. Use the PNNL PreProcessor (*see* **Note 9**) to demultiplex these files. This creates a file (.d.DeMP.d format) per sample.
2. Recalibrate the demultiplexed files for accurate mass using the Agilent MassHunter AgtTofReprocessUi software. Use the reference masses *m/z* 121.050873 and 922.009798 in positive ionization mode and *m/z* 112.985587 and 1033.988109 in negative ionization mode.

3. Use the Agilent MassHunter IM-MS Browser to extract beta and TFix values from the demultiplexed tuning mix data file (*see* **Note 13**). Apply the obtained beta and TFix values to demultiplexed experimental data files to calculate collision cross section values from drift time data (*see* **Note 13**).
4. Extract molecular features from the processed data files (.d. DeMP.d, one ionization mode at a time) using Agilent Mass Profiler software. Set retention time tolerance to ±0.3 min, drift time tolerance to ±1.5% and accurate mass tolerance to ± (5 ppm + 2 mDa). Export the extracted molecular features list as compound exchange format files. This creates a file (.d. DeMP.cef format) per sample.
5. Perform additional peak deconvolution using the High Resolution Demultiplexer (HRdm) [16] algorithm. Load the raw multiplexed data file (.d), the reconstructed demultiplexed data file (.d.DeMP.d) and the molecular features list file (.d.DeMP. cef) to the software and apply default settings (*see* **Note 14**). This creates a file (.d.DeMP.d.HRdm.d) per sample (Fig. 1c).
6. Re-extract molecular features from the processed data files (.d. DeMP.d.HRdm.d, one ionization mode at a time) using Agilent Mass Profiler software (*see* **Note 15**). Set retention time tolerance to ±0.3 min, drift time tolerance to ±1.5%, and accurate mass tolerance to ±(5 ppm + 2 mDa).
7. Annotate features using the ID Browser feature built into the Agilent Mass Profiler software by matching features against the accurate mass and CCS values in the McLean CCS Compendium PCDL [17]. Apply a 5 ppm window for m/z matches, a CCS value tolerance of ±1%, and search for the positive ion species $[M + H]^+$ and $[M + Na]^+$, and the negative ion species $[M - H]^-$. Export the annotated features from Agilent Mass Profiler software in an Excel-compatible format to perform further data analysis.

3.11 Statistical Analysis and Biological Interpretation Using MetaboAnalyst

1. Use the exported feature file to create a comma-separated (.csv format) file containing the names of samples, labels for sample groups, feature annotations, and peak intensities for each feature in each sample.
2. Open MetaboAnalyst (*see* **Note 16**) and begin analysis by clicking on click here to start. Choose statistical analysis (one factor) to proceed to the data upload page.
3. Select peak intensities for data type and sample in columns or rows (unpaired). Then upload the.csv file and submit the data.
4. A data integrity check window will open and summarize the information in the.csv file that was uploaded. This window will contain a checking data content passed message (*see* **Note 17**). Click proceed to continue data analysis.

5. Apply data filtering (*see* **Note 18**) and proceed. A reliability filter of 25% and a variance filter of 10% interquantile range are recommended for large datasets.
6. Perform data normalization (*see* **Note 19**) by selecting log transformation (base 10) and Pareto-scaling. Click normalize and visualize the results in the view result window before proceeding.
7. The univariate, advanced significance, multivariate, and cluster analysis tools available for the dataset are shown in the new window. Visualize overall sample relationships and clustering of the sample groups with heatmap, principal component analysis (PCA) and partial least squares—discriminant analysis (PLS-DA). The corresponding loading plots show the features contributing to the observed clustering in PCA and PLS-DA plots. Use variable importance in projection (VIP) scores from PLS-DA to determine the features that contribute most to the separation between the sample groups.
8. When the dataset has two sample groups, use the univariate tools for fold change, *t*-test, and volcano plot analyses to identify significantly perturbed features contributing to the differences between the groups. When the dataset has three or more sample groups, use analysis of variance (ANOVA) to identify significant features.
9. When testing multiple features, the chance of false positives increases. Apply Bonferroni correction or Benjamini–Hochberg correction for false discovery rate. A value of 0.05 is recommended.
10. Confirm the identity of statistically significant and biologically relevant features using Kyoto Encyclopedia of Genes and Genomes (KEGG) database (*see* **Note 20**).
11. For biological interpretation of the metabolomics data, map the significant metabolites to known metabolic pathways and identify significantly enriched or altered pathways between the experimental groups using the MetaboAnalyst pathway analysis and enrichment analysis tools.

4 Notes

1. The number of cells to be harvested is specific to the trypanosome being investigated. The number specified here is for the bloodstream form of *Trypanosoma brucei* and is provided as general guidance. For most trypanosomes, cell numbers of 1×10^6 or over per sample will provide efficient intracellular metabolome extraction.

2. If multiple samples are being harvested at the same time, leave the harvested samples (cell pellets or spent media) on dry ice until further processing.
3. If samples are being harvested on different days (for instance, longitudinal sampling), store the harvested samples (cell pellets or spent media) in an ultralow-temperature freezer until further processing.
4. The spray shield and related spray chamber components are likely to be hot. Care should be taken while cleaning the instrument.
5. Care should be taken not to hit the tip of the nebulizer against anything. Any damage will have a large negative effect on system performance.
6. If the retrieved parts are visibly dirty with charred residue, prepare a paste of aluminum oxide with MS-grade water and scrub the dirt from the parts using cotton swabs. Once all the dirt has been removed, clean the parts with MS-grade water before proceeding to sonicate in MS-grade isopropanol.
7. A separate instrument tuning should be performed in positive and negative ionization modes using the settings listed in Table 1.
8. To pass the mass calibration check, the corrected residuals value should be under 1 ppm. If these values are above 1 ppm, tune the instrument and check mass calibration again.
9. Use the following PNNL PreProcessor [18] settings to demultiplex the file. Interpolate each drift bin to two drift bins, apply moving average smoothing with the number of drift points set to 3, and apply spike removal and saturation repair above an abundance limit of 40%.
10. If the residual value is over 0.1 ms, tune the instrument and check drift tube calibration again.
11. When connecting the guard column to the running column and the column compartment, ensure all connections are secured and checked for leaks before acquiring data.
12. Make sure there are no trapped air bubbles in the solvent lines. Purge the lines with fresh solvent before setting the flow rate and letting the solvent flow through the column.
13. The beta and TFix values are obtained separately for the positive and negative ionization mode data from the tuning mix data files collected during data acquisition in each mode. The obtained values should be applied to positive and negative ionization data files appropriately.

14. The default settings apply (1) a moving average smoothing to reduce noise in the data, (2) saturation repair to address saturation artefacts, and (3) remove unwanted spikes in the data. In some cases, such as when the noise is higher, these settings will need to be modified to achieve higher resolution.
15. HRdm increases the total feature count as a consequence of improved resolution. Compare the feature count to the number of features obtained in **step 4** of Subheading 3.10 to ensure high-resolution demultiplexing and deconvolution were successful.
16. Access MetaboAnalyst at https://www.metaboanalyst.ca. The steps described here were written for version 6.0. There might be subtle changes to the workflow if the version changes, but basic concepts remain.
17. If the uploaded file was not in the right format or if there were issues reading the content of the file, these will be listed in a red box at the top of the page, and the data integrity window will not open.
18. As a starting point, it is recommended that no data filtering is applied to the dataset. Use an overview of the entire dataset to inform the right filtering parameters to be applied to the dataset.
19. Apply any sample-specific normalization here. Alternative data transformation and data scaling methods are available on this page. The specific methods recommended here work best with the untargeted metabolomics datasets generated on the specific instrumentation described here.
20. To convert common chemical names to KEGG compound identities, use the chemical translation service https://cts.fiehnlab.ucdavis.edu.
21. The settings listed in the table are specific for the Agilent 6560 IM-qTOF instrument (models 6560a and 6560c) with Dual AJS ESI source. These settings will require optimization on mass spectrometers of different makes and models.

References

1. Patti GJ, Yanes O, Siuzdak G (2012) Metabolomics: the apogee of the omic triology. Nat Rev Mol Cell Biol 13:263–269
2. Trivedi DK, Hollywood KA, Goodacre R (2017) Metabolomics for the masses: the future of metabolomics in a personalized world. New Horiz Transl Med 3:294–305
3. Fiehn O (2002) Metabolomics—the link between genotypes and phenotypes. Plant Mol Biol 48:155–171
4. Nicholson JK, Lindon JC, Holmes E (1999) "Metabonomics": understanding the metabolic responses of living systems to pathophysiological stimuli via multivariate statistical analysis of biological NMR spectroscopic data. Xenobiotica 29:1181–1189
5. Resurreccion EP, Fong KW (2022) The integration of metabolomics with other omics: insights into understanding prostate cancer. Meta 12(6):488

6. Akyol S, Ashrafi N, Yilmaz A et al (2023) Metabolomics: an emerging "omics" platform for systems biology and its implications for Huntington disease research. Meta 13:1203
7. Collino S, Martin FPJ, Rezzi S (2013) Clinical metabolomics paves the way towards future healthcare strategies. Br J Clin Pharmacol 75: 619–629
8. Ford L, Mitchell M, Wulff J et al (2022) Clinical metabolomics for inborn errors of metabolism, 1st edn. Elsevier Inc
9. Moses T, Burgess K (2023) Right in two: capabilities of ion mobility spectrometry for untargeted metabolomics. Front Mol Biosci 10:1–9
10. Pičmanová M, Moses T, Cortada-Garcia J et al (2022) Rapid HILIC-Z ion mobility mass spectrometry (RHIMMS) method for untargeted metabolomics of complex biological samples. Metabolomics 18:1–12
11. Nare Z, Moses T, Burgess K et al (2023) Metabolic insights into phosphofructokinase inhibition in bloodstream-form trypanosomes. Front Cell Infect Microbiol 13:1–9
12. Myllymäki H, Astorga Johansson J, Grados Porro E et al (2021) Metabolic alterations in preneoplastic development revealed by untargeted metabolomic analysis. Front Cell Dev Biol 9:1–15
13. Needham N, Campbell IH, Grossi H et al (2023) Pilot study of a ketogenic diet in bipolar disorder. BJPsych Open 9:1–8
14. Gibson SER, Frost I, Hierons SJ et al (2025) Bacteria encode post-mortem protein catabolism that enables altruistic nutrient recycling. Nat Commun 16:1–12
15. Massie F, MacKenzie A, Pandor S et al (2025) An automated metabolite extraction workflow for global metabolomics analysis using the Agilent Bravo liquid handling platform. bioRxiv.
16. Butler KE, Dodds JN, Flick T et al (2022) High-resolution demultiplexing (HRdm) ion mobility spectrometry-mass spectrometry for aspartic and isoaspartic acid determination and screening. Anal Chem 94:6191–6199
17. Picache JA, Rose BS, Balinski A et al (2019) Collision cross section compendium to annotate and predict multi-omic compound identities. Chem Sci 10:983–993
18. Bilbao A, Gibbons BC, Stow SM et al (2022) A preprocessing tool for enhanced ion mobility-mass spectrometry-based omics workflows. J Proteome Res 21:798–807

Chapter 17

Exploring Protein Interactomes Using TurboID-Directed Proximity Labeling and Mass Spectrometry

Elvio Rodríguez Araya, Gonzalo Martínez Peralta, Lucila Attala, Victoria Boselli, Azul de Hernández, Julia Pinheiro Chagas da Cunha, and Esteban Serra

Abstract

Proximity labeling (PL) with TurboID provides a powerful alternative to traditional protein interaction mapping methods, allowing the capture of both stable and transient interactions in living cells. Here, we describe a standardized and optimized protocol for generating high-confidence proximity proteomes in *Trypanosoma cruzi* using a novel vector system, pTcTurboID. This vector allows the stable and regulatable expression of TurboID fusion proteins, ensuring minimal bait expression to preserve physiological function while maintaining sufficient biotinylation activity for protein detection. The protocol comprises eight steps, including the generation of spatial control and bait lineages, optimization of biotinylation conditions, and refinement of the purification process with magnetic streptavidin beads. Key features include the use of compartment-specific spatial controls to minimize nonspecific background and statistical analysis to identify true interactors. Our methodology has been validated with nuclear and cytoplasmic baits, yielding reproducible proximity interactomes. This protocol provides an efficient, reproducible, and robust framework for proximity proteomics in *T. cruzi*.

Key words Proximity labeling, TurboID, *Trypanosoma cruzi*, Protein–protein interactions, Proteomics, Interactomics, Biotinylation, Mass spectrometry

1 Introduction

In the microenvironment of each subcellular compartment, it is unusual for proteins to act in isolation. In yeasts, it was estimated that each protein participates in an average of 16 interactions with other proteins, forming complex networks that are essential for various cellular processes [1]. These interactions, which can be stable or transient, are key to the formation of protein complexes with specific functions [2]. However, traditional methods for identifying these interactions, such as co-immunoprecipitation (Co-IP)

Paul A. M. Michels et al. (eds.), *Euglenozoa: Methods and Protocols, Volume 1*, Methods in Molecular Biology, vol. 3013, https://doi.org/10.1007/978-1-0716-5142-1_17,

followed by mass spectrometry (MS), have limitations. Co-IP relies on specific antibodies to recover a bait protein physically bound to its interactors, often resulting in missing transient interactions during the purification process due to weak or short-lived binding that dissociates during washing steps [2].

Unlike Co-IP, proximity labeling (PL) with nonspecific biotin ligases like TurboID allows capturing transient interactions by biotinylating proximal proteins in live cells without relying on physical stability of complexes [3]. This approach consists of fusing the enzyme to a bait protein, resulting in a localized biotinylation zone where activated biotin covalently binds to proximal molecules. Then, the highly stable streptavidin–biotin interaction enables stringent washing steps for efficient and specific recovery of biotinylated proteins. However, the success of PL depends on several factors, including the concentration of biotin and the expression level of the fusion protein. Excessive biotinylation can saturate proximal proteins, increasing background noise, while insufficient biotinylation may prevent the recovery of relevant proximal proteins [4]. Thus, achieving a balance in these parameters is critical for obtaining reliable proximity proteomes.

To address these challenges, we designed pTcTurboID, a pTcINDEX-derived vector for systematic generation of stable *T. cruzi* cell lineages expressing TurboID fusions in a regulatable manner [5, 6]. We optimized conditions for this system to achieve minimal expression levels that preserve the physiological function of the bait protein while ensuring sufficient biotinylation to detect proximal proteins. Additionally, the entire purification pipeline with magnetic streptavidin beads was refined to obtain high-quality proximity proteomes.

Our protocol consists of eight steps (Fig. 1) and includes the use of spatial control lineages to account for nonspecific biotinylation resulting from the diffusion of proteins before reaching their target complex or interactors [7, 8]. Spatial controls involve fusing an inert protein, such as GFP, to TurboID and targeting it to the same compartment as the bait protein (e.g., via a signal peptide). In step 1, both target protein and spatial control are cloned in frame to TurboID, generating separated plasmids (**step 1**: Cloning target proteins and spatial controls). These plasmids are then integrated into *T. cruzi* epimastigote's genome to generate stable spatial control and bait lineages (**step 2**: Epimastigotes transfection and direct cloning). Once the lineages are stabilized, their optimal expression is normalized, guided by western blot analysis (**step 3**: Normalize TurboID fusion expression and biotinylation levels by blotting), and the localization of the fusions is assessed by immunofluorescence (**step 4**: Verify localization of fusions by immunofluorescence). During the normalization process, different tetracycline

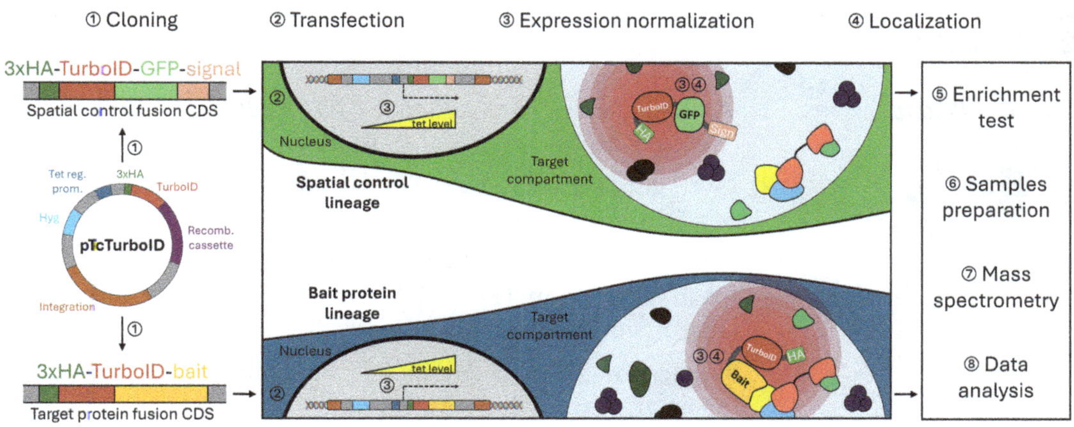

Fig. 1 Schematic representation of the methodology. Each major step is highlighted with a number inside a circle. *Left*: representation of the cloning process to generate the spatial control and bait construct with their corresponding CDS after replacing the recombination cassette of pTcTurboID. *Center*: representation of molecular events happening inside the spatial control (green) and bait protein (blue) expressing lineages. The nuclei contain the constructs inserted in the *T. cruzi* genome, with tetracycline levels regulating the expression of the fusions. The target compartment contains the spatial control fusion (up) and the bait protein fusion (bottom), each with their respective biotinylation radius. *Right*: final steps of the methodology applied on both lineages

levels are tested to identify the conditions that minimize as much as possible the expression of the fusions, while reaching similar biotinylation levels between lineages and enough signal to detect biotinylated proteins. Upon selecting the working conditions, a small-scale enrichment of biotinylated proteins is performed to compare conditions (step 5: Biotinylated proteins enrichment test). If the purification worked as expected, it is scaled up (step 6: Proteomic samples preparation) and replicates are further processed to perform liquid chromatography coupled to mass spectrometry (step 7: Mass spectrometry analysis). Finally, the obtained data are processed to identify proteins enriched during the purification, followed by statistical analysis to distinguish specific interactors from nonspecific background (step 8: Data analysis). Here, detected proteins are compared against the spatial controls to distinguish those that are truly proximal to the target, thereby providing a high-confidence proximity proteome.

Our methodology has been extensively validated using several nuclear and cytoplasmic baits alongside their corresponding spatial controls. These tests successfully yielded a nuclear proteome (by comparing cytoplasmic vs. nuclear spatial controls) and highly reproducible expression patterns and proximity proteomes for several bromodomain factors with known interactors (data to be published). With this tool, we can now generate high-confidence proximity proteomes efficiently and reproducibly using a standardized protocol.

2 Materials

2.1 Cell Culture

1. *T. cruzi* pLew13 (Addgene ID #24007) epimastigotes (we used Dm28c strain).
2. Liver infusion tryptose (LIT) medium: 5 g/L liver infusion, 5 g/L bacto-tryptose, 68 mM NaCl, 3 mM KCl, 22 mM Na_2HPO_4, 0.4% (w/v) glucose, pH 7.3.
3. Fetal Calf Serum (FCS) UV-irradiated and heat-inactivated (Internegocios, S.A., Argentina).
4. Supplemented LIT medium: LIT medium with 0.002% (w/v) hemin and 10% FCS.
5. Culture flasks (T25 and T75).
6. 24-well plates.
7. Biological safety cabinet (Class II).
8. Refrigerated incubator (temperature set at 28 °C).
9. Neubauer counting chamber.
10. Centrifuge.
11. Microcentrifuge.

2.2 Cloning

1. Entry vector pENTR™3C (Invitrogen, USA).
2. Destination vector pTcTurboID-GW (Addgene ID #234714).
3. Optional: pTcTurboID-GW_GFP (Addgene ID #234715) for restriction enzyme cloning.
4. LR Clonase™ II Enzyme Mix (Invitrogen, USA).
5. Plasmid DNA extraction kits for miniprep and midiprep.
6. Luria Broth (LB).
7. LB with agar.
8. *Escherichia coli* DH5α competent cells (stored at −80 °C).
9. Ampicillin stock solution (100 mg/mL).
10. Chloramphenicol stock solution (25 mg/mL).
11. Kanamycin stock solution (50 mg/mL).
12. TE Buffer: 10 mM Tris–HCl, 10 mM EDTA, pH 8.
13. Inner-TurboID Forward Primer: 5′-ATAAACAGGGAGCCC TGC-3′.
14. Reverse primer to detect the inserted gene of interest.
15. Incubator with agitation (temperature set at 37 °C).

2.3 Trypanosoma cruzi *Transfection*

1. *T. cruzi* pLew13 (Addgene ID #24007) epimastigotes (we used Dm28c strain).
2. Materials for cell culture (*see* Cell Culture materials).

3. Spatial control plasmid(s): pTcTurboID-GW_GFP (Addgene ID #234715), pTcTurboID-GW_GFP-NLS (Addgene ID #234716) and/or custom spatial control plasmid(s).
4. Bait plasmid(s): pTcTurboID construct(s) generated during the cloning step.
5. Phosphate buffer saline (PBS): 137 mM NaCl, 2.7 mM KCl,10 mM Na_2HPO_4, 2 mM KH_2PO_4, pH 7.4.
6. Electroporation buffer (TbBSF): 5 mM KCl, 0.15 mM $CaCl_2$, 90 mM Na_2HPO_4, 50 mM HEPES, pH 7.3.
7. Electroporation cuvettes (0.2 cm gap).
8. G418 disulfate stock solution (200 mg/mL).
9. Hygromycin B stock solution (200 mg/mL).
10. Nucleofector™ 2b (Lonza).
11. *Optional:* SpeI restriction enzyme.

2.4 TurboID Labeling Optimization

1. Biotin stock solution: 100 mM in DMSO (Gold Biotechnology, USA).
2. RIPA Lysis Buffer: 50 mM Tris–HCl, 0.1% (w/v) SDS, 0.5% (w/v) sodium deoxycholate, 1% (v/v) Triton X-100, 150 mM NaCl, pH 7.5.
3. HEPES-urea lysis buffer: 50 mM HEPES (4-(2-hydroxyethyl)-1-piperazineethanesulfonic acid), 8 M urea, pH 7.
4. Protease Inhibitor Mix (Cytiva, USA).
5. Phenylmethylsulphonyl fluoride (PMSF) stock solution.
6. 6× Protein Loading Buffer: 12.5% (v/v) β-mercaptoethanol, 45% (v/v) glycerol, 1% (w/v) SDS, 0.04% (w/v) bromophenol blue, 0.5 M Tris–HCl, pH 6.8.
7. SDS-PAGE gel and electrophoresis system.
8. SDS-PAGE protein transfer system.
9. Running buffer: 25 mM Tris–HCl, 190 mM glycine, 0.1% (w/v) SDS, pH 8.3.
10. Transfer buffer: 25 mM Tris–HCl, 190 mM glycine, 20% (v/v) methanol.
11. Nitrocellulose or polyvinylidene fluoride (PVDF) membrane.
12. PBS-T 0.1%: PBS with 0.1% (v/v) Tween 20.
13. Ponceau S: 0.5% Ponceau 4R, 1% glacial acetic acid, in water.
14. Blocking buffer: 5%(w/v) non-fat milk in PBS-T 0.1%.
15. Anti-HA primary antibody: rat Anti-HA monoclonal antibody (Roche, Switzerland).
16. Bovine Serum Albumin (BSA) stock solution: 3% BSA in PBS.

17. HRP-conjugated secondary antibodies: goat IgG anti-rat HRP-conjugated (Roche, Switzerland).
18. Streptavidin-HRP stock solution: 0.5 mg/mL in 50% glycerol (Jackson ImmunoResearch, USA).
19. Rocker or rotator.
20. Protein ladder.
21. Chemiluminescence with an ECL+ western blotting detection kit (Cytiva, USA).
22. Chemiluminescent detection and digital imager system Amersham™ Imager 600.

2.5 Localization Verification and TurboID Expression

1. Tetracycline stock solution: 50 mg/mL in ethanol 70% (Sigma-Aldrich, USA). Homogenize and prepare 1000× solutions in ethanol 70% (working solutions, one for each tested concentration). Store at −20 °C and protect from direct light due to its light sensitivity.
2. 4% (w/v) p-formaldehyde (PFA) in PBS.
3. 0.01% (w/v) Poly-L-Lysine Hydrobromide.
4. PBS with 0.2% (v/v) Triton X-100.
5. PBS with 0.01% (v/v) Tween 20 (PBS-T).
6. Anti-HA primary antibody: rat Anti-HA monoclonal antibody (Roche, Switzerland).
7. Fluorophore-conjugated secondary antibody: goat Anti-rat IgG (H + L), Alexa Fluor® 488 Conjugate (Cell Signaling Technology).
8. 4′,6-diamidino-2-phenylindole (DAPI) (Sigma-Aldrich).
9. Antifade Mounting Medium (Vectashield, Vector Laboratories).
10. Microscope slides with reaction wells and coverslips.
11. Fluorescence microscope.

2.6 Purification of Biotinylated Proteins

1. Streptavidin-conjugated magnetic beads (Thermo Scientific™ Pierce™, USA. Catalog No PI88817).
2. 1 M Potassium chloride (KCl) solution prepared in Milli-Q water.
3. 0.1 M Sodium carbonate (Na_2CO_3) solution prepared in Milli-Q water (made fresh).
4. 50 mM Tris–HCl, pH 7.5 prepared in Milli-Q water.
5. 2 M Urea, 10 mM Tris–HCl, pH 8 prepared in Milli-Q water (made fresh).
6. 2 M Urea, 50 mM Tris–HCl, pH 8 prepared in Milli-Q water (made fresh).
7. 1 M DTT stock solution.

8. Sequencing grade modified trypsin 5 × 20 μg (Promega, USA. Catalog No. V5111).
9. Cracking buffer: 3× Protein loading buffer, 2 mM Biotin, 20 mM DTT prepared in Milli-Q water.
10. On-bead digestion buffer: 0.2 μg/mL trypsin, 50 mM Tris–HCl pH 7.5, 2 M urea, 2 mM DTT, 1 mM $CaCl_2$, diluted with Milli-Q water (made fresh).
11. Magnetic Separation Rack.
12. Vertical rotating mixer.
13. Vortex.
14. Ultrasonicator Bioruptor® (Diagenode, Belgium).

2.7 SDS-PAGE Gel Staining with Silver Nitrate

1. Fixation solution: 12% (v/v) acetic acid, 50% (v/v) methanol, 0.0185% (v/v) formaldehyde (made fresh).
2. Wash solution: 30% (v/v) ethanol in Milli-Q water (made fresh).
3. 0.02% (w/v) sodium thiosulfate ($Na_2S_2O_3$) in Milli-Q water (made fresh).
4. Staining Solution: 0.2% (w/v) silver nitrate ($AgNO_3$), 0.027% (v/v) formaldehyde in Milli-Q water (made fresh).
5. Developing solution: 6% (w/v) sodium carbonate (Na_2CO_3), 0.0004% (w/v) sodium thiosulfate ($Na_2S_2O_3$), 0.0185% (v/v) formaldehyde in Milli-Q water (made fresh).
6. Stop solution: 50% (v/v) methanol, 12% (v/v) acetic acid in Milli-Q water (made fresh).
7. Conservation solution (optional): 25% (v/v) ethanol, 3%(v/v) glycerol in Milli-Q water (made fresh).

2.8 LC-MS Analysis

1. Iodoacetamide (IAA).
2. Formic Acid (FA).
3. Methanol.
4. ZipTip C18 (Millipore, USA).
5. Solvent 1: 0.1% (v/v) formic acid, 50% (v/v) acetonitrile.
6. Solvent 2: 0.1% (v/v) formic acid.
7. Solvent A: 0.1% (v/v) formic acid.
8. Solvent B: 0.1% (v/v) formic acid, 100% (v/v) acetonitrile.
9. Vacuum centrifuge.
10. Chromatographic nanoLC C18 column: EASY-Spray ES902 (25 cm × 75 μm ID, PepMap RSLC C18).
11. Nanoflow HPLC: nanoHPLC Ultimate3000 (Thermo Scientific, USA).
12. High-resolution mass spectrometer: Q-Exactive HF mass spectrometer (Thermo Scientific, USA).

2.9 MS Data Analysis

1. Latest proteome and annotation for *T. cruzi* used strain downloaded from TriTrypDB (we used Dm28c).
2. Sequences of common contaminant proteins such as proteases and human keratins.
3. Proteomic processing software: MaxQuant [9], freeware available from the Max Planck Institute of Biochemistry (download from https://www.maxquant.org and install).
4. Statistical software: Perseus [10], freeware available from the Max Planck Institute of Biochemistry (download from https://maxquant.net/perseus and install).

3 Methods

The numbers of the successive subheading in this section correspond with those of the steps depicted in Fig. 1.

3.1 Cloning Target Proteins and Spatial Controls

pTcTurboID is designed to work with the Gateway® recombination system [11] for easy CDS transferring from pENTR3c entry clones (*see* **Note 1**).

1. Clone your protein of interest (bait) into the entry vector pENTR3c.
2. Clone GFP or any other inert protein (*see* **Note 2**) into the entry vector pENTR3c in frame to a signal peptide targeting the fusion protein to the desired compartment (*see* **Note 3**).
3. Transfer the CDS from the resulting entry vectors to the destination vector pTcTurboID using the LR reaction (*see* **Note 4**).
4. Perform midipreps of the obtained destination vectors and store them at −20 °C.

3.2 Epimastigotes Transfection and Direct Cloning

1. Grow *T. cruzi* epimastigotes Dm28c pLew13 lineage (*see* **Note 5**) in LIT medium supplemented with 200 μg/mL of G418 and maintain them in exponential phase for 2 weeks ($10–30 \times 10^6$ epimastigotes/mL). Expand the culture to reach at least 40×10^6 epimastigotes for each transfection without exceeding 30×10^6 epimastigotes/mL.
2. Centrifuge the cells at 3000× *g* for 10 min. Wash the cells at room temperature with an equal volume of PBS as the culture volume. Centrifuge again and discard the supernatant.
3. Resuspend the cells in 375 μL of TbBSF buffer for every 40×10^6 epimastigotes. In parallel, for each plasmid, resuspend 60–80 μg of alcohol-precipitated circular pTcTurboID, or 15–30 μg of SpeI-linearized plasmid (*see* **Note 6**) in 25 μL of TbBSF buffer.

4. For each transfection, take 375 μL of the cells and mix them gently with the resuspended plasmid. Transfer them to an electroporation cuvette and keep them on ice for 10 min. For each transfection round, include a negative control using 25 μL of TbBSF buffer in replacement of DNA (*see* **Note 7**).
5. Electroporate the cells with a pulse of the program X-014 using the Nucleofector™ 2b [12] and immediately resuspend them in 12 mL of LIT medium without antibiotics inside a T-25 flask. Incubate at 28 °C.
6. After 24 h, add G418 (200 μg/mL) and hygromycin B (200 μg/mL). Homogenize the culture and distribute 500 μL per well in 24-well plates. Incubate the plates at 28 °C in a sealed container with a humidified chamber, regularly monitoring the cultures (*see* **Note 8**).
7. After 4–8 weeks, expand the positive wells by transferring them to fresh supplemented LIT medium with selection antibiotics (*see* **Note 9**).

3.3 Normalize TurboID Fusion Expression and Biotinylation Levels by Blotting

This section aims to determine the optimal concentration of tetracycline, that is, the minimum amount required to induce detectable biotinylation by the TurboID fusion. Different concentrations should be tested to identify the best condition. As a reference, a 75.2-kDa band corresponding to the endogenously biotinylated 3-methylcrotonyl-CoA carboxylase is used to normalize biotinylation levels across samples (Fig. 2).

1. For each transfection, select the clone that performs best according to **Note 9**. Include a negative control lineage that does not express any TurboID (*see* **Note 10**), along with the relevant spatial controls.
2. For each tetracycline concentration to be tested, harvest 15×10^6 log-phase epimastigotes by centrifugation at 3000× *g* for 10 min and resuspend the pellet in 1 mL of pre-warmed supplemented LIT (*see* **Note 11**).
3. Add tetracycline (0, 1, and 5 ng/mL) to induce TurboID fusion expression and initiate biotinylation (*see* **Note 12**). Homogenize and incubate for 24 h at 28 °C.
4. Stop the labeling reaction by placing the cells on ice. Transfer to microcentrifuge tubes and centrifuge at 3000× *g* for 10 min at 4 °C. Discard the supernatant.
5. Wash each pellet twice with 1 mL of ice-cold PBS (4 °C), then resuspend in 80 μL of HEPES-urea lysis buffer supplemented with protease inhibitors and 1× Protein Loading Buffer. Boil samples at 95 °C for 10 min.
6. Load 8 μL of each sample onto 12% SDS-PAGE gels and run the electrophoresis (*see* **Note 13**). Prepare two gels: one for

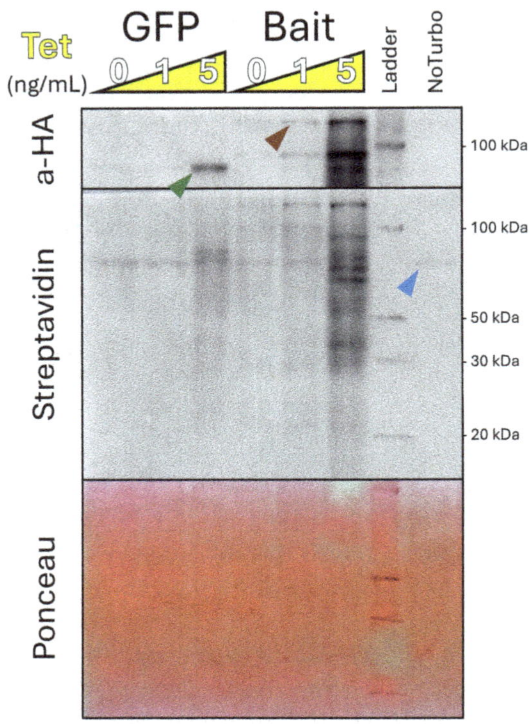

Fig. 2 Example of expression minimization and biotinylation normalization by blotting. Three tetracycline conditions (yellow triangles: 0, 1, and 5 ng/mL) were tested for a bait and a cytoplasmic spatial control (TurboID-GFP). A lineage that does not express TurboID was included (Dm28c pLew13 pTcINDEX-Empty). Parasites were incubated for 24 h at 28 °C in fresh supplemented LIT medium. The upper image shows the TurboID fusion signals (anti-HA), the middle one shows the signals of biotinylated proteins (Streptavidin-HRP), and the lower one shows the loading controls using Ponceau S staining. The green and brown triangles highlight the bands corresponding to the GFP and Bait TurboID fusions, respectively. The blue triangle point to the endogenously biotinylated 3-methylcrotonyl-CoA carboxylase (75.2 kDa)

detecting TurboID fusions using anti-HA antibodies and another for detecting biotinylated proteins using streptavidin (*see* **Note 14**).

7. Transfer proteins to nitrocellulose or PVDF membranes using an electrotransfer system. Stain membranes with Ponceau S to confirm transfer and loading consistency. Rinse with PBS-T.
8. Block membranes with 5% milk in PBS for 1 h at room temperature with rocking. Wash three times with PBS-T 0.1% for 10 min each (*see* **Note 15**).
9. Incubate one membrane with anti-HA primary antibody (0.1 μg/mL in PBS-T 0.1%) for 3 h at room temperature or overnight at 4 °C. Incubate the second membrane with streptavidin-HRP (0.3 μg/mL in PBS-T 0.1% + 3% BSA) for 1 h at room temperature.

10. Wash both membranes three times with PBS-T 0.1% (10 min per wash). The streptavidin-HRP is now ready for detection. For the anti-HA blot, incubate with secondary antibody (diluted in PBS-T 0.1% + 1% BSA) for 1 h at room temperature, then wash the membrane three times with PBS-T 0.1%.
11. Apply ECL substrate following the manufacturer's instructions and capture the signal using a digital imaging system (e.g., Amersham™ 600).
12. For each lineage, select the tetracycline concentration that yields biotinylation bands with intensities similar to the endogenously biotinylated carboxylase. Simultaneously, the selected concentrations must produce biotinylation intensities similar to those seen in other lineages. For example, in Fig. 2, the best combination of tetracycline concentrations will be of 5 ng/mL for the cytoplasmic control (TurboID-GFP) and 1 ng/mL for the bait (*see* **Note 16**).

3.4 Verify Localization of Fusions by Immunofluorescence

In this section, the lineages are tested to see whether TurboID fusion is expressed in the expected compartment using the normalized tetracycline concentration determined in Subheading 3.3.

1. Harvest 10×10^6 log-phase epimastigotes by centrifugation at 3000× *g* for 10 min and resuspend it in 1 mL of supplemented LIT medium in a microcentrifuge tube. Add the optimized tetracycline concentration to each lineage and incubate for 24 h at 28 °C.
2. Centrifuge the cells and wash twice with 1 mL of PBS (1500× *g* for 5 min each cycle). Resuspend each sample in 100 μL of 4% PFA solution. Pause point: The samples can be stored at 4 °C.
3. On microscope slides with reaction wells, place 10 μL of poly-L-lysine (0.1 μg/μL) to each well and incubate for 20–30 min at 37 °C. Remove any excess.
4. Place 10 μL of each sample in the wells and incubate for 15 min at room temperature. Remove any excess.
5. Permeabilize the cells by adding 10 μL of PBS-Triton X-100 0.2% to each well. Incubate for 10 min at room temperature.
6. Wash each well with PBS three times and then place 10 μL in each well of anti-HA primary antibody (1 μg/mL) diluted in PBS-BSA 1%. Incubate for 3 h at room temperature in a humidity chamber or overnight at 4 °C. Remove the liquid with a micropipette.
7. Wash each well three times with PBS-T 0.01% and then place 10 μL in each well of FITC-conjugated secondary antibody (1 μg/mL) and DAPI (2 μg/mL) diluted in PBS-BSA 1% (*see* **Note 17**). Incubate for 1 h at room temperature. Remove the liquid with a micropipette.

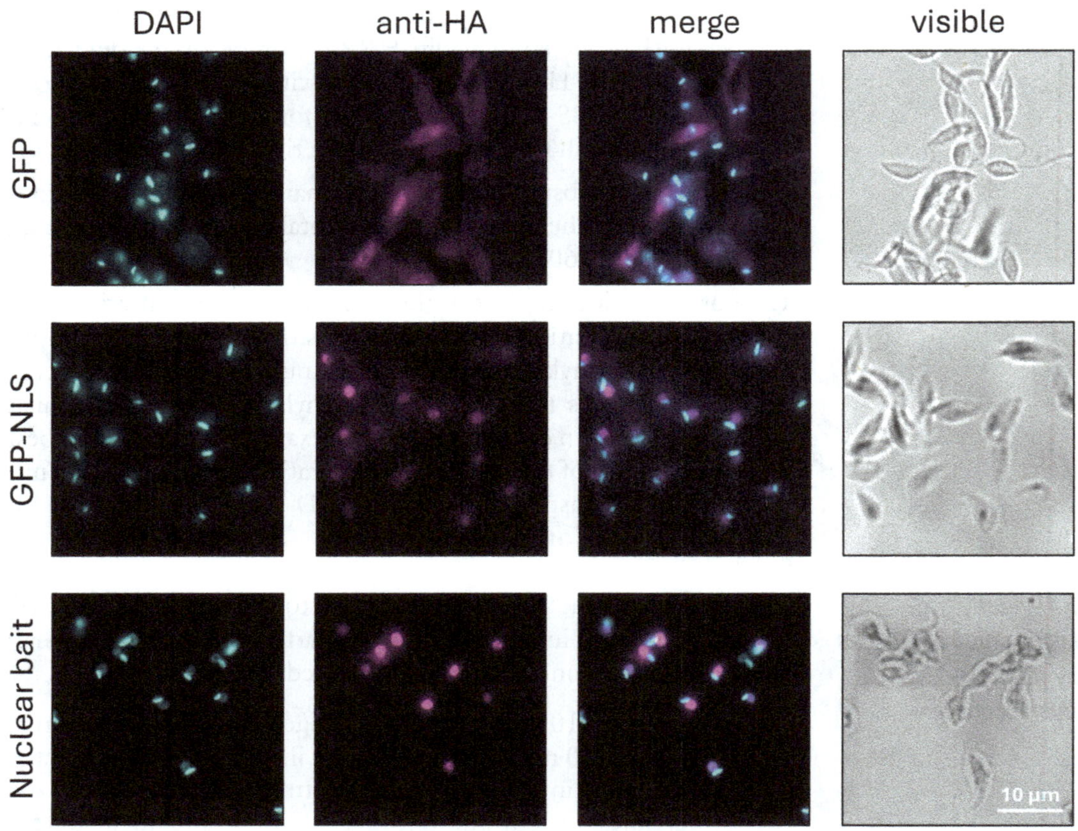

Fig. 3 Expected localizations of the cytoplasmic and nuclear spatial controls compared to a nuclear bait fused to TurboID determined by immunofluorescence. Subcellular localization of TurboID-GFP (cytoplasmic spatial control, top row), TurboID-GFP-NLS (nuclear spatial control, middle row) and a nuclear bait (bottom row). The image shows the DAPI signal (cyan), the anti-HA signal (magenta), the merged fluorescent signals, and the visible channel. Expression conditions were determined as described in **step 3** in Fig. 1

8. Wash each well twice with PBS-T 0.01% and mount with 2 μL of antifade mounting medium per well.
9. Acquire images using a fluorescence microscope with the appropriate filter settings. Compare the immunolocalization patterns of the HA marker with the DNA marker (DAPI) and the visible phase to determine the localization of the constructs. Figure 3 shows the expected localization of the TurboID-GFP (cytoplasmic) and TurboID-GFP-NLS (nuclear) spatial controls, compared with a nuclear bait.

3.5 Biotinylated Proteins Enrichment Test

This section aims to perform a small-scale enrichment of biotinylated proteins to verify the purification process works as expected before scaling up for proteomic analysis. The enrichment test serves as a quality control check to ensure that the TurboID fusion proteins are biotinylating proximal proteins that can be successfully captured using streptavidin beads under the optimized tetracycline induction conditions determined in Subheading 3.3. This validation ensures that when scaled up, the protocol will yield sufficient and specific biotinylated proteins for mass spectrometry analysis.

1. For each lineage, harvest 120 × 10^6 log-phase epimastigotes and resuspend them in a final volume of 8 mL of fresh supplemented LIT medium into a T-25 bottle (final concentration of 15 × 10^6 parasites/mL). Add the appropriate normalized concentration of tetracycline, homogenize and incubate for 24 h at 28 °C. Include a negative control lineage that does not express any TurboID.
2. Stop the biotinylation reaction by placing the flasks on ice and count the number of parasites using a Neubauer chamber to determine the parasite concentration. From now on, maintain everything cold.
3. Centrifuge 150 × 10^6 parasites at 3000× *g* for 5 min at 4 °C and discard the supernatant. Wash the pellet twice using an equal volume of ice-cold PBS as the initial medium. Resuspend the pellet in 750 μL of RIPA buffer with protease inhibitors. Homogenize with a micropipette and vortex and incubate on ice for 30 min.
4. Split the sample into two tubes, one with 500 μL (containing the equivalent to 100 × 10^6 parasites) and the other with 250 μL (50 × 10^6 parasites). The 250 μL will be used later as input.
5. Sonicate the samples in a Ultrasonicator Bioruptor® with two 10 min cycles (settings: 30″ ON, 30″ OFF, Mid power).
6. Centrifuge the samples at 20,000 × *g* for 10 min at 4 °C and recover the supernatant (*see* **Note 18**). Store the 250 μL supernatant (50 × 10^6 parasites) at −20 °C, labeled as "**INPUT (I)**."
7. In parallel to **step 6**, homogenize the streptavidin-conjugated magnetic beads (*see* **Note 19**) and dispense 25 μL per lineage into microcentrifuge tubes. Wash the beads twice with 1 mL of RIPA buffer at room temperature, using a magnetic rack.
8. Add the supernatant from the 500-μL samples (100 × 10^6 parasites) to the tubes with the washed beads. Homogenize by inversion for 1 min at room temperature and then incubate the sample on a vertical rotating mixer for 16 h at 4 °C.
9. From this point onward, work at room temperature. Place the tubes on the magnetic rack with the caps open and allow them to settle for 2 min to ensure complete separation from the solution. Carefully recover the liquid using a micropipette and transfer it to new tubes labeled as "**FLOWTHROUGH (FT)**." Store them at −20 °C. At this stage, the beads should have the biotinylated proteins attached (*see* **Note 20**).

10. Wash the beads sequentially to remove non-specific binders. After each wash, let the beads settle on the magnetic rack for 2 min with the cap open and carefully discarding the supernatant with a micropipette:
    (i) Two washes with 1 mL of RIPA buffer, homogenizing for 2 min each.
    (ii) One wash with 1 mL of 1 M KCl, homogenizing for 2 min.
    (iii) One wash with 1 mL of 0.1 M Na_2CO_3, homogenizing for 10 s (*see* **Note 21**).
    (iv) One wash with 1 mL of 2 M urea in 10 mM Tris–HCl pH 8.0, homogenizing for 10 s (*see* **Note 21**).
    (v) Two final washes with 1 mL of RIPA buffer, homogenizing for 2 min each. At the last washing step, transfer the beads to new tubes and discard the supernatant.
11. Add 30 μL of 3× cracking buffer with 2 mM biotin to the beads and incubate at 95 °C for 10 min to elute the biotinylated proteins.
12. Briefly spin down the sample (no more than 8 s), then place it on the magnetic rack for 2 min with the cap open. Recover the supernatant using a micropipette and transfer it to a new tube labeled as "**ELUTED (E)**." Store at −20 °C. Optionally, the remnant beads can also be resuspended in 30 μL of 3× cracking buffer with 2 mM biotin, labeling them as "**BEADS (B)**" to be included in the SDS-PAGE.
13. Process the E samples to visualize the whole protein enrichment pattern using a 12% SDS-PAGE stained with silver nitrate (Fig. 4a). For this, load 10 μL of each sample and run electrophoresis at 20 mA. Place the gel in 50 mL of fixation solution and heat for 30 s (microwave at max power). Agitate at room temperature for 5 min. Discard and wash with 50 mL of washing solution, heating for 30 s. Agitate at room temperature for 5 min. Discard and incubate with 50 mL of 0.02% (w/v) sodium thiosulfate, heating for 30 s. Agitate at room temperature for 2 min. Discard and wash with Milli-Q H_2O, heating for 30 s. Agitate at room temperature for 2 min. Repeat the washing step. Discard and wash with 30 mL of staining solution, heating for 30 s. Agitate at room temperature for 5 min. Wash with Milli-Q H_2O for 20–60 s. Incubate with the developer solution (enough to cover the gel) for 5–15 min at room temperature (*see* **Note 22**). Discard and stop the reaction with stop solution (enough to cover the gel) when the bands are sufficiently visible. If you want to preserve the gel, use the storage solution.

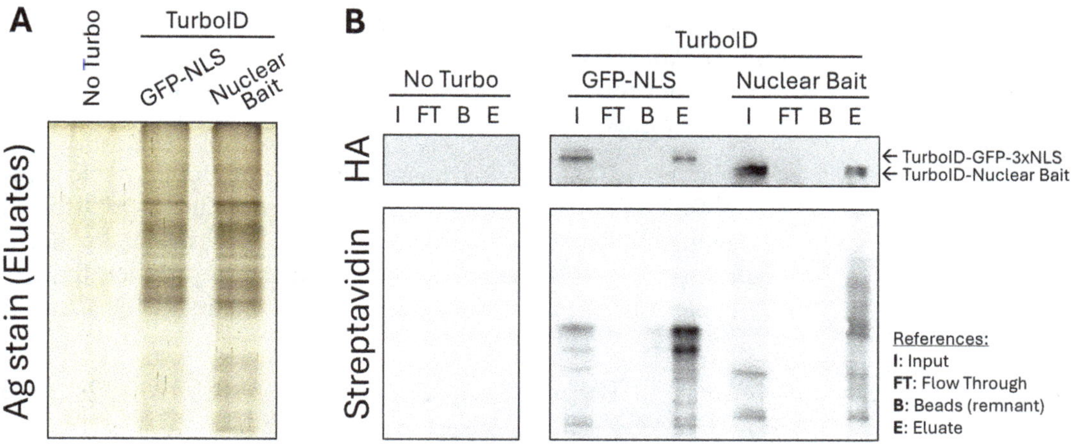

Fig. 4 Example of biotinylated proteins purification test. (**a**) Silver staining of the eluates from a purification test using the parental lineage Dm28c pLew13 as negative control (No Turbo), GFP-NLS as nuclear spatial control lineage and a nuclear bait TurboID fusion lineage. (**b**) Western blots of all fractions of the purification test example using streptavidin-HRP (bottom) and anti-HA (top). The position of the expected bands sizes in the anti-HA blot are highlighted with arrows and a reference for the labels is shown

14. In parallel, process the samples I, FT, B (optional), and E by western blotting as indicated in **steps 6–11** of Subheading 3.3. For this add protein loading buffer to I and FT samples, boil all samples at 98 °C for 10 min, and centrifuge at 12,000× *g* for 3 min. Load 15 μL of I and FT, and 5 μL of B (optional) and E samples (Fig. 4b).
15. Analyze the results of the gel and blots in parallel. An optimal result for TurboID expressing lineages using streptavidin-HRP should show biotinylated proteins present in the I fraction, their absence in the FT and B, and their successful recovery in the E, confirming an efficient purification process (Fig. 4b, *see* **Note 23**). Since the bait protein is also biotinylated, a similar pattern should be observed when probing for the TurboID fusion with anti-HA (*i.e.*, only present in I and E samples). Additionally, the silver staining of the eluates (Fig. 4a) offers complementary evidence of protein enrichment. Lineages expressing TurboID fusions should display enriched protein bands, while the no-TurboID control lineage (e.g., Dm28c pLew13) should show no enrichment (*see* **Note 24**).

3.6 Proteomic Samples Preparation

This section aims to scale up the biotinylation enrichment process for mass spectrometry analysis by preparing high-quality proteomic samples in quadruplicate. Building upon the optimized conditions validated in Subheading 5.5, this procedure involves processing a larger quantity of parasites to obtain sufficient amounts of biotinylated proteins for reliable proteomic identification. The protocol includes on-bead trypsin digestion of captured proteins, followed by peptide recovery and desalting to prepare samples suitable for mass spectrometry analysis. The protocol to follow is similar to the one previously validated. However, this purification requires a higher concentration of parasites (four times more) and is performed by quadruplicate.

1. Harvest 1200×10^6 log-phase epimastigotes for each lineage at $3000\times g$ for 10 min and resuspend the pellet in 80 mL of fresh supplemented LIT medium inside a T-75 flask (final concentration 15×10^6 parasites/mL). Add the normalized tetracycline concentration, homogenize, and incubate for 24 h at 28 °C.
2. Follow **steps 2–10** of Subheading 3.5, with the following modifications (*see* **Note 25**).
   (i) Step 3: Centrifuge 400×10^6 parasites and resuspend them in 500 μL of RIPA with protease inhibitors per replicate.
   (ii) Step 4: Do not split the sample (*see* **Note 26**).
   (iii) Step 7: Use 100 μL of beads per replicate instead of 25 μL.
   (iv) Step 9: Discard the flow-through (*see* **Note 26**).
3. After the last washing step, resuspend the beads in 1 mL of 50 mM Tris–HCl pH 7.5 and transfer them to new tubes. Homogenize for 2 min at room temperature, then leave on the magnetic rack for 2 min with the lid open. Discard the liquid.
4. Add 1 mL of 2 M urea, 50 mM Tris–HCl, pH 7.5, mix for 10 s at room temperature, and leave on the rack for 2 min with the lid open. Discard the liquid and repeat the wash.
5. For the elution, add 200 μL of on-bead digestion buffer (2 M urea Buffer, 50 mM Tris–HCl pH 7.5, 2 mM DTT, 1 mM $CaCl_2$, 0.2 μg/mL trypsin). Incubate overnight at 37 °C with orbital agitation at 1200 rpm.
6. Spin down the sample (8 s), place it on the magnetic rack for 2 min with the cap open. Recover the liquid and transfer it to a new tube. Place it on ice.
7. Resuspend the remaining beads with 100 μL of fresh 2 M urea, 50 mM Tris–HCl, pH 7.5, and 2 mM DTT and vortex for 1 min.
8. Spin down the sample (8 s) and place it on the magnetic rack for 2 min with the cap open. Recover the liquid and mix it with the first elution, obtaining a final volume of 300 μL. Stop point: you can store the samples at −80 °C.
9. Add IAA to a final concentration of 10 mM and incubate at 25 °C for 20 min.

10. Dry the sample with a vacuum centrifuge to a final volume of 100 μL.
11. Acidify the sample by adding FA such that the sample contains approximately 1% (v/v) and is at pH 3.
12. Desalt the sample with C18-ZipTips. Begin by conditioning the C18 ZipTips with 100 μL of methanol, followed by 100 μL of solvent 1, and then twice with 100 μL of solvent 2. Next, load the acidified peptides onto the conditioned ZipTips and wash them twice with 100 μL of solvent 2. Finally, elute the peptides from the ZipTips using 50 μL of solvent 1 and dry the samples completely using a vacuum centrifuge at room temperature. Stop point: store the samples at −80 °C until processing by mass spectrometry.

3.7 Mass Spectrometry Analysis

1. Resuspend each sample in 9 μL of solvent A (0.1% formic acid in water).
2. Inject 3 μL of each sample onto a nano column EASY-Spray ES902 (25 cm × 75 μm ID, PepMap RSLC C18). Adjust the mobile flow rate to 300 nL/min with solvent A and solvent B. Eluate with a gradient of 4–30% solvent B for 64 min, 30–80% solvent B for 7 min and 90% solvent B for 1 min.
3. To perform MS analysis, we used a Q-Exactive HF mass spectrometer (Thermo Scientific) in a data-dependent acquisition manner. For ionization, apply a liquid junction voltage of 1.9 kV and a capillary temperature of 300 °C. Employ a full scan method with a mass range of *m/z* 375–2000, an Orbitrap resolution of 120,000 (at m/z 200), an automatic gain control (AGC) target of 1e6, and a maximum injection time (IT) of 100 ms. Select the 10 most intense precursor ions for MS/MS fragmentation, performed with a normalized collision energy (NCE) of 27 eV. Acquire MS/MS scans with a dynamic first mass, an Orbitrap resolution of 30,000 (at m/z 200), an isolation window of 1.4 *m/z* units, an AGC target of 5e5, and a maximum IT of 55 ms. Set charge state screening enabled to exclude unassigned, singly charged, and equal or more than six protonated ions. Apply a dynamic exclusion time of 25 s to prevent repeated selection of previously analyzed ions.

3.8 Data Analysis

This section focuses on transforming raw mass spectrometry data into a high-confidence proximity proteome by distinguishing true interactors from nonspecific background through statistical comparisons with spatial controls. Key objectives include identifying proteins significantly enriched in bait samples relative to compartment-matched TurboID-GFP fusions, validating results via multidimensional visualization techniques (volcano plots, PCA, heatmaps), and integrating functional annotations to prioritize biologically relevant interactions. The final output provides a curated list of proximal proteins, filtered through significance thresholds (FDR <0.01, log2 difference ≥2), ensuring robust interpretation of interaction networks specific to the bait's subcellular niche.

1. *Prepare Input files*: Download the latest proteome FASTA file of *T. cruzi* strain used from TriTrypDB and their annotations (as mentioned, the examples were done with Dm28c) as TSV [13] (*see* **Note 27**). Include all relevant annotations that can help identify protein function (e.g., gene names, functional descriptions, GO terms, domains). Add the amino acid sequence of TurboID and any additional sequences (e.g., GFP) used in your experiments to the FASTA file (*see* **Note 28**). Save this curated FASTA file for database searching.
2. *Create a MaxQuant project*: Move the FASTA file, RAW files, and annotations into a project directory. Open MaxQuant. In the "Raw data" tab, load all RAW files from your experimental and control samples. Give experimental names to each file using the "Set experiment" button (e.g., GFP_1, GFP_2, ..., Bait_1, Bait_2).
3. *Configure and run MaxQuant*: Go to "Group-specific parameters" tab. Let "Type," "Misc.," "Cross links," "Instrument," and "First search" parameters as default. In "Modifications," set oxidation of methionine and protein N-terminal acetylation as variable modifications (*see* **Note 29**) and carbamidomethylation of cysteine as fixed modification. In "Label-free quantification," select LFQ mode and leave parameters as default. In "Digestion," select Trypsin/P as specific enzyme, with a maximum missed cleavage of 2. Change to "Global parameters" tab. In "Sequences," add the modified FASTA file as database for searching. Ensure that the option "Include contaminants" is selected. In "Identification," add the option "Match between runs." Let the rest of the parameters as default. Hit on "Start" to run MaxQuant.
4. *Create a Perseus project*: Locate the "proteinGroups.txt" file in the combined/txt folder generated by MaxQuant. Open Perseus and load the proteinGroups.txt file using "Generic matrix upload." Set *LFQ intensity* columns as "Main"; annotation columns like *protein IDs*, *protein names*, etc., as "Text"; numerical columns like *molecular weight*, *isoelectric point*, *protein length*, etc., as "Numerical"; and categorical columns like *Reverse*, *Potential contaminant*, etc., as "Categorical." Rename the main columns with "Rearrange>Rename columns" to match the experimental names (e.g., GFP_1, GFP_2, ..., Bait_1, Bait_2) and create a new categorical annotation row named "Condition" to distinguish between conditions with

"Annot. rows>Categorical annotation rows" (e.g., GFP and Bait).

5. *Filtering, transformation, and imputation*: Remove contaminants, reverse hits, proteins with less than two peptides identified and proteins only identified by site using "Filter rows" function. Transform LFQ intensities to Log_2 by going to "Basic>Transform." Select "Filter rows based on valid values" and set criteria to require at least 2 valid values in at least one group (*see* **Note 30**). For missing values, use "Imputation>-Replace missing values from normal distribution" with default parameters (width = 0.3, down shift = 1.8).
6. *Data verification using histograms*: Go to "Analysis>Histogram" to create histograms of the Log_2-transformed LFQ intensities and color the bars with "Selection from imputation" to inspect the distribution of protein abundances and the impact of missing value imputation. In Fig. 5a, an example dataset is shown, with blue bars representing the original data distribution and red bars indicating imputed values added to the lowest intensity range. This placement reflects the assumption that missing values are likely due to low-abundance proteins falling below the detection limit. Histograms allow for identifying differences in data quality between samples and conditions (*see* **Note 31**).
7. *Data verification using PCA*: Go to "Analysis>Principal component analysis" to use Principal Component Analysis (PCA) to explore relationships between samples and conditions. In the example shown in Fig. 5b, samples of cytoplasmic fusions cluster more closely in the principal component 1 (GFP, Bait1, and Bait2) than nuclear proteins (GFP-3xNLS and Bait3), reflecting their different proteomic profiles. Such clustering patterns help validate experimental replicates and assess biological relevance. Deviations, such as a sample clustering far from its group, could indicate technical issues or even sample mislabeling and warrant further inspection.
8. *Data exploration using clustering and heatmaps*: Normalize the Log_2-transformed LFQ matrix using *Z*-score transformation by going to "Normalization>Z-score." With the new matrix selected, go to "Analysis>Hierarchical clustering" to identify clustering patterns at both sample (column-wise) and protein (row-wise) levels. In the example shown in Fig. 5c, samples coming from cytoplasmic fusions cluster together, while nuclear fusion form a separate group, reflecting the different patterns of enrichment of biotinylated proteins during the experiment. Heatmaps also highlight proteins specifically enriched in certain conditions, which can guide functional interpretations.

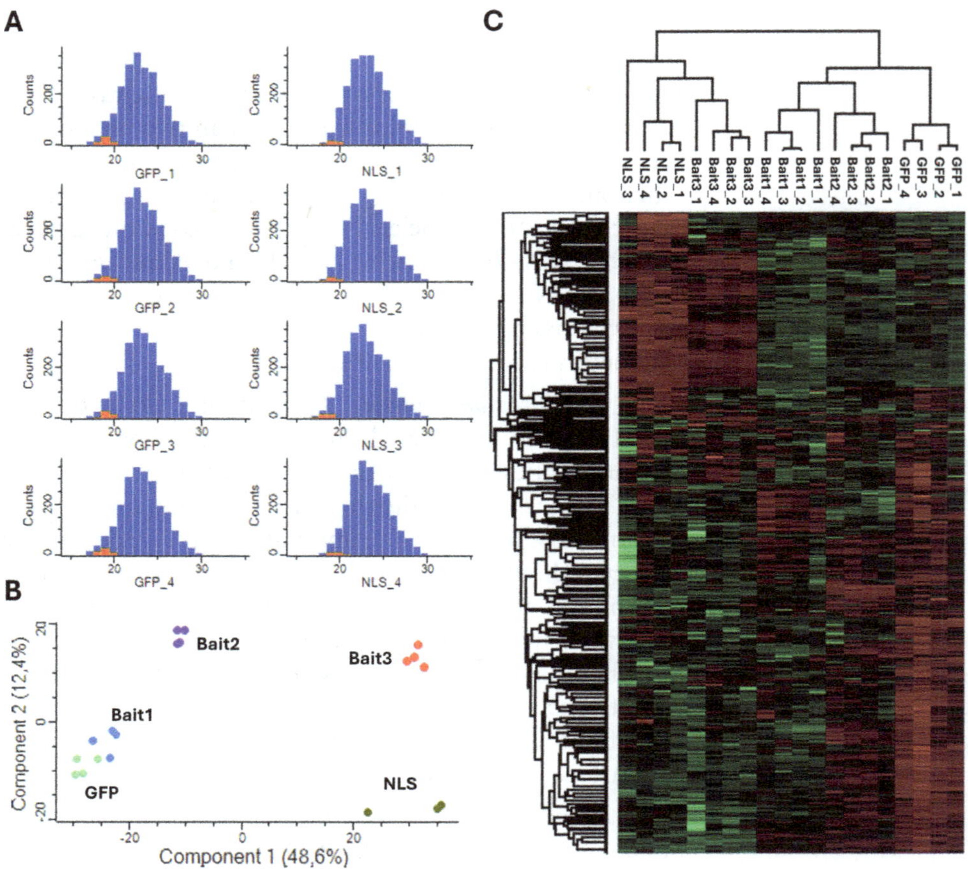

Fig. 5 Data verification and exploratory analysis. The experimental design of the example includes two spatial controls (Turbo-GFP and Turbo-GFP-3xNLS) and three baits (Bait1: cytoplasmic, Bait2: cytoplasmic/cytoskeleton and Bait3: nuclear). (**a**) Histograms of Log2-transformed LFQ intensities for GFP and GFP-3xNLS samples. Detected proteins are colored in blue and imputed in red. (**b**) PCA plot showing sample clustering based on protein abundance profiles. Each point represents an experimental replicate, colored by condition. The percentage of variance explained by each principal component is displayed along the axes, highlighting the separation of conditions based on their proteomic signatures and compartmentalization. (**c**) Heatmap and hierarchical clustering of *Z*-score normalized Log2-transformed LFQ intensities. Rows correspond to detected proteins, and columns represent experimental replicates, with clustering performed to reveal similarities among conditions. The color scale represents relative protein abundance, with red indicating higher abundance and green indicating lower abundance. Dendrograms illustrate the relatedness of replicates and proteins, enabling the identification of condition-specific enrichment patterns

9. *Statistical analysis*: Go back to the Log_2-transformed LFQ matrix and normalize the values by subtracting the median going to "Normalization>Subtract" (*see* **Note 32**). Then, perform a two-sample *t*-test by going to "Tests>Two-sample test," selecting the desired bait condition (right), the desired spatial control condition (left), an FDR of 0.01 and the $-Log_{10}$ *p*-value to be informed (*see* **Notes 33** and **34**). Repeat

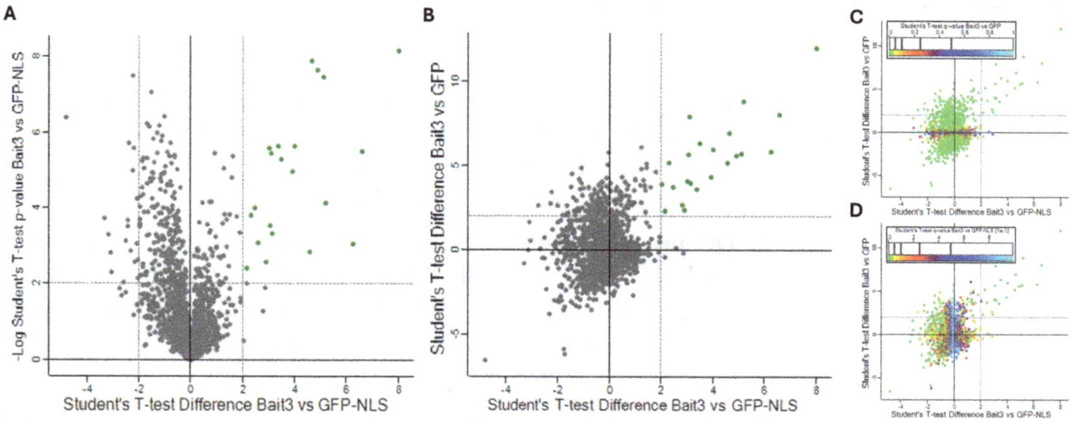

Fig. 6 Differential enrichment visualizations. Example visualizations for differential enrichment of a nuclear bait (Bait3). (**a**) Volcano plot displaying differential enrichment analysis. Each point represents a protein, with the x-axis showing Student's *t*-test Difference (Bait3 vs. Turbo-GFP-3xNLS) and the y-axis showing the −Log10 of Student's *T*-test *p*-value. Dashed lines mark the significance thresholds for enrichment differences (±2 on the x-axis) and FDR (y = −Log10[0.01] = 2). Proteins in green are those that surpass the thresholds. (**b**) Scatter plot comparing simultaneous differential enrichment against Turbo-GFP-3xNLS (x-axis) and Turbo-GFP (y-axis). This plot visualizes enrichment patterns across multiple spatial controls. (**c**) Same scatter plot as panel B, colored by *q*-values for the *t*-test comparison of Bait3 against Turbo-GFP. The color scale reflects statistical significance, enabling easier identification of proteins with strong differential enrichment. (**d**) Same scatter plot as panel B, colored by *q*-values for the *t*-test comparison of Bait3 against Turbo-GFP-3xNLS

this for each pairwise condition you want to explore to determine significant differences in protein abundance.

10. *Differential enrichment visualization*: To identify proteins enriched in the bait compared to the spatial control samples, generate a volcano plot by going to "Analysis>Scatter plot" and access the matrix data using columns. Plot Student's *T*-test Difference on the x-axis and the −Log Student's *T*-test *p*-value on the y-axis and add curves to mark the FDR threshold of 0.01 ($y = -\text{Log}_{10}[0.01] = 2$) and enrichment differences of ±2 (*see* **Note 35**). Enriched proteins will have values above both thresholds (*see* **Note 36**). Explore different annotations, such as GO terms, domains, protein names, and highlight them based on biological relationships (Fig. 6a). If the experiment includes multiple controls, visualize simultaneous differential enrichment by plotting Student's *T*-test Difference against one control on the x-axis and the other on the y-axis (Fig. 6b), coloring the points using the *q*-value (Fig. 6c, d). Such visualizations enable an intuitive assessment of patterns across conditions and help prioritize candidate proteins for follow-up studies.

4 Notes

1. Traditional cloning can be achieved by replacing the GFP cassette from pTcTurboID-GW_GFP with the desired CDS using BamHI and EcoRV restriction enzymes. Also, any CDS can be cloned in frame to GFP using only EcoRV.
2. Do not skip the inclusion of an inert protein in frame to TurboID for the spatial control. We saw that TurboID activity in isolation is significantly higher than when fused to an inert protein, which overestimates the non-specific biotinylation background, significantly affecting sensitivity.
3. We have designed and tested two plasmids for use as spatial controls for cytoplasmic (pTcTurboID-GW_GFP: Addgene ID #234715) and nuclear baits (pTcTurboID-GW_GFP-3xNLS: Addgene ID #234716). If the bait is known to be in either compartment, we recommend using these controls instead. For mitochondrial, glycosomal, or any other compartment, use appropriate signal peptides fused to the desired inert protein (e.g., GFP). Consider that some signal peptides need to be at the N-terminal end of the fusion to work properly [14]. For these cases, NotI and NaeI restriction sites are available upstream of the start codon to fuse the desired signal peptide or bait protein directly using pTcTurboID-GW_GFP.
4. We use 75 μg of each vector and 1 μL of the LR Clonase™ II mix in a final volume of 5 μL inside 0.2 mL PCR tubes (25 °C, overnight). Recombination efficiency can be increased by cutting the entry vector outside the gateway cassette with any of the following enzymes: PvuII, PstI, NruI, or NheI.
5. The *T. cruzi* Dm28c pLew13 lineage constitutively expresses the bacteriophage T7 RNA polymerase, the tetracycline repressor (TetR), and a neomycin resistance gene (conferring resistance to G418), all from a multicopy episomal vector (pLew13: Addgene plasmid #24007). This system enables inducible gene expression using an integrative vector (pTcTurboID in this case) containing a tetracycline-regulated T7 promoter to drive expression of the gene of interest. Prior to using this system, the parental strain must be transfected with pLew13, as described by Taylor et al. (2006) [5].
6. Chromosomal integration is more efficient if the plasmid is linearized. Make sure that the insert does not contain the recognition site for SpeI. We typically perform an overnight isopropanol precipitation with the desired amount of plasmid (previously linearized or not).
7. The negative control typically dies at the end of week 3.
8. To maintain humidity and minimize evaporation during incubation, place a paper towel moistened with sterile distilled

water at the bottom of the sealed container and position the 24-well plates above it. If the container is properly sealed, evaporation should be minimal. However, if necessary, an additional 500 μL of supplemented LIT medium with selection antibiotics can be added to the wells at any time.

9. Using this direct clonogenic assay, between 1 and 6 wells typically show active epimastigote growth after 4–8 weeks. However, some of these clonal populations may exhibit undesirable traits, such as leaky constitutive expression of the TurboID fusion protein. Therefore, the clonogenic selection step is critical. Clones can be screened by western blot by culturing them in the presence or absence of tetracycline (500 ng/mL) for 24 h, followed by detection using anti-HA antibodies. For each condition, we use 10×10^6 epimastigotes in 1 mL of supplemented LIT medium. Clones showing significant signal in the absence of tetracycline should be excluded, while a strong band should be detected in the induced condition. However, if the leak is minimal, a clone may still be suitable for use without tetracycline induction.
10. Suitable negative controls include the parental Dm28c pLew13 lineage or a lineage transfected with the empty pTcINDEX vector. These are used to estimate the baseline intensity of the 75.2-kDa endogenous carboxylase band. At least one lane of every SDS-PAGE gel should include this control to use as reference.
11. From this point on, use a single batch of supplemented LIT medium (e.g., 2 L) throughout the protocol to avoid variability once normalized. Also, there is no need to add biotin, since the FCS already contains sufficient levels. Due to this, before starting every proximity labeling experiment, the centrifugation and resuspension of the parasites in fresh medium ensures the same amount of starting free biotin. However, if your construct fails to produce enough biotinylation, you can try adding 50–500 μM biotin for 10–15 min prior the harvesting step.
12. We have observed that most constructs begin to show induction at tetracycline concentrations between 1 and 5 ng/mL. The goal is to determine the minimum concentration that yields detectable activity. In rare cases, some constructs or clones may require slightly higher concentrations (e.g., 5–20 ng/mL). In some cases, a small basal expression may result in detectable biotinylation, enough to reach the desired minimal expression (0 ng/mL).
13. Always include all relevant controls—negative (no TurboID) and spatial controls—in each gel for normalization and comparison purposes.
14. If you plan to run only one gel, start with the anti-HA blot and perform a mild stripping step before probing with

streptavidin-HRP. Do not start with streptavidin-HRP, as the biotin–streptavidin interaction is too strong and may interfere with subsequent antibody binding.

15. Before incubating with streptavidin-HRP, make sure all milk-based blocking reagent has been thoroughly removed. Milk contains free biotin, which can reduce signal intensity by competing for streptavidin binding.
16. In the example shown in Fig. 2, 0 and 1 ng/mL of tetracycline were insufficient to induce TurboID-GFP in that clone. Conversely, 5 ng/mL was too high for the bait, while 1 ng/mL can be considered borderline. This assay can be repeated with additional tetracycline concentrations to pinpoint the optimal level. In practice, if the biotinylation levels between baits and controls are similar enough, and do not deviate markedly from the endogenous carboxylase band, that condition is acceptable.
17. From this step on, make sure to work in darkness in order to avoid quenching the fluorescent dyes.
18. Here we are working with soluble proteins, so the supernatant is what contains the biotinylated proteins of interest and must be carefully recovered without disturbing the pellet. If you are planning to work with membrane proteins, consider using a stronger lysis buffer (e.g., containing higher SDS concentration or additional detergents) and potentially applying harsher sonication settings to ensure efficient solubilization. In such cases, it may also be useful to include an additional clarification step to remove insoluble debris before proceeding with the enrichment.
19. If the beads show significant signs of agglutination, try dispersing them by placing the magnetic bead flask stock in a vertical rotating mixer at 4 °C overnight.
20. Avoid leaving the beads without liquid for extended periods.
21. Try not to delay more than 10 s in these two steps, otherwise, the beads will clump together.
22. Follow the incubation time rigorously to avoid overdevelopment or underdevelopment of the silver stain. Excessive incubation can lead to high background noise, while insufficient incubation may result in weak or undetectable protein bands. If no bands are visible after 20 min, try by adding more formaldehyde or heating by 10 s the solution.
23. Depending on the exposure time, sometimes you will also see the 75.2 kDa band of the 3-methylcrotonyl-CoA carboxylase in the input sample and very dimly in the eluate. In the example of Fig. 6b, it has not appeared. On the other hand, for some baits, we did not observe as marked a difference with the No Turbo lineage as seen in Fig. 6a, although the methodology still yielded highly specific proximity proteomes.

24. Some common issues are listed next. (1) Low (biotinylated) protein enrichment in the eluates. If the beads are working properly, it can be related to insufficient biotinylation conditions. Try increasing the tetracycline levels or by adding 50–500 μM biotin to the medium for 10–15 min before cooling down the cells on ice. (2) Significant signal in the flow-through. This indicates that the streptavidin beads are being saturated. It is possible that the biotinylation levels in the tested condition are too elevated. Try to reduce the tetracycline levels. If the clone has a high leaky expression and no tetracycline was added, try with another clone or transfect again. If you proceed under these high biotinylation conditions, the results might not be biologically meaningful. (3) Significant signal in beads samples. It is possible that the elution process has not worked properly. Try by adding twice as much biotin to the cracking buffer and extend the boiling time of the beads.
25. If you are processing many conditions (lineages) simultaneously, it is recommended to perform the purification protocol in blocks. Each block will consist of one replicate of each condition. For example, if you have a "No TurboID" control, two different spatial controls and 6 baits, process one replicate of each condition at a time. The purification protocol is reproducible enough that no statistical block effect was detected in two independent experiments performed by two different operators.
26. Here we skip all the fractionation steps. If you want, you can separate a small amount of input and keep the flow-through to validate the replicates using western blot and silver staining.
27. If you are using another parental strain, select yours instead. If your strain proteome is not available, choose one that belongs to the same Discrete Typing Unit (DTU) as yours.
28. For example, if you are working with a nuclear bait and pTcTurboID-GW_GFP-3xNLS vector as spatial control, add the following lines to the FASTA file:

```
>TurboID
KDNTVPLKLIALLANGEFHSGEQLGETLGMSRAAINKHIQTL
RDWGVDVFTVPGKGYSLPEPIPLLNAKQILGQLDGGSVAVLP
VVDSTNQYLLDRIGELKSGDACIAEYQQAGRGSRGRKWFSPF
GANLYLSMFWRLKRGPAAIGLGPVIGIVMAEALRKLGADKVR
VKWPNDLYLQDRKLAGILVELAGITGDAAQIVIGAGINVAMR
RVEESVVNQGWITLQEAGINLDRNTLAATLIRELRAALELFE
QEGLAPYLPRWEKLDNFINRPVKLIIGDKEIFGISRGIDKQG
ALLLEQDGVIKPWMGGEISLRSAEKEFSRADESIRVDPSITS
LYKKAGSLKEPIQST
>GFP
MSKGEELFTGVVPILVELDGDVNGHKFSVSGEGEGDATYGKL
```

```
TLKFICTTGKLPVPWPTLVTTFTYGVQCFSRYPDHMKRHDFF
KSAMPEGYVQERTIFFNDDGNYKTRAEVKFEGDTLVNRIELK
GIDFKEDGNILGHKLEYNYNSHNVYIMADKQKNGIKANFKTR
HNIEDGGVQLADHYQQNTPIGDGPVLLPDNHYLSTQSALSKD
PNEKRDHMVLLEFVTAAGITHGMDELYK.
>3xNLS
ADPKKKRKVDPKKKRKVDPKKKRKV
>3xHA_tag
MYPYDVPDYAGYPYDVPDYAGYPYDVPDYAASK
```

29. Add any other variable modification that you want to explore. Have in mind that every variable modification added will significantly increase the search space.
30. The filtering criteria for valid values can be adjusted based on your experimental design and stringency requirements. Requiring at least two valid values in at least one group is a moderate stringency approach. More stringent filtering (e.g., requiring valid values in all replicates) will reduce the number of proteins identified but increase confidence. Less stringent filtering will retain more proteins but may increase false positives. Consider your experimental goals when setting these parameters.
31. There are many ways to explore the quality and structure of proteomics data. The methods presented here—histograms, PCA, clustering analysis, and heatmaps—provide a broad overview of the dataset and its features. These visualizations help assess variability between samples, identify potential biases, and uncover biological patterns. Additional exploratory tools include scatter plots between samples, which are particularly useful for checking correlations and spotting outliers. For instance, one sample in a condition might differ significantly from the others, for example, showing almost no protein detection or requiring extensive imputation across the entire distribution. In such cases, it may be prudent to exclude the sample from further analysis to avoid introducing artifacts or biases into downstream results. Always tailor your choice of visualizations to the specific questions you aim to address and the characteristics of your data.
32. Normalizing by subtracting the median is a straightforward approach to adjust for any systemic bias in the dataset, such as variations in sample loading or technical artifacts. This method centers the data around zero, making it easier to detect relative changes in protein abundance across conditions.

Depending on the experimental design, other normalization methods like quantile normalization or total intensity normalization may also be used. These techniques ensure that technical variation is minimized and that the biological differences between conditions are emphasized. Always choose a normalization method based on your dataset characteristics and analysis goals.

33. Depending on the number of conditions and experimental design, you can perform different statistical tests. For single baits versus single control, use a two-sample *t*-test as in the example. For experiments with multiple baits, spatial controls, or temporal conditions, consider using statistical approaches that account for these complexities. One-way or two-way ANOVA, followed by post hoc testing (e.g., Tukey's test), can be applied to examine differences across multiple conditions. Similarly, mixed-effects models may be suitable for data with hierarchical structures, such as replicates nested within conditions. Always ensure the statistical model aligns with the experimental design to accurately interpret your results.
34. When interpreting FDR-adjusted *p*-values, it is important to balance statistical rigor with biological relevance. An FDR threshold of 0.01 is stringent, reducing the likelihood of false positives but possibly excluding biologically meaningful hits. For exploratory studies, consider also reviewing proteins with FDR thresholds of 0.05 ($-\text{Log}_{10} = 1.30$) or even 0.1 ($-\text{Log}_{10} = 1.00$) to identify trends or potential candidates for follow-up validation. Always integrate statistical findings with prior biological knowledge to prioritize meaningful results.
35. Volcano plots are powerful visualizations for identifying differentially enriched proteins and their biological significance. Thresholds for enrichment differences (e.g., ± 2 on the x-axis) and FDR ($-\text{Log10}[0.01] = 2$ on the y-axis) are commonly used to define significance. These thresholds serve as practical guides but should not be interpreted as absolute boundaries. Biologically relevant proteins may fall below these thresholds due to experimental noise, low abundance, or transient interactions that are harder to detect. Therefore, when interpreting volcano plots, consider the broader biological context and integrate findings with complementary data, such as functional annotations, interaction networks, or pathway enrichment analyses. This approach ensures that proteins near the threshold are not prematurely dismissed and that potentially important discoveries are not overlooked.
36. Proteins with higher difference values are often those in closer proximity to the bait or those that form more stable interactions, spending more time nearby. In contrast, transient interactions may be near the threshold of 2.

Acknowledgments

This work was funded by the United Nations University—Programme for Biotechnology in Latin America and the Caribbean, UNU-BIOLAC; the National Research Council, CONICET (PIP 2021-0848); the National Agency for Science, ANPCyT (PICT 2020-01704); and Universidad Nacional de Rosario (80020230200062UR). We want to thank VEuPathDB for providing access to essential bioinformatics resources and databases that facilitated our research. We want to thank Alejo Cantoia and Celina Bessone for technical assistance.

References

1. Michaelis AC, Brunner AD, Zwiebel M et al (2023) The social and structural architecture of the yeast protein interactome. Nature 624: 192–200. https://doi.org/10.1038/s41586-023-06739-5
2. Budayeva HG, Cristea IM (2014) A mass spectrometry view of stable and transient protein interactions. Adv Exp Med Biol 806:263–282. https://doi.org/10.1007/978-3-319-06068-2_11
3. Qin W, Cho KF, Cavanagh PE, Ting AY (2021) Deciphering molecular interactions by proximity labeling. Nat Methods 18:133–143. https://doi.org/10.1038/s41592-020-01010-5
4. May DG, Scott KL, Campos AR, Roux KJ (2020) Comparative application of BioID and TurboID for protein-proximity biotinylation. Cells 9. https://doi.org/10.3390/cells9051070
5. Taylor MC, Kelly JM (2006) pTcINDEX: a stable tetracycline-regulated expression vector for Trypanosoma cruzi. BMC Biotechnol 6:1–18. https://doi.org/10.1186/1472-6750-6-32
6. Alonso VL, Ritagliati C, Cribb P, Serra EC (2014) Construction of three new gateway® expression plasmids for Trypanosoma cruzi. Mem Inst Oswaldo Cruz 109:1081–1085. https://doi.org/10.1590/0074-0276140238
7. Geoghegan V, Mottram JC, Jones NG (2022) Tag thy neighbour: nanometre-scale insights into Kinetoplastid parasites with proximity dependent biotinylation. Front Cell Infect Microbiol 12:1–14. https://doi.org/10.3389/fcimb.2022.894213
8. Cho KF, Branon TC, Udeshi ND et al (2020) Proximity labeling in mammalian cells with TurboID and split-TurboID. Nat Protoc 15: 3971–3999. https://doi.org/10.1038/s41596-020-0399-0
9. Cox J, Mann M (2008) MaxQuant enables high peptide identification rates, individualized p.p.b.-range mass accuracies and proteome-wide protein quantification. Nat Biotechnol 26:1367–1372. https://doi.org/10.1038/nbt.1511
10. Tyanova S, Temu T, Sinitcyn P et al (2016) The Perseus computational platform for comprehensive analysis of (prote)omics data. Nat Methods 13:731–740. https://doi.org/10.1038/nmeth.3901
11. Reece-Hoyes JS, Walhout AJM (2018) Gateway recombinational cloning. Cold Spring Harb Protoc 2018:1–6. https://doi.org/10.1101/pdb.top094912
12. Pacheco-Lugo L, Díaz-Olmos Y, Sáenz-García J et al (2017) Effective gene delivery to Trypanosoma cruzi epimastigotes through nucleofection. Parasitol Int 66:236–239. https://doi.org/10.1016/j.parint.2017.01.019
13. Amos B, Aurrecoechea C, Barba M et al (2021) VEuPathDB: the eukaryotic pathogen, vector and host bioinformatics resource center:1–14. https://doi.org/10.1093/nar/gkab929
14. Mach J, Poliak P, Matušková A et al (2013) An advanced system of the mitochondrial processing peptidase and core protein family in Trypanosoma brucei and multiple origins of the Core I subunit in eukaryotes. Genome Biol Evol 5:860–875. https://doi.org/10.1093/gbe/evt056

Chapter 18

Mapping *Trypanosoma* Protein Interactome by Proximity-Dependent Biotinylation

Chethan K. Krishna, Ralf Erdmann, and Vishal C. Kalel

Abstract

Proximity labeling techniques, such as BioID and TurboID, have emerged as modern tools for studying protein–protein interactions (PPIs) in cells. TurboID is a highly efficient biotin ligase that biotinylates proximal proteins in living cells when fused to a bait protein. These biotinylated proteins are enriched using a pull-down assay with streptavidin beads and identified through proteomic analysis, revealing the interactome of the target protein. Here, we describe proximity labeling using TurboID fused to *Trypanosoma brucei* PEX19, an essential cytosolic protein and a receptor for newly synthesized glycosomal membrane proteins, to identify its interactome. Understanding these interactions can be crucial for identifying new drug targets to combat these parasites. This powerful approach for mapping interactomes can be applied broadly to other organisms, including plants, yeast, and mammalian systems.

Key words *Trypanosoma*, *Leishmania*, Protein–protein interaction, Proximity labeling, BioID, TurboID, Biotinylation, Proteomics

1 Introduction

Trypanosomatids, a subgroup of kinetoplastids, share the characteristic features of harboring a kinetoplast and a single flagellum [1, 2]. Glycosomes, peroxisome-related organelles that compartmentalize the majority of the glycolytic pathway, are found across kinetoplastids and diplonemids. However, their essential role in parasite survival has so far been demonstrated only in certain trypanosomatid species [3, 4]. Understanding the biogenesis of glycosomes or other organelles requires insight into their protein interactome, formed through protein–protein interactions (PPIs), which can be assessed by biochemical methods. Investigating PPIs is instrumental for deciphering protein function and, in turn, understanding their biological roles. However, detecting weak, transient, or indirect interactions remains a significant challenge in the field.

Paul A. M. Michels et al. (eds.), *Euglenozoa: Methods and Protocols, Volume 1*, Methods in Molecular Biology, vol. 3013, https://doi.org/10.1007/978-1-0716-5142-1_18,

Affinity purification (in vitro pull-down) is a widely used and well-established technique for mapping protein interactions. This approach involves the isolation of a target protein using an immobilized matrix that captures both the bait and its binding partners. Typically, this method employs either antibodies against the bait protein or tag-specific beads that bind to genetically fused tags on the bait protein, enabling the purification of protein complexes [5, 6]. However, a major drawback of this method is that the interactions are determined from the cell extract rather than the native cellular environment [7]. Additionally, this method is limited in its ability to detect transient or weak PPIs and is poorly suited for proteins that are low in abundance or highly hydrophobic, particularly membrane or membrane-associated proteins [8].

Proximity labeling has emerged as a modern tool for studying PPIs, effectively overcoming the limitations of conventional methods, such as affinity purification, by eliminating the need to retain interactions during cell lysis and enrichment [5]. In this approach, a catalytic enzyme is fused to a protein of interest (bait protein), which then covalently labels the proteins proximal to or interacting with the bait protein within live cells upon the addition of corresponding substrates [9]. Unlike affinity purification, which requires gentle lysis conditions, labeled proteins can be purified under harsh denaturing conditions, enabling effective enrichment of less abundant proteins that are often challenging to isolate using conventional approaches [9]. Furthermore, this approach allows the detection of weak, transient, or hydrophobic PPIs in their native cellular environment. Thus, proximity labeling followed by affinity purification has been widely used in recent years to study PPIs, and protein–nucleic acid interactions in various organisms, with subsequent identification through proteomic analysis [10]. There are two main classes of enzymes used for proximity-dependent biotinylation. One is ascorbate peroxidase (APEX, APEX2), which rapidly biotinylates proximal proteins within about 1 min. However, its use in live cells is limited by potential oxidative stress from the required hydrogen peroxide (H_2O_2) treatment and the poor membrane permeability of biotin-phenol. The other class comprises non-toxic promiscuous biotin ligases, with engineered enzyme variants such as BioID, TurboID, and miniTurbo, which are designed to improve labeling efficiency and require biotin as a substrate [5, 10–12]. These methods vary in their labeling radius, ranging from ~30 nm for APEX2, ~10 nm for BioID to ~4 nm for recently reported light-triggered labeling technique, MicroMap, which utilizes iridium-based photocatalysts (μMap) [13].

BioID, a promiscuous mutant of *Escherichia coli* biotin ligase (BirA), typically requires 18–24 h or longer for effective labeling. This extended time frame can lead to non-specific labeling and reflects the enzyme's low catalytic activity [12, 14, 15]. To address the limitations of APEX and BioID-based proximity labeling, two enhanced variants of BirA were developed to improve the efficacy of proximity labeling: TurboID, with 15 mutations, and miniTurbo, a smaller version of TurboID with the N-terminal domain deleted and harboring 13 mutations, both relative to the wildtype BirA [12]. However, miniTurbo exhibits lower stability than TurboID, likely due to the loss of its N-terminal domain [12, 16]. TurboID exhibits high activity that enables rapid biotinylation within 10 min at room temperature [12]. Comparative analyses have consistently shown that TurboID offers advantages over BioID for proximity labeling applications across diverse organisms. In recent years, TurboID has been the most active proximity labeling enzyme, making it the preferred choice when maximizing biotinylation yield and sensitivity is the priority [12].

This chapter provides a protocol with detailed information for the use of TurboID-based proximity labeling in *Trypanosoma brucei*. Here, we focus on identifying proximal and potentially interacting proteins of the protein PEX19, a cytosolic receptor for newly synthesized glycosomal membrane proteins, followed by the enrichment of biotinylated proteins using streptavidin beads. This method can be applied to investigate PPIs in trypanosomatids and other euglenozoans, as well as in other systems, such as yeast and mammalian cells.

2 Materials

2.1 Cell Culture

1. *T. brucei brucei* procyclic form (PCF) 29–13 strain, which co-expresses T7 RNA polymerase and tetracycline repressor [17]. This strain is derived from *T. b. brucei* Lister 427 and will hereafter be referred to as wildtype (WT).
2. SDM-79 medium [18], supplemented with 7.5μg/mL hemin, 10 mM D-glucose, 10% heat-inactivated fetal bovine serum, and 1% penicillin-streptomycin.

2.2 Reagents, Consumables

1. Chemicals: DMSO; Tetracycline, Biotin; Digitonin, High Purity (*see* **Note 1**).
2. Antibiotics: Blasticidin, phleomycin, G418, hygromycin B.
3. Phosphate-Buffered Saline (PBS): 137 mM NaCl, 2.7 mM KCl, 10 mM Na_2HPO_4, 1.8 mM KH_2PO_4, pH 7.4 (*see* **Note 2**).
4. Penicillin-Streptomycin (10,000 U/mL).
5. Human T Cell Nucleofector Kit (Lonza).
6. 3 M sodium acetate (pH 5.2).
7. Pierce™ Bradford Plus Protein Assay Reagent (Thermo Fisher Scientific).
8. Protease Inhibitor Cocktail (PIC): 25× PIC prepared using cOmplete™, EDTA-free Protease Inhibitor Cocktail: Roche (*see* **Note 3**).

9. Mini spin columns: Mobicol Spin Classic with screw cap, bottom plug, without filters (MoBiTec).
10. Filter (small) 35-μm pore size, for use with Mobicol mini spin columns (MoBiTec).
11. Pierce™ Streptavidin Agarose (Thermo Fisher Scientific).
12. 5× Laemmli Sample Buffer: 250 mM Tris–HCl, pH 6.8, 10% SDS (w/v), 50% glycerol (v/v), 25% 2-mercaptoethanol (v/v), 0.1% Bromophenol Blue (w/v).

2.3 Antibodies

1. Anti FLAG tag.
2. IRDye® 800CW Streptavidin (LI-COR).
3. *Tb*PEX15 [19], *Tb*PEX19 (Prof. Ralf Erdmann Lab).
4. *Tb*Enolase (provided by Prof. Paul Michels).
5. *Tb*MtHSP70 (provided by Prof. James D. Bangs Lab).

2.4 Equipment

Use the following equipment or equivalent models with similar specifications and performance for centrifugation, cell culturing, electrophoresis, blotting, and imaging:

1. Allegra® X-30R Series Benchtop Centrifuge, BioCertified Swinging-bucket Rotor, Model SX4400 (Beckman Coulter, Brea, USA).
2. Biofuge Fresco, Rotor PP 2/98 #3324 (Heraeus, Hanau, DE).
3. Centrifuge 5417 R, Rotor F45-30-11 (Eppendorf, Hamburg, DE).
4. Test Tube Rotator.
5. Biosafety laminar flow cabinet for sterile cell culture.
6. Lonza™ Nucleofector™ Transfection 2b Device.
7. Cell culture incubator.
8. SDS-PAGE electrophoresis system.
9. Immunoblot transfer system.
10. Power supply for SDS-PAGE and blot transfer systems.
11. Immunoblot imaging System: Odyssey® DLx (LI-COR Biosciences, Lincoln, USA).

3 Methods

3.1 Overview of the Method

The method provides a detailed workflow for TurboID-based proximity labeling in *T. brucei* (also *see* Fig. 1). The workflow comprises the following key steps:

1. Clone a construct of the gene of interest fused to TurboID (*see* Subheading 3.2). TurboID-*Tb*PEX19 is shown here as a representative example.

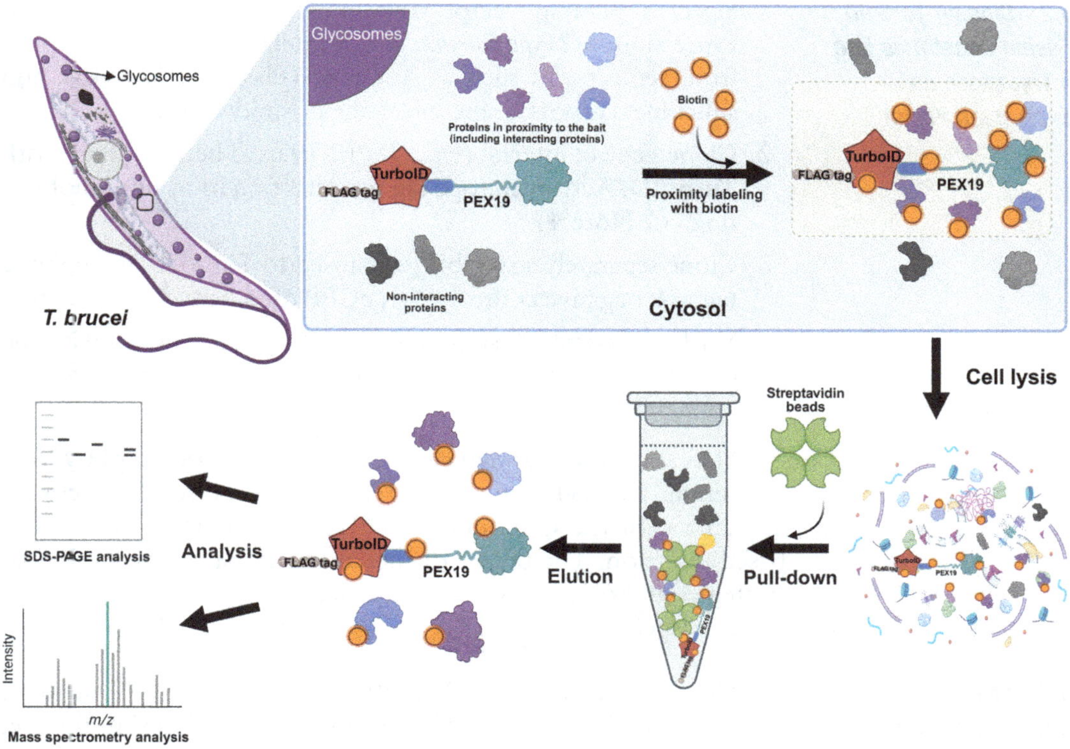

Fig. 1 TurboID-based proximity labeling to identify proximal and potentially interacting proteins of the PEX19 protein in *Trypanosoma brucei* (scheme). FLAG-tagged TurboID, a modified biotin ligase fused N-terminally to the target protein PEX19, is expressed under a tetracycline-inducible system. Upon exogenous addition of biotin, TurboID catalyzes the biotinylation of proteins within its labeling radius (outlined in an orange-shaded area within the dotted line), including both direct interactors and other proximal proteins surrounding the PEX19 fusion protein. Following the labeling, cells are lysed, and biotinylated proteins are enriched through in vitro pull-down using streptavidin agarose beads. Purified proteins are subsequently analyzed by SDS-PAGE and identified by mass spectrometry, revealing the proximity interactome of the target protein

2. Transfect the TurboID fusion construct into *T. brucei* and select clones (*see* Subheading 3.3).
3. Induce the expression of TurboID fusion protein with tetracycline in the presence of exogenous biotin, resulting in proximity-dependent biotinylation of neighboring proteins (*see* Subheading 3.4).
4. Extract biotinylated proteins from detergent-lysed cells, followed by enrichment with streptavidin agarose beads (*see* Subheading 3.5).
5. Analyze the enriched biotinylated proteins by streptavidin blotting, immunoblotting and/or mass spectrometry to reveal the proximal proteome (*see* Subheading 3.6).

3.2 Cloning TurboID Fusion Constructs into a *Trypanosoma* Expression Vector

1. Select a plasmid vector containing the TurboID gene for expression in *Trypanosoma*. If not available, obtain a template from repositories such as Addgene (Plasmid #124646), and subclone TurboID gene into your plasmid vector.
2. Clone gene of interest (e.g., *Tb*PEX19 used here) in frame with TurboID. Add an epitope tag for detection in by immunoblotting (*see* **Note 4**).
3. Clone separately a suitable gene fused to TurboID and epitope tag as a negative control (e.g., eGFP used here) (*see* **Note 5**).
4. Verify the correct fusion of TurboID and epitope tag with the desired gene of interest or control by automated Sanger sequencing.

Here, we used *Trypanosoma* expression plasmid pHD1336 containing *Tb*PEX19 (gene of interest) or eGFP (control) cloned in frame with the C-terminus of FLAG-TurboID by restriction-based insertion. The details of the constructs, along with cloning strategies and oligonucleotide sequences, are provided in Tables 1 and 2, respectively.

3.3 Cell Maintenance and Generation of TurboID-Expressing *T. brucei* Lines

1. Grow the *T. b. brucei* PCF 29–13 cells in SDM-79 medium [18], supplemented with 7.5μg/mL hemin, 10 mM D-glucose, 10% heat-inactivated fetal bovine serum, and 1% penicillin-streptomycin (*see* **Note 6**). In addition to this, G418 (15μg/mL) and hygromycin (50μg/mL) should be used for the maintenance of all PCF cell lines, wildtype and derivatives.
2. Maintain the culture at 28 °C and split them regularly to keep trypanosomes within the exponential growth phase, generally maintained between 1×10^6 and 30×10^6 cells/mL.
3. Once cells are in exponential growth phase, proceed with transfection using plasmid constructs encoding the gene of interest or a negative control (e.g., TurboID-*Tb*PEX19 or TurboID-eGFP).
4. Linearize plasmid constructs prior to transfection using appropriate restriction enzymes (e.g., NotI for pHD1336) and purify the DNA for transfection (**steps 5** and **6**).
5. After restriction digestion, purify the linearized plasmid DNA by ethanol precipitation. Add 1/10th volume of 3 M sodium acetate (pH 5.2) and 2.5 volumes of ice-cold absolute ethanol to the digested DNA. Invert the tube gently to mix, and incubate at −20 °C for 30 min.
6. Centrifuge at 18,000× *g* (Rotor F45-30-11, Eppendorf) for 10 min at 4 °C, discard the supernatant, wash the pellet with 70% ethanol, centrifuge again, remove residual ethanol, air-dry the pellet, and resuspend the DNA in sterile water (e.g., 30μg plasmid DNA in 15μL). Use 5μL of this DNA solution per transfection.

Table 1
***Trypanosoma brucei* expression plasmids used in this protocol**

Sl no.	Construct	Primer pair	Restriction sites	Cloned in vector
1	Myc-BioID-*Tb*PEX19 (pML03-Erdmann Lab collection)	RE6302—RE6169	XhoI/BamHI	pLew100_myc_BirA* (AddGene Cat No. 41716)
2	FLAG-TurboID	RE7159—RE7160	HindIII/MfeI	pHD1336
3	*Tb*PEX19	RE7163—RE7164	MfeI/BamHI	pHD1336-FLAG-TurboID
4	eGFP	RE7309—RE7310	MfeI/BamHI	pHD1336-FLAG-TurboID

Table 2
Oligonucleotides

Primer	Sequence 5′–3′
RE6302	ATACTCGAGATGTCTCATCCCGACAATGAC
RE6169	ATAGGATCCCTACACTGATGGTTGCACAT
RE7159	AAGTCTAAGCTTATGGATTACAAGGATGACGACGATAAGGGAAGCAAAGACAATAC TGTGCCTCTGAAG
RE7160	AATAGACAATTGCTCGAGTGCGGCGGCAAGTTTCTTTTC
RE7163	AAGGACCAATTGAGCATGTCTCATCCCGACAATGAC
RE7164	AAGATCGGATCCTACACTGATGGTTGCACATCGGCA
RE7309	AATAGCAATTGAGCATGGTGAGCAAGGGCGAGGAGCTG
RE7310	AGATAGGATCCTTACTTGTACAGCTCGTCCATGCC

7. Transfect wildtype PCF cells with the linearized plasmid DNA using the Amaxa Nucleofector 2b device with program X-001 and the Human T-cell Nucleofector Kit (Lonza) (**Steps 8–12**). Subsequently, the transfected plasmids will integrate into the rRNA locus of the wildtype cells.
8. Harvest 30×10^6 cells by centrifugation at $1500 \times g$ (SX4400 Rotor, Beckman Coulter) for 10 min at room temperature. Resuspend the pellet in 100 μL Nucleofector solution mix (82.8 μL Nucleofector Solution + 18.2 μL Supplement 1). Mix the cells with 10 μg linearized plasmid DNA in a cuvette and electroporate using program X-001.
9. Transfer the electroporated mixture immediately to 4.5 mL of complete SDM-79 medium without selection antibiotics in a

T25 flask and incubate overnight (O/N, 16 h) at 28 °C in a cell culture incubator.

10. In parallel, prepare conditioned medium by inoculating 1–2 × 10^6 cells/mL in complete SDM-79 medium and incubate O/N. Following the incubation, centrifuge the culture at 1500 × *g* (SX4400 rotor, Beckman Coulter) for 10 min to remove cells, then transfer the supernatant (conditioned medium) to a fresh tube.
11. Add 4.5 mL of conditioned medium, supplemented with 2× appropriate selective antibiotics to the transfected culture, bringing the total volume to 9 mL and yielding final concentrations of 10 μg/mL blasticidin for TurboID-fused constructs and 5 μg/mL phleomycin for BioID-fused constructs.
12. For clonal selection, prepare two 24-well plates by adding 600 μL of conditioned medium containing the appropriate selection antibiotics to wells in rows B and C, and 300 μL to wells in row D. Add 1 mL of the transfected culture to wells A1–A9, then perform a serial 1:3 dilution by transferring 300 μL of culture from wells in row A to B, then B to C, and C to D, ensuring all wells in rows B–D reach a final volume of 600 μL. Incubate the plates at 28 °C in a cell culture incubator.
13. Monitor the plates for transfectant outgrowth, typically visible by day 9–10. Once outgrowth is observed, transfer clones from row D to T25 flasks containing SDM-79 medium, select individual clones, and confirm target protein expression by immunoblotting following induction.

3.4 Induction of Proximity Labeling in Trypanosomes

To demonstrate the superiority of the TurboID method, we included BioID-*Tb*PEX19 for comparison of the biotinylation efficiency.

1. Inoculate SDM-79 medium with wildtype, TurboID-*Tb*PEX19, or BioID-*Tb*PEX19 (*see* **Note 7**) cell lines in a 12-well plate at a density of 6 × 10^6 cells/mL, with 3 mL total volume per well.
2. Induce protein expression by treating cells with either DMSO (control) or tetracycline (Tet; 5 mg/mL stock in DMSO) at a final concentration of 0.1μg/mL for 6 h (*see* **Notes 8** and **9**).
3. Following tetracycline induction for 6 h, add biotin to a final concentration of 50μM (*see* **Note 10**).
4. To determine the optimal biotinylation efficiency, incubate cells with 50μM biotin (*see* **Note 11**) for different periods of time (1, 30, or 90 min). After incubation, harvest cells by centrifugation at 1500× *g* (SX4400 Rotor-Beckman Coulter) for 10 min at 4 °C in a 15-mL Falcon tube.

5. Wash the resulting cell pellet twice with PBS, centrifuging at 2400× *g* (Rotor PP 2/98 #3324, Heraeus) for 3 min at 4 °C each time.
6. Resuspend the final pellet in 1× Laemmli buffer diluted in PBS and heat denature.
7. Perform immunoblot analysis with the denatured samples. Detect biotinylated proteins with Infrared dye-coupled streptavidin (IRDye® 800CW Streptavidin, LI-COR 1:15,000 dilution). Wildtype cells serve as a negative control to assess background biotinylation. Consistent with previous studies, TurboID demonstrated rapid biotinylation, with detectable labeling even within 1 min (*see* **Note 12**), in contrast to the slower catalytic activity of BioID (*see* Fig. 2).
8. Also, confirm the expression of BioID or TurboID-tagged *Tb*PEX19 by immunoblotting with the *Tb*PEX19 antibody. The results show that 0.1μg/mL tetracycline for 6 h is sufficient for optimal induction of the fusion proteins (*see* Fig. 2).

3.5 Enrichment of Biotinylated Proteins Using Streptavidin Agarose Beads

1. Following optimization of induction and biotinylation conditions in small-scale assays, perform large-scale enrichment of biotinylated proteins using streptavidin agarose beads. The protocol described below, using 40 mL of PCF culture grown in a single T75 flask (corresponding to ~2.5×10^8 cells), yields an eluate sufficient for 3–4 SDS-PAGE and immunoblot analyses.
2. *Tb*PEX19, a cytosolic receptor for newly synthesized glycosomal membrane proteins, was used as a representative example for proximity labeling. To assess non-specific labeling, a cytosolic eGFP construct was included as a negative control.
3. Both *Tb*PEX19 and the eGFP control were fused at the N-terminus with TurboID and an additional FLAG tag to ensure consistent tagging and detection (*see* **Notes 5** and **13**). A schematic representation of these constructs is shown in Fig. 3a.
4. For the enrichment of biotinylated proteins, inoculate wildtype and cells expressing TurboID fused to eGFP or *Tb*PEX19 in 50-mL flasks at a density of 6×10^6 cells/mL, with a total volume of 40 mL per flask.
5. Induce protein expression by adding tetracycline to a final concentration of 0.1μg/mL and incubate for 6 h.
6. Following the 6 h induction, add biotin to a final concentration of 50μM, and incubate for the duration that was found optimal in **steps 4–7** of the small-scale assay described in Subheading 3.4, that is, 90 min in our example with PEX19.

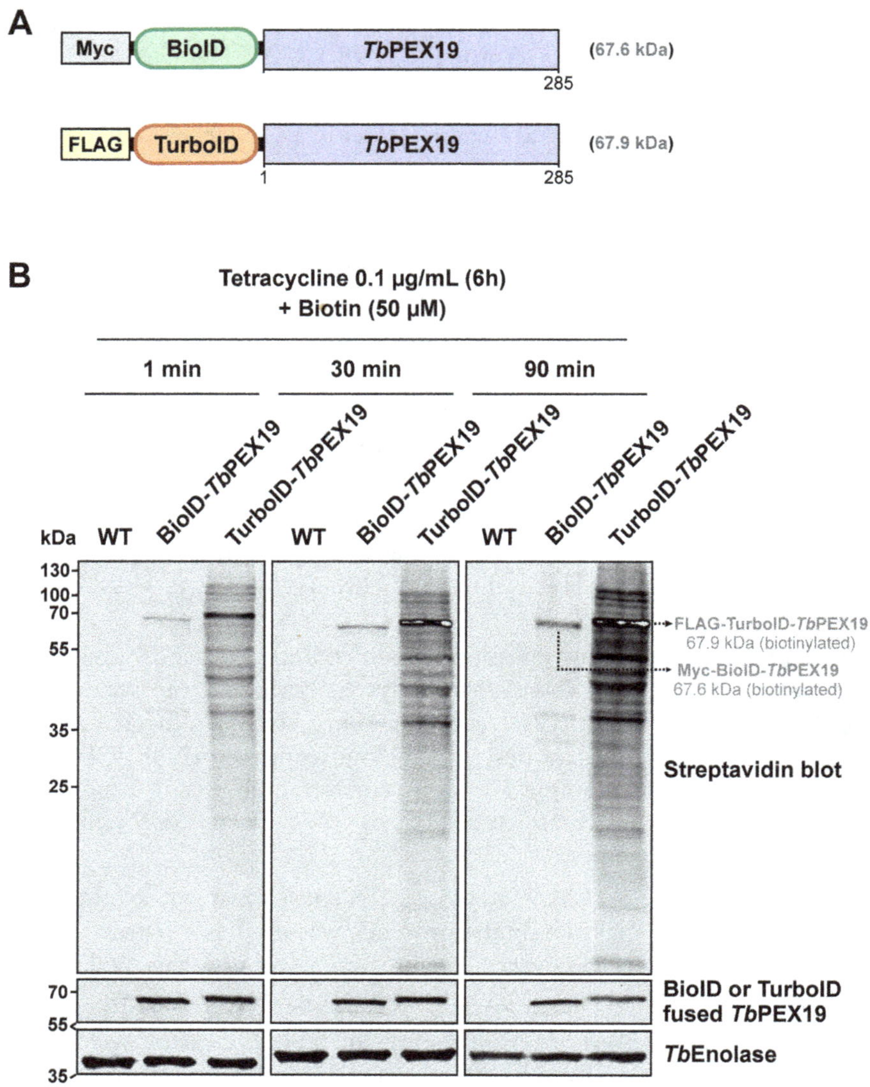

Fig. 2 Comparison of biotinylation efficiency between BioID and TurboID in *T. brucei* cells. (**a**) Schematic representation of the BioID and TurboID fusion constructs of *Tb*PEX19 used for the comparison of biotinylation efficiency. (**b**) Expression of BioID- or TurboID-tagged *Tb*PEX19 is tetracycline-inducible (0.1 mg/mL for 6 h). Following induction, cells were treated with 50µM biotin and incubated for different time points (1, 30, or 90 min). Immunoblot analysis was performed on total cell lysates of the harvested samples. Biotinylated proteins were detected using infra-red dye labelled streptavidin (top panel). Wildtype cells (lacking ligase expression) were included as a negative control to assess background biotinylation. The correct expression of the BioID or TurboID fusion constructs of *Tb*PEX19 (running at ~68 kDa size) was detected in the respective part of the immunoblot using *Tb*PEX19 antibody (middle panel), which also shows that their relative expression levels were comparable. Enolase was used as a loading control (bottom panel)

7. Harvest cells by centrifugation at 1500× *g* (SX4400 rotor, Beckman Coulter) for 10 min at 4 °C in 50-mL Falcon tubes.
8. Wash the resulting cell pellets twice with PBS buffer, with centrifugation at 2400× *g* (Rotor PP 2/98 #3324-Heraeus) for 3 min at 4 °C. Freeze the cell pellets rapidly in liquid nitrogen and store at −80 °C.

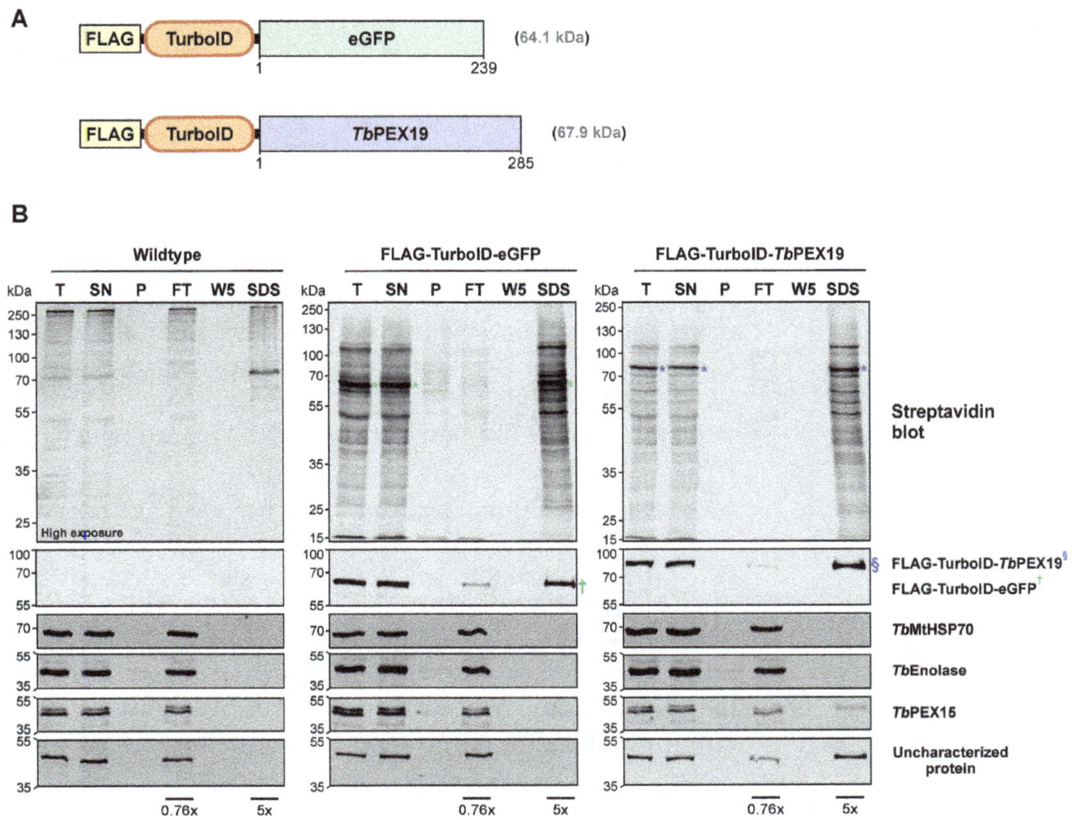

Fig. 3 In vitro pull-down of biotinylated proteins using streptavidin agarose beads. (**a**) Schematic representation of the control eGFP and test *Tb*PEX19, both tagged N-terminally with FLAG-TurboID, used for the study to identify proximal and interacting proteins. (**b**) Immunoblot of streptavidin pulldown profiles from wildtype and TurboID-eGFP or *Tb*PEX19 expressing cells. Biotinylated proteins were detected using infra-red dye labelled streptavidin (top panel). Wildtype cells, which lack ligase expression, served as a negative control to assess background biotinylation (high exposure). Self-biotinylation of eGFP and *Tb*PEX19 is shown by the green and blue asterisks, respectively. Expression of the biotin ligases (TurboID fused to eGFP or *Tb*PEX19) was detected using FLAG antibody. MtHSP70 and enolase were included as negative controls, and the known glycosomal membrane protein (PMP) PEX15 was included as a positive control. An uncharacterized protein, identified as a prominent cytosol-binding partner of PEX19 (from unpublished work), was also detected. PEX15 and the uncharacterized protein were present at significantly higher levels in the eluate from *Tb*PEX19 than eGFP, indicating potential *Tb*PEX19-binding partners. The profile includes total cell lysate (T), supernatant (SN), pellet (P), flow-through (FT), W5 wash, and SDS-denatured samples (SDS). Concentration factors relative to the total are depicted at the bottom

9. The following day, thaw the frozen pellets, and resuspend the samples in PBS buffer with 1× PIC.
10. Estimate protein concentration using the Bradford assay. Equal amounts of total protein (1.8 mg in 800μL) were used for each cell line (wildtype, TurboID-eGFP, and TurboID-*Tb*PEX19). Samples (Total, T) were taken before processing for purification for immunoblot analysis.

11. Subsequently, treat the cells with digitonin at a final concentration of 3 mg per mg of protein for 30 min at 4 °C to permeabilize cellular membranes. Centrifuge the samples at 16,000× *g* (Rotor F45-30-11, Eppendorf) for 30 min to separate the soluble and insoluble protein fractions (*see* **Notes 14** and **15**).
12. Preparation of streptavidin beads: In parallel with the centrifugation of cell lysates above, centrifuge streptavidin agarose beads at 500× *g* (Rotor F45-30-11, Eppendorf) for 1 min in Mobicol Spin Classic tubes to remove the storage buffer. Wash the beads three times with PBS directly in the Mobicol Spin Classic tubes with screw caps to equilibrate. Prepare 600μL of a 50% slurry (i.e., 300μL of settled beads) and aliquot 100μL of settled beads into separate Mobicol tubes for each sample (*see* **Note 16**).
13. Meanwhile, the 30-min centrifugation from **step 11** results in the separation of solubilized proteins in the supernatant (SN) and an insoluble pellet (P). Transfer the supernatant from wildtype (WT), eGFP control, and *Tb*PEX19-expressing cells, which contain the biotinylated proteins, to the respective tubes containing 100μL of equilibrated streptavidin beads. Incubate the samples at 4 °C for 1 h with rotation at 10 rpm using a test tube rotator to allow binding of the beads.
14. After incubation, collect the flow-through (FT), and wash the beads five times with 700μL of PBS, with each wash followed by centrifugation at 500× *g* (Rotor F45-30-11, Eppendorf) for 1 min.
15. Seal the Mobicol Spin Classic tubes containing the beads with bound biotinylated proteins and add 1× Laemmli buffer. Denature the samples by heating at 95 °C for 10 min. After denaturation, remove the seals, and place the content of the tubes into new 1.5-mL microfuge tubes. Centrifuge the samples at 500× *g* (Rotor F45-30-11, Eppendorf) for 2 min. The resulting eluate (E) contains the bait protein along with its proximal interacting proteins (*see* **Note 17**).

3.6 Analysis of the Enriched Biotinylated Proteins to Reveal Proximal Proteome

1. Separate the various samples collected from each step of the enrichment process (from Subheading 3.5), including the eluate by SDS-PAGE and analyze using Streptavidin blot (or immunoblot, *see* **step 2** below). Biotinylated proteins are detected with IR-Dye labeled streptavidin (LI-COR, 1:15,000 dilution), and wildtype cells are used as a negative control to assess background biotinylation. Consistent with previous studies (*see* Fig. 2), TurboID demonstrated rapid biotinylation, with detectable labeling even within 1 min, in contrast to the slower catalytic activity of BioID.

2. Perform immunoblot analysis of samples from each step of the enrichment process, including the eluate, using various relevant markers, as shown in Fig. 3b.

Our analysis included multiple validation steps. First, to confirm successful biotinylation and capture of the bait protein and identify proximal proteins, we analyzed the eluate by immunoblot using a FLAG antibody (to detect the tagged bait protein) and streptavidin blot (to detect biotinylated proteins) (*see* Fig. 3b). Second, the enrichment of PEX15, a known glycosomal membrane protein [19], in the *Tb*PEX19 eluate compared to the eGFP control, along with the specific identification of an uncharacterized protein in the *Tb*PEX19 eluate, validated the method's reliability in identifying both established and novel interaction partners. Additionally, the absence of enolase and MtHSP70 in eluates from both *Tb*PEX19 and eGFP samples served as negative controls (*see* Fig. 3b), further validating the TurboID-based labeling as a reliable approach for identifying proximal interactors.

4 Notes

1. Dissolve 1 g of digitonin powder in 20 mL of water and boil for 10 min under a fume hood. Allow the solution to cool slowly to room temperature, then store at 4 °C. Before use, centrifuge the solution to sediment any precipitates and use only the clear supernatant. Refer to the supplier's safety datasheet for handling instructions, as digitonin is toxic.
2. Prepare the buffer 1 day before use, filter sterilize it, and store at 4 °C. Use the buffer at room temperature when washing the cells to avoid thermal shock.
3. Prepare 25× PIC by dissolving one tablet of cOmplete™, EDTA-free Protease Inhibitor Cocktail (Roche) in 2 mL of PBS buffer.
4. In this study, a FLAG tag was used as the epitope tag for detection via immunoblotting. However, other epitope tags such as Myc, HA, or Strep II can also be used, depending on the availability of specific antibodies or compatibility with downstream applications.
5. For studies aimed to identify the cytosolic proximal interactome, eGFP was used as a control. In contrast, for studies focused on the membrane protein interactome, it is advisable to use a membrane protein control from a different organism, such as a protein containing a transmembrane domain (TMD) that is correctly integrated into the organelle of interest. For instance, in the case of *T. brucei* glycosomal membranes, the human ALDP (amino acids 66–164) has been demonstrated to

integrate into the glycosomal membrane [20] and could serve as a suitable control for studying the membrane protein interactome. However, if the control protein is highly conserved, it may introduce non-specific interactions, which should be carefully considered during data interpretation.

6. The SDM-79 complete medium, supplemented with G418 (15μg/mL) and hygromycin (50μg/mL) for maintaining PCF cell lines, is stored at 4 °C and then warmed to 28 °C using a water bath before use.
7. The BioID-*Tb*PEX19 expressing cell line used in this study was previously generated in the Prof. Ralf Erdmann lab. This cell line was established using the pML03 construct, which expresses an N-terminal Myc-BirA(R118G)-*Tb*PEX19 fusion protein cloned into the pLEW100 plasmid.
8. Control cultures were treated with the same volume of vehicle as the tetracycline stock solution (5 mg/mL) to maintain consistent DMSO concentrations, given that tetracycline was dissolved in DMSO.
9. Optimal inducing conditions were determined by treating cells with varying tetracycline concentrations and incubation times. While higher concentrations (e.g., 1μg/mL) and prolonged induction (up to 24 h) resulted in extensive biotinylation, including self-biotinylation, these conditions could lead to increased non-specific labeling. Consequently, an induction time of 6 h with 0.1μg/mL tetracycline was chosen for optimal expression with minimal background biotinylation. It is advised to determine the optimal expression level of the target protein experimentally in each case.
10. In this study, a 90-min biotinylation time with 50μM biotin was sufficient for identifying proximal and binding partners of *Tb*PEX19. However, the biotinylation time can be extended, especially with lower biotin concentrations, depending on experimental requirements. For studies focusing on organelle biogenesis and more stable interactions, particularly when using low or no supplemental biotin, extending the biotinylation time to at least 12 h may be beneficial. This aligns with the approximate doubling time of PCF *T. brucei* cells, facilitating the characterization of interactions occurring in a full cell cycle and providing insights into dynamic processes like glycosome biogenesis.
11. Prepare a 1-mM biotin stock solution in SDM-79 medium and sterile-filter through a 0.2-μm filter. Add the filtered stock solution to the induced cells to achieve a final concentration of 50 μM.

12. Studies have shown a low level of labeling even before the addition of exogenous biotin, suggesting that TurboID can utilize the trace amounts of biotin present in cells or organisms grown in standard biotin-containing media. For instance, a study in *T. brucei* showed that SDM-79 medium, containing approximately 827 nM biotin, is sufficient for high-level biotinylation [9].
13. The choice of N-terminal and C-terminal tagging is determined by the specific functional requirements of the protein. In this study, the TurboID tag was fused to its N-terminus to maintain the C-terminal region of PEX19 free of any tag as it serves as the PMP binding site.
14. Mild detergents, such as digitonin, can be effective; the use of harsher detergents/denaturants like urea is recommended for improved protein solubilization, enhanced extraction of biotinylated proteins, or removal of non-biotinylated contaminants from streptavidin beads post pull-down [5, 21].
15. If the goal is to identify interacting proteins in the cytosol, digitonin can be used at lower concentrations as long as it is sufficient to permeabilize the plasma membrane and release cytosolic content. This allows for the enrichment of the cytosolic fraction for the detection of biotinylated proteins. In contrast, the organelle-enriched pellet can be treated with harsher detergents or higher concentrations of digitonin to enable the detection of biotinylated proteins in organelle fractions.
16. Before preparing the beads, it is important to first assemble the column by inserting the 35-μm pore-size filter into the Mobicol Spin Classic column (MoBiTec). Once the column is assembled, the required volume of beads should be transferred, including the storage buffer provided by the manufacturer. This storage buffer must be removed by centrifugation, and the beads should be equilibrated by washing with PBS to prepare them for subsequent binding steps.
17. In this protocol, denaturation with Laemmli buffer was used to prepare samples for immunoblot analysis. For downstream applications such as mass spectrometry, it is recommended to perform a non-denaturing elution followed by desalting or dialysis of the relevant fractions prior to proteomic analysis.

Acknowledgments

This project work has received funding from the European Union's Horizon 2020 research and innovation program under the Marie Skłodowska-Curie grant agreement no. 812968 (R.E.). C.K. was recipient of the Marie-Curie ITN Fellowship (grant agreement no. 812968). The work received funding support from Ruhr University Bochum, InnovationsFoRUM (Host Microbe Interactions: IF-009N-22 and IF-018N-22 to R.E.). The work was supported by DFG grant ER178/17-1 to R.E. We thank Bettina Warscheid for kindly providing TurboID templates used to generate our expression constructs, and Paul Michels and James D. Bangs for providing the antibodies against Enolase and MtHSP70, respectively. The overview of the method was created using BioRender.com.

References

1. Kaufer A, Ellis J, Stark D et al (2017) The evolution of trypanosomatid taxonomy. Parasit Vectors 10(1):287. https://doi.org/10.1186/s13071-017-2204-7
2. Stuart K, Brun R, Croft S et al (2008) Kinetoplastids: related protozoan pathogens, different diseases. J Clin Invest 118(4):1301–1310. https://doi.org/10.1172/JCI33945
3. Kalel VC, Maser P, Sattler M et al (2018) Come, sweet death: targeting glycosomal protein import for antitrypanosomal drug development. Curr Opin Microbiol 46:116–122. https://doi.org/10.1016/j.mib.2018.11.003
4. Opperdoes FR, Baudhuin P, Coppens I et al (1984) Purification, morphometric analysis, and characterization of the glycosomes (microbodies) of the protozoan hemoflagellate Trypanosoma brucei. J Cell Biol 98(4): 1178–1184. https://doi.org/10.1083/jcb.98.4.1178
5. Geoghegan V, Mottram JC, Jones NG (2022) Tag thy neighbour: nanometre-scale insights into kinetoplastid parasites with proximity dependent biotinylation. Front Cell Infect Microbiol 12:894213. https://doi.org/10.3389/fcimb.2022.894213
6. Gingras AC, Gstaiger M, Raught B et al (2007) Analysis of protein complexes using mass spectrometry. Nat Rev Mol Cell Biol 8(8): 645–654. https://doi.org/10.1038/nrm2208
7. Branon T, Han S, Ting A (2017) Beyond immunoprecipitation: exploring new interaction spaces with proximity Biotinylation. Biochemistry 56(26):3297–3298. https://doi.org/10.1021/acs.biochem.7b00466
8. Zhang Y, Li Y, Yang X et al (2020) TurboID-based proximity labeling for in planta identification of protein-protein interaction networks. J Vis Exp (159). https://doi.org/10.3791/60728
9. Moreira C, Kelemen CD, Obado SO et al (2023) Impact of inherent biases built into proteomic techniques: proximity labeling and affinity capture compared. J Biol Chem 299(1): 102726. https://doi.org/10.1016/j.jbc.2022.102726
10. Yang X, Wen Z, Zhang D et al (2021) Proximity labeling: an emerging tool for probing in planta molecular interactions. Plant Commun 2(2):100137. https://doi.org/10.1016/j.xplc.2020.100137
11. Samavarchi-Tehrani P, Samson R, Gingras AC (2020) Proximity dependent biotinylation: key enzymes and adaptation to proteomics approaches. Mol Cell Proteomics 19(5): 757–773. https://doi.org/10.1074/mcp.R120.001941
12. Branon TC, Bosch JA, Sanchez AD et al (2018) Efficient proximity labeling in living cells and organisms with TurboID. Nat Biotechnol 36(9):880–887. https://doi.org/10.1038/nbt.4201
13. Geri JB, Oakley JV, Reyes-Robles T et al (2020) Microenvironment mapping via Dexter energy transfer on immune cells. Science 367(6482):1091–1097. https://doi.org/10.1126/science.aay4106
14. Choi-Rhee E, Schulman H, Cronan JE (2004) Promiscuous protein biotinylation by Escherichia coli biotin protein ligase. Protein Sci 13(11):3043–3050. https://doi.org/10.1110/ps.04911804
15. Roux KJ, Kim DI, Raida M et al (2012) A promiscuous biotin ligase fusion protein identifies proximal and interacting proteins in mammalian cells. J Cell Biol 196(6):801–810. https://doi.org/10.1083/jcb.201112098
16. Rahmati S, Emili A (2025) Proximity labeling: precise proteomics technology for mapping receptor protein neighborhoods at the cancer cell surface. Cancers (Basel) 17(2):179. https://doi.org/10.3390/cancers17020179

17. Wirtz E, Clayton C (1995) Inducible gene expression in trypanosomes mediated by a prokaryotic repressor. Science 268(5214): 1179–1183. https://doi.org/10.1126/science.7761835
18. Brun R, Schonenberger (1979) Cultivation and in vitro cloning or procyclic culture forms of Trypanosoma brucei in a semi-defined medium. Short communication. Acta Trop 36(3):289–292
19. Krishna CK, Das H, Hohnen L et al (2025) High-confidence glycosomal membrane protein inventory unveils trypanosomal peroxin PEX15. Cell Rep 44(5):115614. https://doi.org/10.1016/j.celrep.2025.115614
20. Saveria T, Halbach A, Erdmann R et al (2007) Conservation of PEX19-binding motifs required for protein targeting to mammalian peroxisomal and trypanosome glycosomal membranes. Eukaryot Cell 6(8):1439–1449. https://doi.org/10.1128/EC.00084-07
21. Peach M, Marsh N, Macphee DJ (2012) Protein solubilisation: attend to the choice of lysis buffer. Methods Mol Biol 869:37–47. https://doi.org/10.1007/978-1-61779-821-4_4

Chapter 19

From Genome to Function: Metabolism-Wide Models of Lifecycle Stages of Parasites

Bruno R. Pinto, Gabriela T. Montanaro, Mayke B. Alencar, Milou Hoving, Jurgen R. Haanstra, Ariel M. Silber, and Pranas Grigaitis

Abstract

Metabolism is a complex network of biochemical reactions that cells use to obtain free energy transduced from nutrients and synthesize new cellular components. Understanding the wiring of metabolism can aid in metabolic engineering to boost certain metabolic functions, or, alternatively, inspire interventions to block undesired metabolism, for example, of parasites. The challenge is that interventions to metabolic networks should be studied in the context of the entire, extensive cellular metabolic networks. Incoming and outgoing fluxes may be readily measured but how fluxes are distributed within the metabolic network often remains a black box. Systems biology offers a toolset to shine light inside this black box: constraint-based metabolic models can predict intracellular flux distributions. But how to construct such a model? In this chapter, we will guide you through the steps to make a constraint-based model out of the genome sequence and demonstrate that it may be easier than it sounds. We provide a walkthrough of model construction, curation and validation—with a specific focus on protistan parasites and their lifecycle stages. While genomes can help to infer the entirety of metabolic potential encoded in the organism's genome (a so-called genome-scale metabolic model), cells do not necessarily express all metabolic enzymes simultaneously. This is particularly relevant for parasites as the different lifecycle stages rely on different sets of metabolic enzymes and host-imposed constraints. We discuss approaches to restrict the genome-wide network to become lifecycle stage specific and how you can test and use the resulting models.

Key words Genome-scale metabolic models, Systems biology, Biomass equation, Dry weight, Growth-associated ATP maintenance (GAM), Non-growth-associated ATP maintenance (NGAM)

1 Introduction

To grow and function, cells need energy and building blocks for new cells. The precursors of the new cell material can be either taken up from the surroundings or made by the cell from other

Bruno R. Pinto and Gabriela T. Montanaro contributed to this work equally.

Paul A. M. Michels et al. (eds.), *Euglenozoa: Methods and Protocols, Volume 1*, Methods in Molecular Biology, vol. 3013, https://doi.org/10.1007/978-1-0716-5142-1_19,

molecules. In parallel, the energy is harvested from energy-rich nutrients (e.g., sugars, amino acids, or fats) and stored in the high-energy chemical bonds of adenosine triphosphate (ATP). Metabolic waste that is generated in these processes needs to be excreted from the cell. Thus, the chemistry of life forms large networks of conversion reactions, collectively called metabolism [1].

Molecular (reductionist) approaches in biology typically focus on deciphering the function of a single or small group of genes or proteins and thus require extensive experimental work. Although this foundational knowledge is essential, it provides an incomplete picture as proteins do not function in isolation, but rather are embedded in large networks. Systems biology is an interdisciplinary field focused on understanding biological systems as a totality: an integrative approach to life. Systems biology considers organisms as complex networks made of interconnected components, with interactions and feedback loops essential to their operation. These interconnections give rise to emergent behaviors and so complicate assessing the behavior of the system as a whole.

The logic of such networks (or significant parts of them) can be encoded into computer models that greatly facilitate understanding of their function. Broadly speaking, modeling approaches to cellular networks fall into three large categories (and of course combinations thereof): interaction-based, mechanistic, and constraint-based models [2, 3]. Interaction-based models catalogue the interactions between network components (e.g., proteins) without explicitly considering network parameters, such as interaction stoichiometry. Such models can, for example, reveal functional modules within a network. However, when focusing on more dynamic networks, such as metabolism, the other two types of models are more relevant.

Mechanistic (or kinetic) models of metabolism are formulated as a system of ordinary differential equations for metabolite concentrations over time. The individual reactions are described by rate equations that can follow either simple mass action relations or more elaborate kinetics. For enzymes, these rate equations are often Michaelis–Menten type of equations with parameters for maximal velocities of the reaction (v_{max}) and affinities of the enzymes for substrates (and products where applicable; also called Michaelis constant, or K_M). Each rate equation often has multiple parameters that need to be experimentally determined or fitted based on pathway-wide flux data: the larger the modelled network, the more parameter values these models require. Therefore, kinetic models are often made for specific, well-studied pathways rather than the entirety of metabolism. Examples for parasites are the glycolysis models of bloodstream-form (BSF) *Trypanosoma brucei* (originally made in 1997 [4]; most recent version in [5]), *Entamoeba histolytica* [6], *Schistosoma mansoni* [7] and *Plasmodium falciparum*-infected red blood cells [8]. These models have been used to predict drug targets and to understand the relevance of allosteric feedforward loops in the regulation of glycolytic flux.

Yet, in cells, including those of parasites, distinct pathways like glycolysis do not operate in isolation. Constraint-based models offer a broader, metabolic network-wide view. These models aim to include all metabolic reactions in an organism (and are therefore often named genome-scale metabolic models (GEMs)); this comes at a trade-off of detail: these models generally only include stoichiometry and reaction rate bounds rather than detailed kinetics. Constraint-based models predict reaction rates (fluxes) in the network, usually subject to optimization of a pre-selected objective function (e.g., maximization of growth rate) dictated by the constraints put on the model (e.g., substrate availability). GEMs are particularly useful for predicting the network-wide effects of perturbations that affect individual metabolic reactions, such as changes in enzyme availability, nutrient limitation, gene knockout, or inhibition of enzymatic function. Capitalizing on that, GEMs have been successfully applied to address a wide range of biological questions, such as development of metabolic strategies for synthesizing valuable compounds, discovering gene functions, media design, and identifying new drug targets [9].

Parasites often have different hosts and have host-adjusted lifecycle stages with distinct metabolic networks. As in each lifecycle stage, only a subset of the genome-encoded metabolic proteins is expressed, the most relevant models are a subset of the metabolic potential encoded in the genome. In this chapter, we will describe step-by-step how metabolism-wide constraint-based models can be made for parasites. We will focus on constructing lifecycle stage-specific metabolic networks, although the modeling framework is transferrable to reflect different sorts of metabolic plasticity (e.g., euglenozoans that periodically alternate between autotrophic and heterotrophic metabolism).

We will first briefly describe how an initial model is constructed based on information from the genome. The construction can partly be automated but will also require manual curation. This genome-scale model, however, will have limited relevance for parasites with lifecycle stages. We will discuss the types of data that can be used to generate lifecycle stage–specific models. In each lifecycle stage, a different objective might be needed, and we will discuss different options. We will build, and reflect on, successful examples from the Kinetoplastea, where such models were used to compare the metabolic networks of different life stages and/or identify drug targets; namely, for the red blood cell and liver stages of *Plasmodium* species (see for a review [10]) and *Trypanosoma cruzi* [11]. At the end, we will describe typical strategies to validate the lifecycle stage–specific models and how you can use them.

2 Materials

2.1 Computational Tools for GEM Reconstruction

1. A desktop- or laptop computer.
2. Genome sequencing data (either de novo sequencing or obtained from public databases like NCBI Genome [12]).
3. Public databases like KEGG [13], BRENDA [14], UniProt [15], ModelSEED [16].
4. Software packages referred to in the text (*see* **Note 1**).

2.2 Experimental Materials

2.2.1 Dry Weight Measurements

1. Filters: Nucleopore™ Track- Etch membrane; 0.4µm; diameter 25 mm; Whatman™.
2. Petri dish (⌀ 92 mm).
3. Small petri dishes (⌀ 35 mm).
4. Tweezers.
5. Vacuum manifold (e.g., VACUUBRAND diaphragm vacuum pump and Satorius manifold).
6. 60 °C incubator.
7. Analytical scale (e.g., Satorius Cubis® II Semi-Micro Configurable Lab Balances).
8. Desiccator.

2.2.2 Oxygen Consumption Measurements

1. A standard cultivation medium (for *T. cruzi*: Liver infusion tryptose (LIT) medium supplemented with 10% fetal calf serum).
2. Oligomycin A (5 mg/mL and 0.5 mg/mL stocks in ethanol).
3. High-resolution respirometer (e.g., Oroboros-O2K).
4. Hamilton microliter syringes.

2.2.3 Metabolite Quantification Using NMR

1. Cell culture supernatant.
2. 3 kDa MWCO filter.
3. Internal NMR standard (e.g., TMSP-$_{d4}$).
4. NMR spectrometer (we typically operate a 600 MHz instrument).

3 Methods

3.1 GEM Reconstruction Pipeline

The construction of a metabolic model begins with an annotated genome (and detailed data on the biochemical reactions carried out by the organism, where applicable) (Fig. 1a, b). This should also include non-enzymatic reactions. Essential information must be specified for each reaction: substrate- and product stoichiometry,

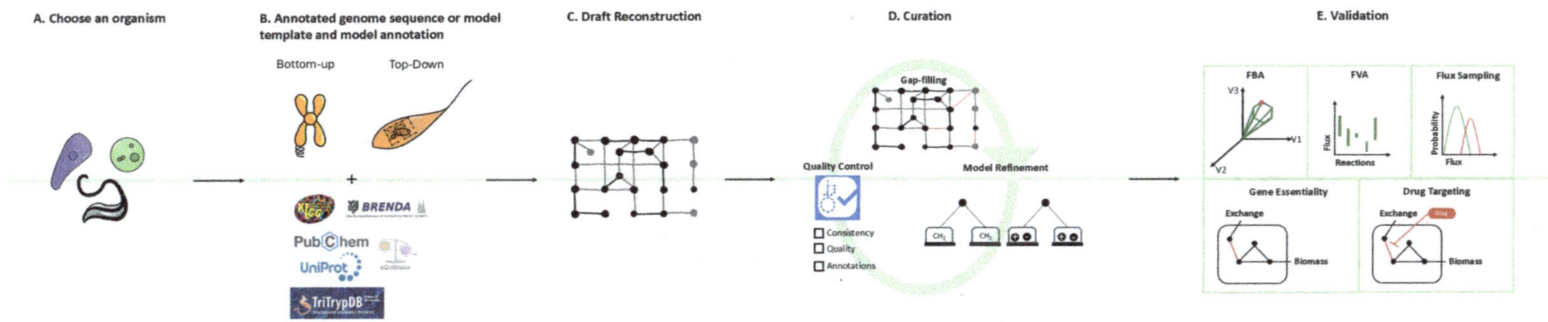

Fig. 1 Schematic pipeline of GEM reconstruction. (**a**) The first step to reconstruct a genomescale model is to choose an organism of interest. (**b**) A draft is made based on the organism of choice, using the annotated genome sequence or an already existing model phylogenetically similar. (**c**) The draft reconstruction is done automatically by different software. (**d**) Once the draft is done, a curation step is necessary, involving gap-filling, quality control and refinement. (**e**) Once the curation is over, the model can be validated using different techniques such as Flux Balance Analysis (FBA), Flux Variability Analysis (FVA), flux sampling, gene essentiality and drug targeting

cofactors, reaction reversibility, and cellular compartmentalization. Mathematically, metabolic models are often expressed as a stoichiometric matrix S, with dimensions $m \times n$, where each row corresponds to a metabolite (in a system with m metabolites), and each column represents a reaction (n reactions). The elements in this matrix are stoichiometric coefficients: in a column, positive values indicate production of that metabolite, negative values indicate consumption, and zero means that the metabolite is not part of the reaction [17].

Within the network, reactions (and reaction fluxes [18]) are categorized as either internal reactions, which take place within the system of interest, or exchange reactions that represent influx/efflux of metabolites into the system. Transport reactions connect the extracellular environment with the cell by representing uptake or excretion of metabolites into and from the simulated cell.

Next, known enzymatic reactions in the metabolic model are assigned one or more enzymes inferred from the genome sequence via so-called Gene–Protein–Reaction (GPR) associations. GPR associations frequently are not a unique 1:1 match; many enzymes contain multiple subunits that function together, and some enzymes catalyze more than one reaction. Consequently, a single GPR may involve multiple genes (and their combinations), or one gene appears in multiple GPRs [19, 20].

The initial model (Fig. 1c) must undergo multiple refinements using data from various databases, experimental results, and literature to minimize artifacts and biases. This iterative process, more commonly known as model curation (Fig. 1d), is an essential part of generating a high-quality reconstruction. The resulting model can then be validated with experimental data (Fig. 1e).

3.1.1 Reconstruction Strategies

We typically distinguish between bottom-up and top-down approaches of GEM reconstruction. The bottom-up reconstruction is an approach to build GEMs by assembling metabolic networks systematically assessing the organism's genomic and biochemical data. This method meticulously incorporates individual enzymatic reactions based on the genome annotation. Later, metabolite transport and exchange reactions are added to create an organism-specific representation of metabolism. The process relies heavily on manual, iterative curation and access to experimental data. For these reasons, it is highly accurate but labor-intensive [21].

In brief, the bottom-up approach consists of the following steps:

1. Genome annotation: Identify metabolic genes via KEGG [13], UniProt [15], ModelSEED [16], or another database.
2. Draft model construction: Convert annotated genes into reactions and pathways. For this it is possible to use freely available tools, for example, CarveMe [22].

Table 1
Comparison between GEM reconstruction approaches

	Bottom-up	Top-down
Input	Genome annotation	Template model
Effort	High (manual curation)	Faster (automated refinement)
Accuracy	High (organism-specific)	Depends on template model quality

3. Manual curation and gap-filling: Resolve dead-ends and missing reactions via literature/biochemical evidence [23].
4. Validation: Compare model predictions with experimental data.

The alternative, top-down approach is to generate a model from a template GEM, ideally of a biologically related organism. With this approach, metabolic networks are reconstructed based on homology; to achieve accurate mapping, the ideal template should be as phylogenetically close as possible to the organism of interest. Several freely available software packages do this [24], for example, the RAVEN Toolbox [25].

The top-down approach works as follows:

1. Template selection: Select template model of another organism.
2. Homology detection: Use a software to automate homology reconstruction.
3. Validation: Compare model predictions with experimental data.

The choice between approaches depends on the research objectives and available resources (Table 1). The bottom-up reconstruction is the preferred method when constructing de novo models for well-studied organisms or when maximal accuracy is required. A top-down reconstruction, on the other hand, is very useful when the parts and interactions are not known in advance, making it more suitable for high-throughput studies, personalized medicine applications, or investigations of systems lacking comprehensive biochemical data [26]. A drawback of the top-down approach is, however, that it will never capture the reactions that are unique to the organism of interest but absent in the template model.

A hybrid methodology that combines aspects of both approaches has become widely used over the years. A typical reconstruction pipeline begins with a template-based draft generation, followed by extensive manual curation (and extensions) to ensure biological accuracy. In particular, hybrid models help to strike a balance between biological accuracy and investments in time and labor [27].

Table 2
Interpretation of flux bounds for exchange reactions

Lower bound	Upper bound	Interpretation
0	0	No exchange
0	>0	Only excretion allowed
<0	0	Only uptake allowed
<0	>0	Both uptake and excretion allowed

3.1.2 Constraining the Solution Space: Exchanges and Virtual Medium

The structure of the metabolic network (stoichiometric ratios, reaction reversibility) defines the so-called solution space, or the set of possible flux distributions in the network that satisfy stoichiometric or thermodynamic relationships. In an *unbounded* solution space, fluxes in a flux distribution can, theoretically, trend towards infinity. To create a more defined solution space, we use biological reasoning to assign values to upper and lower flux bounds, or to *constrain* the model. The common practice in the field is to set these bounds to 1000 units (i.e., [0, 1000] or [−1000, 1000] for irreversible and reversible reactions, respectively).

Cells, cellular compartments, and the extracellular environment are explicitly distinguished in GEMs. Exchange reactions connect the system with its boundary and typically take form of $metabolite_{extracellular} \leftrightarrow \emptyset$. Manipulating the flux bounds of these reactions will define what nutrients can be taken up and/or excreted. Setting the flux bounds of exchange reactions to either zero or non-zero value (as introduced above, ±1000) will have different consequences as constraints (Table 2). As the first approximation, the flux bounds for all the exchange reactions are set to [−1000, 1000], that is, uptake and excretion are allowed for every metabolite defined in the extracellular compartment. The resulting scenario is a virtual medium: an environment with high concentrations of ready-to-use nutrients.

The virtual medium can be used to troubleshoot missing transport reactions (i.e., whether the intracellular metabolic network can take up all compounds from its environment). Flux bounds of exchange reactions are the primary "knob" to constrain the GEM; typically, a limited set of nutrients is available (e.g., defined growth medium) and typical metabolic by-products are a rather restricted repertoire of molecules. This property also allows another use of GEMs—as tools for growth media design [28].

3.1.3 Objectives, Flux Sampling, and Flux Balance Analysis (FBA)

In constraint-based modeling, we typically assume the metabolic network to operate in a steady state (i.e., the metabolite concentrations do not change), and the resulting solution space can be already sampled to obtain ensembles of potential flux distributions.

However, GEMs are often used in combination with optimization-based approaches like Flux Balance Analysis (FBA) [29] to reduce the solution space even further.

In FBA, we define an objective to be optimized (minimized/maximized): in most general form, a weighted sum of fluxes v. For instance, in a system consisting of three fluxes (v_1, v_2, v_3), one such objective could be max($2 \times v_1 + 1.2 \times v_3$). The output of such a linear program (optimization problem) is in a set of steady-state fluxes (distribution) in the network with the optimal value of the objective function. To obtain biologically meaningful solutions, a key decision is to choose a representative objective function. As highlighted above, the objective function might be a combination of flux values, yet typically, can be defined as a single "biologically relevant objective" of the cell. What could these objectives be in parasites? There are a few "usual suspects" in the field; however, keep in mind that the inferred objectives might be lifecycle stage-specific: here we will only briefly describe some options (to be elaborated on in Subheading 3.2.3).

From the historical perspective, as the first GEMs have been reconstructed for microbes like *Escherichia coli* and *Saccharomyces cerevisiae*, the most frequently used objective function is biomass formation (production of new cells). This objective function represents the hypothesis that cells try to grow and divide as fast as possible: that is, the metabolic network will "convert" simpler molecules into building blocks needed for new cells in the way that supports the fastest production of new cells. Other popular objective functions for microorganisms are related to production of high-value metabolites (e.g., vitamins [30]) or undesired end-products (e.g., off-flavors in food production [31]).

Furthermore, production of new cells is not a representative objective function for many cell types (e.g., cells from terminally differentiated human tissues or non-replicative stages of parasites); in this case, alternative objectives should be considered, such as maximization of uptake flux for a certain nutrient (e.g., glucose). In some cases, maximization of glucose uptake flux corresponds to maximization of biomass production, but they are not interchangeable stricto sensu.

Finally, a typical objective function for non-growing cells is ATP hydrolysis (steady-state ATP production). In non-growing cells, the ATP consumption is typically attributed to maintaining the cell integrity, a.k.a. "ATP maintenance" (*see* **Note 2**). In non-growing cells, the "non-growth-associated ATP maintenance" (NGAM) value (i.e., ATP production and hydrolysis flux under no growth) is a value that can be obtained using a combination of modeling and experiments—we will return to discussing how to determine the ATP maintenance parameters in Subheading 3.2.5.

Optimal solutions are usually non-unique, and the optimal solution space (sets of flux distributions resulting in an optimal value of the objective function) should ideally be inspected further. Two principal approaches to this problem exist: optimization- or flux sampling-based methods. Optimization-based methods, namely, Flux Variability Analysis (FVA), is used to infer the interval (span) of the individual flux values corresponding to the optimal solution, and is generally a fast approach to get a feeling of the size of the optimal solution space. On the contrary, sampling of the solution space allows unbiased exploration of the solution space, including suboptimal solutions. However, a brute-force sampling approach is computationally extremely intensive, therefore, many methods have been developed to facilitate this task [32].

3.1.4 Quality Control of the Reconstruction

To produce meaningful predictions, the newly generated GEM should first be checked for inconsistencies. Here, we will highlight a few types of potential issues frequently present in draft reconstructions: (1) mass or charge imbalance and (2) energy- or matter-generating cycles. Such checks are likely to be done more than once during the curation process, and it is critical to perform an additional check on the final version of the model to ensure its consistency.

First, the GEM should be checked for violations of mass and charge conservation, that is, the net sum of masses and charges of reaction substrates and of products should be equal in each reaction. Most software packages that handle GEMs contain routines to perform the consistency check automatically; for these methods to return insightful results, every metabolite in the GEM should ideally be annotated with both the elemental formula and the charge of the molecule (reactions where at least one metabolite is not annotated will be flagged as inconsistent) (*see* **Note 3**). Also, note that the charge of (bio-)molecules is typically pH-dependent, and some of them may have a different charge based on their ionization state. It is therefore important to assume a certain pH value and remain adherent to it during the curation process (*see* **Note 4**).

Defining the elemental formulae (and charges) right generally greatly helps to detect model inconsistencies leading to further issues, for example, proton imbalance (missing/extra H^+ in the reaction). Such imbalances may give rise to energy- or matter-generating cycles (GCs), that is, sets of reactions that may fuel the in silico representation of metabolism without external inflow of nutrients and/or oxygen. Typical GCs are, for example, sets of transport reactions that generate proton-motive force across membranes (eventually re-phosphorylating ADP to ATP), or intracellular cycles that recover not only their starting products but also additional moieties, typically, hydrocarbon fragments.

A good sanity check to prune typical GCs is to create a new reaction representing ATP hydrolysis (ATP + H_2O→ADP + P_i + H^+) and set all exchange flux bounds to [0, 1000] (only excretion allowed; see above for more details). Then, maximize the flux through the ATP hydrolysis reaction. Normally, the optimization result should be 0.0; this would mean no "complete" energy-GCs present. For identifying carbon-GCs, typically it is useful to repeat the same procedure allowing oxygen uptake to the model (i.e., oxygen exchange set to [−1000, 1000]). In case the predicted ATP hydrolysis flux does not equal zero, the reactions that carry flux should be inspected in detail to identify the culprit(s) for creating the GC. Imbalances (and GCs) related to other chemical elements are rare; nonetheless, ideally, all inconsistencies highlighted by the mass and charge balance check-up should be resolved in a meaningful way.

At the end of the curation process, it might be useful to evaluate the quality (and completeness) of the annotations, and the stoichiometric consistency of the GEM. To facilitate this process, established software packages like MEMOTE [33] and/or a recent package FROG [34] may be used to assess and benchmark the quality of GEMs.

MEMOTE also scores GEMs using a numeric scale (in between 0% and 100%), which is a combination of assessing the stoichiometric consistency of the model and the coverage of its annotations. However, the scheme for scoring model annotations in MEMOTE is rather rigid. For instance, completeness of metabolite annotations is scored across 11 naming schemas/databases, and it is virtually impossible to acquire close-to-complete mappings for all 11 databases. In our view, the modeler should aim for >90% coverage for one to two databases to achieve a reasonable interoperability of model identifiers. Overall, MEMOTE is a diagnostic tool, and the interpretation of a low MEMOTE score is dependent on the actual culprit: stoichiometric inconsistencies are an important issue that can be highlighted by MEMOTE, and should be mended at the best capacity to ensure model consistency. However, a low MEMOTE score coming from the lack of comprehensive annotations should not discourage from using the GEM further, if the information is sufficient to ensure a reasonable degree of interoperability.

3.1.5 Naming the Genome-Scale Metabolic Model

The historical naming convention for the GEMs focused on three components: (1) flagging the reconstruction of metabolism as an "in silico" representation, (2) acknowledging the main author(s) of the GEM, and (3) specifying a single metric of the model, typically, number of genes present. This convention has been exercised, among others, for all recent *E. coli* GEMs; as an example, two reconstructions published in 2012 and 2017, respectively, were named *i*JO1366 [35] and *i*ML1515 [27].

Despite the proposed nomenclature, approaches to naming the GEMs are subject to personal taste. Variations on the theme include instead of genes, modelers choose to include the number of reactions, or use the organism's scientific name instead of the initials of the modelers (e.g., the *Bordetella pertussis* GEM *i*BP1780 [36]). Moving from the convention even further, major iterations of the *S. cerevisiae* GEM were named just "yeast": the latest version "Yeast9" was released in 2024 [37]. Similarly, the latest iteration of the *Homo sapiens* GEM is called Human1 [38], which served as the template for the suite of GEMs representing other metazoan model organisms [39].

Recently, there was a call to standardize the meta-data collection for GEMs to improve the FAIRness (*see* **Note 5**) of GEMs: the "standard-GEM" [40] specification. There, the naming convention relates more to the "yeast" or "human" GEM, suggesting that either a trivial name or the species identifier (e.g., human-GEM, or sapiens-GEM) is used. However, we discourage from using the naming convention used in the "standard-GEM" specification: the "canonical" system generates GEM identifiers with a high likelihood to be unique, an important consideration for FAIRness. The historical naming conventions allow for a wide range of alternatives if a name is reserved and do not suggest a historical sequence of GEMs.

3.2 From a Genome-Scale Reconstruction to Stage-Specific Models

A genome-scale reconstruction of metabolism is generally a start—not the endpoint. While the GEM represents the overall metabolic *capacity* of your organism of interest, at every given moment, the *factual capacity* will always be a subset of the reactions in the network. Here we have to acknowledge the elephant in the room: the chemistry of life requires catalysts, namely, enzymes—which are a great expenditure of resources for the cell [41], including taking up physical volume of the cell [42]. Given these costs, even the unicellular, free-living microbes are highly unlikely to express the entire (metabolic) genome at once. In multicellular organisms, such expectation is, in fact, the norm: differentiated tissues execute highly specific gene expression programs. Somewhere in between these two extremes between free-living bacteria and well-differentiated cells, many unicellular parasites exhibit multi-stage lifecycles: it is well known for many parasites that the metabolic map usually changes drastically upon transition to another lifecycle stage [43].

To obtain a good representation of lifestage-specific metabolism, it is critical to extensively curate the organism-specific GEM (which we will call the "pan-GEM," referring to the theoretical metabolic capacity as a union ["pan"] of capacity at different lifecycle stages) prior to crafting lifestage-specific GEMs (Fig. 2a). The model reduction process shall, in most cases, generate a lot of gaps in the pan-GEM. These gaps will be a mix of true- and false-

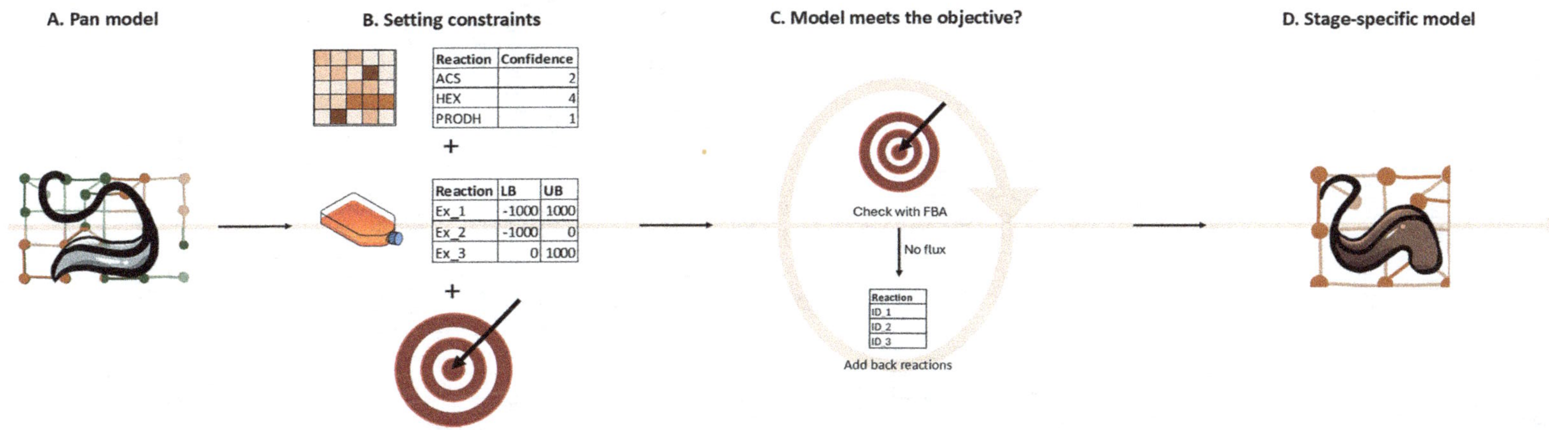

Fig. 2 Schematic pipeline of stage-specific GEM. (**a**) To build a pan model stage-specific, it is necessary to include a set of constraints. (**b**) It is possible to obtain a list of reactions present in the life stage of interest, by using OMICs datasets or information already published in known databases. Other constraints, such as an objective function and medium components are necessary to match with the biological situations that the life stage of interest actually faces. Once the information is gathered, it is necessary to attribute to the reactions high, medium, and low confidence scores, depending on the algorithm of use. (**c**) In the end, the algorithm has to check if the objective function has flux to maintain or remove reactions. If the objective function does not have flux, there is a need to add back reactions and do the FBA again. (**d**) If the objective function has flux, the stage-specific model is complete. The network is smaller than that of the pan-model

negatives (e.g., enzyme-encoding gene not expressed during the lifecycle stage vs. true expression, but below detection limit) due to incomplete datasets. Acknowledging these limitations, the pan-GEM should guide the critical assessment of the impact of losing certain metabolic functions to the lifestage-specific metabolic network.

To make the lifestage-specific metabolic network, three parts (Fig. 2b) are needed:

1. Information to reduce the pan-GEM to the stage-specific network.
2. Definition of the environment in which the lifestage dwells. The availability of specific nutrients is a determining factor for metabolic reprogramming and, consequently, for metabolic flux simulations.
3. An objective function that can steer the algorithm to make a functional model. The selected objective function(s) (e.g., biomass production or ATP hydrolysis) should be included into the pan-GEM to represent the variety of metabolic objectives the organism of interest might adhere to. In different lifecycle-stage GEMs, different reactions should be flagged as objectives.

We will first discuss the data to reduce the network, setting lifecycle stage–specific context, and the choice of the objective function. Then, we will introduce the general idea for (a) model reduction algorithm(s) (Fig. 2c), and the software for creating lifecycle stage–specific GEMs (Fig. 2d). Note that even though these lifecycle stage–specific models are by definition not genome-wide anymore, we will keep referring to them as GEMs.

3.2.1 The Data for Network Reduction

The process of making lifecycle stage–specific models begins with compiling the information on the makeup of the lifestage of interest, usually in the form of large biological datasets (-omics). Two kinds of -omics data are typically used, transcriptomics and proteomics (and a combination of both, if available). Both types of data come with trade-offs: currently, transcriptomics techniques, like RNA-sequencing (RNA-seq), are often still more sensitive than proteomics (e.g., mass spectrometry-based proteomics), generating more data-rich datasets. However, detection of protein makes a stronger case for the enzymatic reaction to take place: mRNA levels frequently are not a good indicator of the protein level [44] due to post-transcriptional regulation; in the end, frequently there is no protein translated at all. This is especially the case for trypanosomatids (as e.g., reviewed in [45]).

Furthermore, protein expression alone does not necessarily imply enzymatic activity in cases when post-translational modifications or compartmentalization are required. In such cases, profiling of post-translational modifications (e.g., phosphoproteomics) or use of compartment-specific proteomics datasets (e.g., mitochondrial proteome [46]) may help to infer the enzymatic activity. Furthermore, most of the computational methods to reduce GEMs from -omics data rely on thresholding, so it is critical to place the cut-off right. A threshold too low may preserve metabolic functions that should not be present in the stage-specific model, while the opposite may remove pathways that are actually operational.

When using a single type of data (and/or single dataset), some data exploration (e.g., plotting the values, computing percentiles) might help to get a feeling for numbers and select the threshold value. This decision becomes non-trivial when blending multiple datasets coming from different techniques (e.g., transcriptomics and proteomics, or protein expression vs. phosphorylation degree), since different values have to be harmonized to form a weighing scheme.

For some (usually biomedically or biotechnologically relevant) organisms, decades of research have yielded a vast amount of biochemical knowledge from experiments. Keeping in mind that -omics datasets will likely be incomplete, other streams of knowledge should be harnessed when possible. Databases which systemize evidence from multiple data sources (e.g., evidence for the presence of activity of specific enzymes) are available for many model organisms such as *E. coli*, or budding yeast *S. cerevisiae*. The TrypanoCyc [47] (http://vm-trypanocyc.toulouse.inra.fr/) and LeishCyc [48] (http://vm-trypanocyc.toulouse.inra.fr/LEISH/organism-summary) databases store information for *T. brucei* and *Leishmania major*. For instance, TrypanoCyc contains a massive amount of manually curated and evidence-scored entries regarding known and predicted enzyme activities (inspired by the schema introduced by Thiele and Palsson [21]). The predicted activities in this case mean that the reactions take place in cell lysates, yet the responsible enzyme(s) has not been identified.

Maintaining organism-specific databases is a laborious task, and many databases have been unfortunately discontinued and/or are not updated anymore. In such cases, one might resort to more generic, well-established resources, like KEGG (https://www.genome.jp/kegg/) or BioCyc (https://biocyc.org/), in order to aggregate information from, for example, closely related organisms.

3.2.2 Constraints for the Stage-Specific Context

Next to a different network topology, also the environment of the lifecycle stage should be taken into account. In many cases, the "environment" could be interpreted synonymously as the growth medium: cells can only take up what is provided in their growth medium. Metabolite uptake and excretion rates can be computed from time-resolved metabolic measurements and can be used to constrain (=set the bounds for) exchange fluxes. These fluxes can be used as quantitative constraints on the exchange flux in the model.

If metabolite measurements are done during a growing lifecycle stage, it is important to also track the increase in biomass (or cell numbers) to compute specific (biomass or cell number-normalized) fluxes. The change in the metabolite over time is then a combination of a specific flux (e.g., per unit biomass) and the increase of cell numbers: flux calculations should take this into account. Thus, the specific flux q for a metabolite can be calculated with the formula [49]:

$$q_{metabolite} = \frac{\mu}{\Upsilon} \tag{1}$$

in which the μ is the specific growth rate in (cells/cells)/time and Υ is the biomass yield on the metabolite [cells/mol metabolite]. The μ can be obtained from an exponential fit to biomass data over time. The Υ can be obtained by plotting the biomass (e.g., culture weight, or number of cells) over the metabolite amount (from time-resolved measurements). If there is a constant flux, all points will be on a line. The slope of the linear fit through the points will give the Υ on the metabolite. This method has been used to calculate the glucose consumption flux in batch cultures of *T. brucei* [50] and that paper shows the plots and the calculations for μ, Υ, and $q_{glucose}$.

In cases where biomass is constant, flux calculations are relatively simple. Given no increase in biomass, a plot of the metabolite concentration over time will show a linear relationship, indicative of a constant flux. The slope of a linear fit through the points will yield concentration/time. Correction for cell culture volume and division by the (average) amount of biomass in the incubation will give the specific flux q in mol/time/cell number. This method is applicable to lifecycle stages that are stationary (e.g., *stumpy* form *T. brucei*), and for proliferative lifecycle stages if the measurements are done in a timeframe where growth can be neglected. An example for glucose and glycerol fluxes from a 12-h incubation of *T. brucei* can be found in [51] (*see* **Note 6**).

Sometimes, and especially when working with organisms growing on rich media (such as parasites or mammalian cells), it is complicated to quantify the consumption and excretion of all nutrients and byproducts. Information from the medium composition can be still useful to constrain nutrient exchange fluxes even without flux quantification. In practice, this allows to impose qualitative constraints (a binary decision whether to allow uptake of the metabolites or not) based on the media composition. If the component is absent from the media, the flux bound is set to [0, 1000] (*see* Subheading 3.1.2). Qualitative constraints could be used to test whether cells could grow on different media, to identify nutritional dependencies (auxotrophies) or to test whether cells can grow anaerobically (without oxygen).

Furthermore, in the absence of flux quantification, we could resort to some simplification to derive reasonable exchange flux bounds. For such cases, we could assume that after a certain time (e.g., 24 or 48 h), the cells will halt growth because one (or more) medium component has been depleted. Following this assumption, we could infer the maximal possible uptake flux over from the concentration of the component X (denoted as x in [mmol/L]) of the starting medium, cell dry mass (m in [gDW/cell]) and initial cell seeding density (D in [cell/L]):

$$LB_x = -1.0 \times \frac{x\,[mmol/L]}{D\,[cell/L] \times m\,[gDW/cell]} \times \frac{1}{\tau}\,[h^{-1}] \quad (2)$$

This represents the maximal theoretical uptake rate, which likely will be considerably lower in real life, and is highly dependent on the growth rate of the cells. However, this can be a useful first approximation to, for example, narrow down media components of interest to be quantified more precisely.

3.2.3 Selection of an Objective Function for the Different Life Stages

Next, the stage-specific models will also need an objective function that is maximized or minimized (*see* Subheading 3.1.3). Here we would like to emphasize that the objective will likely be stage specific too. Biomass production is a logical objective for the growing lifecycle stages of parasites, and we will discuss this in Subheading 3.2.4.

However, non-growing lifecycle stages should be represented by a different objective function. Non-replicative stages require energy and nutrients to maintain cell viability, enable the parasite to evade immune responses, infect new hosts, and continue its life cycle. In that light, the objective function for these non-replicative stages can be rendered as the set of maintenance processes required to maintain cellular integrity and viability. Identifying cellular and metabolic processes that are active during the non-replicative phases of these organisms can help in the formulating such an objective function.

Typically, minimizing the flux through the non-growth-associated ATP maintenance (NGAM) reaction is rendered as a sufficient objective for the non-replicative states. Then, a practical approach is to set the lower bound of the NGAM reaction to the inferred ATP production flux in non-replicative cells (*see* Subheading 3.2.5) and minimize the flux through the reaction—so the flux corresponds to that value. In this way, we essentially obtain a flux distribution corresponding to the producing ATP to maintain cell integrity. Overall, the ATP maintenance parameters are an essential part of any GEM, and we will discuss how to obtain the ATP maintenance parameters in more depth in Subheading 3.2.5.

3.2.4 Objective Function for Growing Life Stages: Biomass Objective Function (BOF)

The build-up of new cells is not explicitly encoded in the genome: a virtual reaction needs to be included that produces "cell biomass." The biomass objective function (BOF), or, colloquially, the biomass equation, represents the fractions of (macro-)molecules that are parts of the cell in the ratios they are present in dry cell biomass. Generally, a biomass equation will include DNA, RNA, protein, lipids, and carbohydrates. If quantities are known, minor components like cofactors and metal ions may be included as well.

The major components of biomass are largely the same among different cells, but their fractional contribution to biomass may differ. Therefore, the stoichiometric coefficients (quantities of components per dry weight) need to be tailored to the organism of interest. Proteins, nucleotides, and lipids make up the vast majority of biomass, so it is important to get a good estimate or measurements for these.

To date, there is no standardized protocol for experimentally determining *all* cellular components. These values either need to be experimentally measured or inferred from available data—we will briefly discuss combining both approaches. For the latter, a good starting point is to check whether there is a related organism for which a BOF already exists. Then, consider which fractions may differ in (the lifecycle stages of) your organism—variations in lipid composition, protein and nucleic acid content, as well as cofactors and other essential metabolites, must be incorporated into the biomass equation.

Defining a BOF In all reactions of the GEM, the stoichiometric coefficients are, in fact, in *mol* (in the field, we typically operate in a space where expressing molar amounts in *mmol* is more convenient). The biomass equation is a special case, because for the dimensions to work out right (broader discussion on this available in [52]), the biomass constituents should be expressed as their abundance in the biomass, that is, *mmol/gDW*. Multiple strategies to encoding the biomass equation in the GEM exist, and the critical sanity check in formulating the representation of the biomass is to make sure that the abundance of the biomass components is expressed in the correct dimensions.

Here we will briefly describe our preferred approach, where we will create a simplified BOF which represents the bulk components in their mass fractions—this is typically what is experimentally measured. Then, for constituents formed of simpler components, we will create reactions representing production of these biomass precursors. This will end up in a set of reactions which are relatively easy for the modeler to interpret and are not cluttered as if all known biomass components were described in a single, very long reaction equation.

We will start with the simplified BOF. There, we describe the composition of the cells in terms of mass ($m, g/gDW$) of high-level components (nucleic acids, protein, etc.). Moreover, we include the estimated GAM value in molar units (n, $mmol/gDW$). The general representation of this equation will be as follows:

$$m_{DNA}\text{DNA} + m_{RNA}\ \text{RNA} + m_{prot}\ \text{protein} + m_{carb}\ \text{carbohydrates} + m_{lip}\ \text{lipids} + \text{minor components} + n_{GAM}\ \text{ATP} + n_{GAM}\ \text{H2O} \rightarrow 1\ \text{biomass} + n_{GAM}\ \text{ADP} + n_{GAM}\ \text{P}_i + n_{GAM}\ \text{H}^+ \quad (3)$$

The term "minor components" refers to any compounds (cofactors, vitamins, etc.) with minor mass contributions to the biomass.

The next step is to add more detail, that is, describe the molar composition of the bulk constituents, that is, their molar contribution to the constituent mass n_m (in $[mmol/g\ constituent]$). This way, the "coupling," or multiplication of, for example, the moles of nucleoside triphosphates used to make 1 g of RNA ($mmol/g\ RNA$) and the mass of the RNA in the BOF ($g\ RNA/gDW$), represents the demand of nucleoside triphosphates in $mmol/gDW$. We will take polymerizing bulk RNA from individual nucleoside triphosphates as an example:

$$n_{m,A,RNA}\ \text{ATP} + n_{m,C,RNA} CTP + n_{m,G,RNA}\ \text{GTP} + n_{m,U,RNA}\ \text{UTP} + n_{m,RNA}\text{H2O} \rightarrow 1\text{RNA} + n_{m,RNA}\ PP_i \quad (4)$$

where

$$n_{m,RNA} = \left(n_{m,A,RNA} + n_{m,C,RNA} + n_{m,G,RNA} + n_{m,U,RNA}\right)$$

This procedure is repeated for each biomass constituent. Below, we will describe how to obtain the stoichiometric coefficients to formulate the final BOF.

Weight fractions for the biomass equation can be obtained from existing data in literature or from new measurements. It is important to note that each weight fraction needs to be normalized to the dry weight of the organism (i.e., g/gDW). The dry weight is not something that is often determined for parasites. Many studies will report the biomass used in assays as (1) the number of cells, (2) optical density, or (3) protein amounts, and it will be important to know how much dry weight these measures of biomass represent. Therefore, we will here first provide a protocol for dry weight measurements, before we discuss the composition of the biomass.

Measuring the Dry Weight of Cells For dry weight measurements intact cells are needed. As parasites do not have a cell wall, they are more vulnerable to lysis than organisms like bacteria and yeast.

To highlight the adjustments needed for parasites, here we present an entire protocol that has successfully been used for the sleeping-sickness parasite *T. brucei* (data unpublished; personal communication, J.R. Haanstra). For materials *see* Subheading 2.2.1.

The steps for dry weight measurements are:

1. Dry filters by placing them in a Petri dish in a 60 °C stove to remove water.
2. Take filters from the stove and keep them in a desiccator.
3. Weigh the filters (shiny side down) after at least 24 h: measure weights on a milligram scale immediately after placement on the scale ($t = 0$) and after 30 s of the filter placement ($t = 30$). The relevant measurement for dry weight determination is after 30 s, since scales need some time to adjust and to provide the most accurate measurement.
4. Place-weighed filters in individual small Petri dishes and write down their measurements. Weighed filters can be stored at room temperature until use.
5. Grow cells in the exponential phase (*see* **Notes 7** and **8**).
6. Count the cells to know the number of cells that will be in the pellet.
7. Pellet the cells by centrifugation (*see* **Note 9**).
8. Carefully resuspend the pellet in 1 mL of the growth medium, adjust volume if necessary.
9. Place the dried filters (shiny side down) on the manifold holder and secure the filter with a clamp.
10. Turn on the vacuum pump and wet the filters with growth 1 mL medium (*see* **Note 10**).
11. Pipet the cell suspension slowly on a filter (*see* **Note 11**).
12. Flush the filter with 2 mL of growth medium.
13. Turn off the vacuum pump.
14. Place each filter in their corresponding petri dish and place this in a 60 °C incubator for at least 24 h [you need to identify the exact filter as they will have slightly different weights].
15. Weigh the filter on the same scale and in the same manner as in the preparation phase.
16. As a control perform the same procedure for medium without cells [use the same volumes as for the cell suspension].
17. Subtract the filter weight before the addition of the sample at $t = 30$ s from the filter weight after the addition of the sample at $t = 30$ s.

18. If the medium control dry weight was not equal to 0, it has to be subtracted from each dry weight measurement.
19. By dividing the dry weight by the number of cells in the sample (determined at **step 6** above) you get to the dry weight/cell (or per number of cells—generally 10^6 or 10^8 cells).

Measurement of Bulk Components Many of the bulk components of the cells can be measured using rather simple, typically spectrophotometric assays (also, ready-to-use kits are available from many vendors). A successful example of covering most of the components is the biomass quantification for *E. coli* [53]. Here we will highlight some straightforward measurement options for some of the macromolecules:

- DNA and RNA: can be isolated from cells separately and quantified by UV absorbance spectrophotometry according to standard biochemical protocols [e.g., RNA [54]].
- Protein: Several well-established protocols exist to measure proteins in lysates, such as BCA [55], Bradford [56]. In any case it will be extremely important to remove protein components from the culture medium in the assay.
- Lipids: Total amount of lipids in the biomass can be quantified using colorimetric vanillin-phosphoric acid reagent assay [57]. However, determining the composition of the lipids (ratios of different phospholipid classes, and ratios of different fatty acids incorporated into lipids) requires technologically advanced methods (e.g., gas chromatography, mass spectrometry [58]).
- Carbohydrates: the content of typical storage carbohydrates (e.g., the typical eukaryote storage carbohydrates glycogen and trehalose; although mannogen and paramylon are more prevalent in euglenozoans/trypanosomatids [59]), as well as total carbohydrate pool, can be measured using anthrone [60] or phenol-sulfuric acid assays [61]. Alternatively, enzymatic assays can be used for, for example, trehalose quantification [62].

Finally, specific structural metabolites specific or unique to the organism of interest must be incorporated into the biomass equation. Many organisms contain specific [macro-]molecules (glycans and lipids, such as GPI anchors, N-glycans linked to proteins) and structures like the cell wall (in prokaryotes). These compounds and structures are often involved in host–parasite interactions processes, making their biosynthesis essential for the parasite.

Cells require vitamins and cofactors that are involved in enzymatic reactions and redox processes. Many of these compounds are critical to the parasite's survival but are not synthesized by the organism's metabolic network and must be acquired from the host or extracellular environment. Identifying and quantifying these compounds for inclusion in the biomass equation is a challenging task. Experimentally, auxotrophies are identified using defined media [63]; essentiality studies can help guide the construction of the biomass equation. As an entry point, one might resort to reviewing the literature for information on characterized enzymes in the organism of interest that utilize specific cofactors. Typically, the BOF only includes redox cofactors such as NADH, NADPH, glutathione, and trypanothione (for e.g., Kinetoplastea) due to their role in the bioenergetics and mitigating oxidative stress.

Inferring the Fine-Grained Composition of Biomass Constituents Macromolecules are composed of simpler specific building blocks: the different nucleotides of the genomic DNA and in the RNA, the amino acids in the proteins and different carbon-chain fatty acids incorporated into the specific lipid classes. The final step in formulating a BOF is to determine the ratios of different building blocks coming together into the macromolecules. In brief, one has to count all the different building blocks in those datasets and determine their fractional contributions to the total.

For DNA, RNA, and protein, the subfractions can be obtained from the genome composition and -omics datasets. Semi-quantitative lipidomics can be used to infer the composition of lipids. We refer the reader to the description of these computations [64] accompanying the semi-automated tool called BOFdat. Note that it is important to compute the macromolecule composition for different life-stages, as the fractional contributions will likely change during the lifecycle of your organism.

3.2.5 Inferring the ATP Maintenance Parameters

It is important to realize that the ATP demand for biosynthesis arises from the biomass equation implicitly, as in the synthesis of the individual cellular building blocks that are needed for the biomass formation (DNA, RNA, proteins, set of membrane lipids, and other macromolecules), the energy costs are accounted for. This consideration has formed the basis for the theoretical calculation of energy requirements to make a new cell (cf. the calculations for the bloodstream form of the trypanosomatid *T. brucei* [65]).

However, if we tracked the bioenergetics of trypanosomatids at different lifecycle stages, we would realize that biosynthesis and maintenance do not entirely cover the ATP production in these cells: there is a substantial fraction of energy investment that is not explicitly accounted for. We name the additional ATP requirement for biomass production the growth-associated ATP maintenance (GAM): it is represented in the BOF as hydrolysis of ATP to ADP and inorganic phosphate (*see* Eq. 3). We will describe the process of computing the value of GAM, as well as for the non-growth-associated ATP maintenance (NGAM), introduced in Subheading 3.1.3. The NGAM value will be used to set the flux bounds for an ATP hydrolysis reaction, *separate* from the BOF.

How to obtain these parameters? Say, we grew cells in a defined medium and we quantified the growth of the cells and consumption of major nutrients (e.g., glucose). Then, we can use the flux measurements to constrain the GEM we constructed, in this way inferring the specific production of ATP (q_{ATP}) in these cells. The inconsistency between what the GEM perceives as "ATP demand" versus actual ATP demand based on the experimental data (=fraction of unexplained q_{ATP} from the experiments) will allow us to estimate the ATP maintenance parameters to be put into the model: the n_{GAM} in the biomass equation (Eq. 3) and the value for $q_{ATP,\ NGAM}$. Below we describe the calculations needed to quantify growth-associated maintenance and non-growth-associated maintenance (*see* **Notes 12** and **13**).

Theoretical Basis for Inferring Energy Parameters We start by decomposing the ATP demand (specific consumption flux, which, at steady-state, is equal to production) into

$$q_{ATP} = q_{ATP,GS} + q_{ATP,NGAM} \tag{5}$$

where $q_{ATP,\ GS}$ ("GS" for "growth-scaling") represents the ATP demand that scales with growth, that is, the known ATP demand (demand for producing biomass precursors) and unspecified energy demands that are associated with growth (growth-associated ATP maintenance, GAM). The $q_{ATP,\ NGAM}$ (non-growth-associated ATP maintenance, NGAM) represents the energy demand for the processes in non-proliferative cells, for example, maintaining membrane potential, DNA repair, proteome and transcriptome turnover, motility, and other processes related to cellular organization and maintenance. This specific flux will set the lower bound through an ATP hydrolysis reaction (separate from the BOF!) in the model.

However, the $q_{ATP,\ GS}$ has no meaning in the GEM by itself; instead, it is a sum of two growth-related terms:

$$q_{ATP,GS} = (n_{Biomass} + n_{GAM})\mu, \tag{6}$$

The $n_{Biomass} \times \mu$ is the fraction of the q_{ATP} that is accounted for by the GEM by stoichiometric description of biosynthesis of the biomass precursors at a given growth rate μ. The parameter n_{GAM} is the coefficient for ATP hydrolysis part of the BOF (Eq. 3), for which we want to get a value.

With this relation, (Eq. 5) can be rewritten as:

$$q_{ATP} = q_{ATP,NGAM} + (n_{Biomass} + n_{GAM})\mu \tag{7}$$

The q_{ATP} can also be defined in more metabolic terms as a sum of the ATP production fluxes through the two types of ATP production:

$$q_{ATP} = q_{ATP,OxPhos} + q_{ATP,SubPhos} \quad (8)$$

This equation reflects that there are two principal ways to rephosphorylate ADP into ATP in live cells: in substrate-level phosphorylation (SubPhos), the phosphate group of a high-energy organic molecule is directly transferred to ADP to form ATP; in oxidative phosphorylation (OxPhos), the energy from the transfer of electrons from a reduced to a more oxidized molecule is used to create a transmembrane H^+ gradient and the resulting proton-motive force drives the ATP synthase. In the latter process, the electrons need a terminal acceptor, which is often molecular oxygen, but could also be organic molecules such as fumarate, or nitrogen compounds in case of some bacteria.

Combining (Eq. 7) and (Eq. 8) gives:

$$q_{ATP,NGAM} + (n_{Biomass} + n_{GAM})\mu = q_{ATP,OxPhos} + q_{ATP,SubPhos} \quad (9)$$

Below we will discuss measurements and computational routines to infer $q_{ATP,\ NGAM}$ and n_{GAM}. At first, we perform measurements which would allow to estimate $q_{ATP,\ OxPhos}$ and $q_{ATP,\ SubPhos}$ (Eq. 9) and feed them into the GEM in order to obtain the ATP maintenance parameters from the left part of (Eq. 9).

Decomposing the q_{ATP} Terms for OxPhos and SubPhos In trypanosomatids with a "canonical" mitochondrion (e.g., those retaining most respiratory complex activities), most ATP is generated through OxPhos. Thus, we will first address the $q_{ATP,\ OxPhos}$ term from the (Eq. 9).

For a correct estimate of this term, it is important to collect the information about the function of the electron transport chain (ETC) to account for diverse functional capacity of this system. For instance, some *Leishmania* species possess a fully functional (proton-pumping) Complex I of the ETC, whereas *T. cruzi* and *T. brucei* lack critical subunits required for H^+ transfer across the membrane [66], and thus, reduced ATP yield on oxygen (P/O ratio). Electron entry via subsequent complexes (Complex II, glycerol-3-phosphate shuttle, electron transfer flavoprotein etc.) affects the P/O ratio in a similar way (*see* **Note 14**). However, it should be noted that in some organisms (and/or lifecycle stages), electrons may be diverted from entering the ETC towards an alternative oxidase (AOX), as in the bloodstream form of *T. brucei* [67] where the ETC is not even expressed.

Knowing the functional parameters of the ETC, we can infer the value of $q_{ATP,\ OxPhos}$ experimentally from measuring the specific oxygen consumption rate q_{O2}. Most of the total oxygen consumption is covered by OxPhos, yet other contributions exist that do not contribute to ATP production (proton leak, alternative oxidase, or non-mitochondrial oxygen consumption). To correct for the latter, we decompose q_{O_2} into:

$$q_{O_2} = q_{O_2,OxPhos} + q_{O_2,non-OxPhos} \tag{10}$$

The $q_{O_2,OxPhos}$ flux will allow us to estimate the $q_{ATP,\ OxPhos}$; since the $q_{O_2,non-OxPhos}$ does not bear any mechanistic explanation in itself, the most straightforward way to obtain $q_{O_2,OxPhos}$ is to subtract $q_{O_2,non-OxPhos}$ (i.e., residual oxygen consumption when OxPhos is inhibited) from the total oxygen consumption flux q_{O_2}. The q_{O_2} can be robustly quantified through high-resolution respirometry assays conducted on intact cells in both growing (mid-log) and non-growing (early-stationary phase). If we inhibit the F_oF_1-ATP synthase using, for example, oligomycin [68], then the $q_{O_2,OxPhos}$ becomes 0 and the $q_{O_2} = q_{O_2,non-OxPhos}$.

Measuring Oxygen Consumption Below, we describe the steps to measure the oxygen consumption flux in both untreated and OxPhos-inhibited *T. cruzi* epimastigote cells (in the latter, $q_{O_2} = q_{O_2,non-OxPhos}$). For materials, *see* Subheading 2.2.2.

1. Grow *T. cruzi* epimastigotes to the early-stationary phase in LIT medium (~10^8 cells mL^{-1}) (*see* **Note 8**).
2. Centrifuge the cells at 1000× *g* for 10 min and save the supernatant (*see* **Note 15**). Resuspend the cells in the saved supernatant (spent medium) to maintain physiological conditions and adjust to the required density for each assay.
3. Adjust cell concentration to 1×10^9 cells/mL in the saved supernatant.
4. Set the equipment to the appropriate temperature for each cell type (28 °C for *T. cruzi* epimastigotes).
5. Transfer 2 mL of the retained supernatant medium into each respirometer chamber and wait 15 min for signal stabilization.
6. Add 50μL of cell suspension to oxygraph chambers (final concentration: 2.5×10^7 cells/mL).
7. Wait 10 min for signal stabilization.
8. Determine optimal oligomycin concentration for respiratory inhibition, performing stepwise titration by adding 1μL oligomycin (0.5 mg/mL stock).
9. Allow 5 min between additions (leak respiration is established when additional oligomycin produces no further inhibitory effect). For subsequent replicates, use the minimum effective final concentration (use 5 mg/mL stock for this purpose).
10. Allow 5 min for signal stabilization.
11. Compute the OxPhos-associated contribution of the oxygen consumption flux:

$$q_{O_2,OxPhos} = q_{O_2} - q_{O_2,non-OxPhos} \tag{11}$$

Oxygen flux rates derived from respirometric assays are expressed in pmol O_2 per second per cell number. To normalize the value for computational analyses, normalize cell counts to dry weight (*see* Subheading 3.2.4). Although the oxygen consumption flux is the key input to the model, optionally, you can compute the ATP produced by OxPhos ($q_{ATP,\ OxPhos}$) manually (*see* **steps 12–14** below).

12. Calculate the number of protons pumped per O_2 consumed.

    Example: If $q_{O_2,OxPhos} = 150\,\text{pmol}\ O_2/s\ per\ 2.5 \times 10^7$ cells and the electrons enter the ETC via the Q-junction, 6 H^+ are pumped per oxygen atom reduced ($2e^-$) (4 from Complex III and 2 from Complex IV) so this results in 1800 pmol $H^+\ s^{-1}$ per 2.5×10^7 cells. Since this is a measure of $q_{ATP,\ OxPhos}$, this value corresponds to the flux of protons re-entering the mitochondrial matrix through the F_1F_O-ATP synthase.
13. Convert H^+ to ATP using the P/O Ratio.

    The P/O ratio (ATP produced per oxygen atom reduced) (*see* **Note 16**) depends on the H^+ required per ATP synthase: 4 H^+/ATP (standard in mammalian mitochondria) (*see* **Note 17**).

$$P/O = \frac{H/O}{H/ATP} \tag{12}$$

    More on the P/O ratio and how to calculate them can be found in [69].
14. Calculate ATP Flux (*see* **Note 18**)

$$q_{ATP,OxPhos}\ (pmol\ ATP\ s - 1\ \text{per}\ 2.5 \times 107\ \text{cells}) = q_{O_2,OxPhos} \times P/O \times 2 \tag{13}$$

    Example: If $q_{O_2,OxPhos} = 150\,\text{pmol}\ O_2/s$ per 2.5×10^7 cells, and the electrons entry point is the Q-junction, P/O is 1.5, so:

$$q_{ATP,OxPhos} = 150 \times 1.5 \times 2 = 450\ pmol\ ATP\ s - 1\ per\ 2.5 \times 107\ \text{cells} \tag{14}$$

In general form (Eq. 13) becomes:

$$q_{ATP,OxPhos} = \left(q_{O_2} - q_{O_2,non-OxPhos}\right) \times \left(\frac{H/O}{H/ATP}\right) \times 2 \tag{15}$$

In addition to ATP production via OxPhos, an important fraction of q_{ATP} in *T. cruzi* epimastigotes comes from glycolytic- and mitochondrial substrate-level phosphorylation (collectively:

SubPhos, $q_{ATP,\ SubPhos}$) (*see* Eq. 8). Glycolytic SubPhos typically yields two ATP molecules per hexose. Equivalently, the glycolytic SubPhos can be quantified as the excretion of pyruvate and alanine (product of glutamate:pyruvate transamination), with one ATP per Pyr/Ala. This might be a more accurate measurement if multiple hexoses are present in the medium. The mitochondrial SubPhos refers to the production of ATP in the ASCT/SCS cycle (*see* **Note 19**), with one ATP per acetate or propionate produced. Major sources of mitochondrial acetyl-CoA and propionyl-CoA are glycolysis (pyruvate), amino acid catabolism, and the beta-oxidation of fatty acids [70–73]. The contribution of $q_{ATP,\ SubPhos}$ to the total q_{ATP} can be inferred using the GEM, constrained with the fluxes of the excreted products (e.g., by NMR).

Measuring Metabolite Concentrations with NMR One of the most robust ways to quantify fluxes in complex systems is by tracking metabolite concentrations in time using ^{1}H-NMR. Here, we provide a protocol for monitoring these fluxes by NMR using *T. cruzi* epimastigotes as an example. For materials, *see* Subheading 2.2.3.

1. Grow epimastigote culture to early-stationary phase (10^8 cells/mL) in LIT medium (*see* **Note 8**).
2. Collect 1 mL sample from the culture on the first once the cells reach stationary density (10^8 cells/mL) and 24 h after this point. Register the cell density of each sample for control.
3. Centrifuge the culture (3000× *g*, 10 min, 4 °C) to remove cells and cell debris.
4. Deproteinize: Filter supernatant (3 kDa MWCO filter).
5. Freeze the filtered supernatant at −80 °C until the day of data acquisition (*see* **Note 20**).
6. Collect 300μL of sample, and mix it with 240μL deuterium oxide, 60μL of 5 mM the internal standard (TMSP-$_{d4}$), and transfer to a 5-mm glass tube for ^{1}H-NMR acquisition (*see* **Note 21**).
7. Input for determining concentrations: ^{1}H-NMR peak integrals (from specialized software) for glucose, fructose, acetate, propionate, and the internal standard.
8. Normalize the concentration against the internal standard:

$$[metabolite]\ (mM) = \frac{metabolite\ peak\ area \times [TMSP]}{TMSP\ peak\ area} \tag{16}$$

9. For flux calculations, *see* Subheading 3.2.2

Calculation of the ATP Maintenance Parameters
To get the values for $q_{ATP,\ NGAM}$ and n_{GAM}, we will need to use the flux measurements (and growth rate for growing lifecycle stages) to constrain the GEM. The basic principle is as follows: the constraints will dictate the maximal ATP production flux (q_{ATP}), and with using multiple sets of experimental measurements, we will be able to decompose the q_{ATP} into its respective parts from (Eq. 7). Keep in mind that we need to perform the calculations sequentially to obtain these values.

Using the consumption and production fluxes computed from data for *non-proliferative* lifecycle stages, we can infer the value of $q_{ATP,\ NGAM}$: from (Eq. 7). In non-growing conditions (when $\mu = 0$), $q_{ATP} = q_{ATP,\ NGAM}$,

1. Create the reaction of ATP hydrolysis that we will call ATPM (*see* **Note 22**), where ATP is hydrolyzed into ADP and inorganic phosphate ($ATP + H_2O \rightarrow ADP + P_i + H^+$). Set this reaction as an objective flux (maximize).
2. Set the lower and upper flux bounds of the biomass equation to [0, 0].
3. Set the exchange fluxes bounds for known consumption and excretion fluxes (oxygen, glucose consumption; pyruvate, acetate excretion, etc.).
4. Optimize the model: the flux through the ATPM reaction will be $q_{ATP,\ NGAM}$.
5. Set the $q_{ATP,\ NGAM}$ as the lower flux bound for the ATPM reaction. Proceed to estimating the n_{GAM}.

With this computation, we have determined the $q_{ATP,\ NGAM}$ term of (Eq. 7). In a similar vein, to obtain the n_{GAM} value to be input into the BOF, we will need to apply a similar routine as described above, but this time with the data coming from *growing* lifecycle stage. *Warning:* the lower flux bound set in **step 5** of the protocol above should remain set!

1. Set the growth rate of the cells: constrain the BOF to the experimentally observed growth rate (i.e., [μ, μ]). Do *not* set the BOF as the objective; make sure your initial BOF does *not* have ATP hydrolysis terms as indicated in (Eq. 3).
2. Set the exchange fluxes bounds for known consumption and excretion fluxes.
3. Optimize the model: the resulting flux through the NGAM reaction will be $q_{ATP} - n_{Biomass} \times \mu$, because the $n_{Biomass} \times \mu$ is already set by the flux bounds of the BOF.

4. To extract the $n_{GAM} \times \mu$ (Eq. 7), subtract the $q_{ATP,\ NGAM}$ from the obtained flux value. Divide the resulting value by μ to obtain the n_{GAM} value.
5. Add the ATP hydrolysis terms to the BOF with the coefficient of n_{GAM} or $-n_{GAM}$, depending on whether they are substrates or products. The final BOF should adhere to the format in (Eq. 3)

3.2.6 Algorithms and Software to Obtain Lifestage-Specific Models

Data pre-processing usually is the most labor-intensive task to obtain the lifestage-specific models. The actual process is conceptually rather straightforward: in the end, we would like to have a lifecycle stage–specific GEM (more broadly, a context-specific GEM) that contains only reactions for which an [active] enzyme is present at the said lifecycle stage. In other words, metabolic functionalities without the support from the data should be removed from the model (Fig. 2c).

You might consider coding your own algorithm, or leveraging a software package that is publicly available. It is best to make a (semi-)automated pipeline for the stage-specific models that can be redone easily: in case you modify the pan-GEM, the stage-specific models have to be rebuilt as well. Depending on the exact formulation of your algorithm (or the choice of a publicly available one), the process might become computationally heavy because of memory- and computing power-demanding tasks (e.g., solving mixed-integer linear programs [MILPs]). In either case, the process should translate the expression (or enzyme activity) data into an "action plan" to remove reactions from the GEM to obtain a data-informed, reduced model.

Due to multiple confounded factors, like detection limits in -omics data, incomplete annotations of metabolic enzymes, and moonlighting functions, the initial context-specific GEM will probably lack essential reactions to perform most of the metabolic tasks/objectives. This is the usual outcome regardless of the method (custom or publicly available) used, and threshold applied to discriminate between "present" and "absent" enzymes. Thus, algorithms usually also include a "mitigation" step, where reactions, critical to performing metabolic tasks/objectives, are added back to the GEM.

To summarize the algorithm of turning a GEM into a lifecycle stage–specific model:

1. Match the GPRs with the entries in the expression dataset (of the lifestage of interest) in order to score the GPRs (levels of enzymes) in the GEM.
2. Remove reactions for which GPRs are scored below the predefined threshold value.

3. Check if the model still can perform the metabolic tasks/ objectives (e.g., produce biomass, consume, and/or produce specific metabolites).
   If yes: **go to 5**. Else: **go to 4**.
4. Determine a reaction (or a reaction set) to add back to the model, and **return to 3**. Note that different algorithms take different (computational) strategies to identify the reaction (s) to add back.
5. Obtain the final lifecycle stage–specific model.

In the last 15 years, multiple software packages have been developed to automate this task and get to models that can perform desired metabolic functions. Two popular algorithms for generating context-specific models are GIMME [74] and tINIT [75] (and the recently-published "fast tINIT", or ftINIT [76]), and we refer the reader to a recent review/benchmark of some of these methods [77] for a more in-depth discussion.

3.2.7 Naming the Lifestage-Specific Models

The naming of the lifestage-specific models should follow the naming of the PAN-model (*see* Subheading 3.1.5). We propose to append that name with an underscore and a three-letter abbreviation that reflects the lifestage (e.g., **_try** for trypomastigote or **_bsf** for bloodstream form).

3.3 Validation of the Lifecycle Stage Models Using Experimental Data

The lifecycle stage model inherently should function with constraints not violated, and objective(s) satisfied, for example, experimentally determined growth rate set as the lower bound of the BOF. However, the model may predict a flux distribution through the network that, albeit satisfies the constraints, is not necessarily realistic. To determine whether the predicted flux distributions make sense, the stage-specific models can be further validated with experimental data from that life stage. The experimental validation of GEMs must be designed according to the specific objective function that was defined in its construction. In most cases, the objective function is a simulation of proliferation, or biomass production as the growing lifecycle stages have been more extensively studied. Below, we will describe some common experimental data (sets) that can be used to validate the model.

3.3.1 Known Auxotrophies

Many protistan parasites are auxotrophic for certain essential metabolites, such as amino acids and nitrogenous bases, that is, they are unable to synthesize them (instead, they rely on uptake from the host or extracellular environment). Refined metabolic models must accurately reproduce these auxotrophies in silico as part of the validation routine. For this, one should restrict the lower exchange bound for the essential metabolite to zero (i.e., do not allow uptake); the model should report no growth [78].

3.3.2 Known Metabolic End Products

For many parasites there will be known end products that they produce and/or metabolic maps for the specific lifecycle stage (e.g., [79]). In addition, quantitative exo-metabolomic data are a source of information about the organism's metabolism. The model should be able to reproduce these findings. To evaluate this, flux distributions for the model can be calculated, with the medium (and potentially measured uptake fluxes) as constraints (*see* Subheading 3.2.2). With FBA, a flux distribution can be found at the minimal or maximal value of the objective function, for example, at the maximal growth rate. In addition, FVA can predict the flux ranges that are possible for all reactions at a certain value of the objective function, for example, for the growth rate that is experimentally observed. These simulated metabolite profiles can then be compared to the experimental data.

3.3.3 Gene Essentiality Screens

The information from gene essentiality screens can be used to benchmark the model, that is, measure the consistency of the reconstruction. In some cases, they might be more useful for model quality control than for validation.

One can simulate gene knockdowns by lowering the upper flux bound of a reaction; in a similar vein, setting the flux bounds of a reaction to zero will mimic a knock-out. The effect on the objective function value can be then compared to datasets from genetically modified cells (e.g., by using methods such as CRISPR/Cas9 (see, for example, [80] for *T. brucei* and [81] for *T. cruzi*) or RNAi (for *T. brucei*: [82]). The agreement between the model prediction and experimental data will reflect the accuracy of the model.

In a gene-essentiality analysis, the genes fall into one of four groups: True positive (TP), True negative (TN), False positive (FP), and False negative (FN). A "true" (TP and TN) means that computational prediction matches the experimental data: the gene is either essential for growth (TP) or not (TN). In the FP situation, when non-essential genes are predicted to be essential, it is important to check the model for missing reactions to identify potential metabolic routes that bypass the reaction using the enzyme encoded by the FP gene. Lastly, the FNs—the genes that are known to be essential, but are predicted to be non-essential—must be evaluated with caution. This suggests that a bypass might exist in the model unless the model is sufficiently constrained.

There are several points to take into account when comparing the model predictions to the experimental dataset. First, it is important how essentiality is defined. Is a gene essential for growth when upon depletion the cells die or when growth is reduced below a threshold (say, below 5% of the original growth rate)? Model predictions should be evaluated with the same threshold in mind. Furthermore, be aware of the size of the experimental dataset when reporting the fraction of correct predictions (and comparing this to other models). If the experimental dataset is small, the percentage of correct predictions may be artificially high.

3.4 Scientific Questions for the Lifecycle Stage–Specific Models

After the construction and validation of the lifecycle stage specific model, the model can be used for in silico studies. Here we name a few options—often inspired from outside the parasite field.

3.4.1 Comparing Network Structures between Lifecycle Stages

The different lifecycle stage models for a parasite (and between parasites) can be compared. This can both be done for the entire network and for subsystems (pathways). The *T. cruzi* lifecycle stage models showed significant differences among metabolic pathways [11].

3.4.2 Prediction of Intracellular Fluxes

The model can predict which (parts of) pathways are operational within cells. The more constraints on the extracellular fluxes, the more reliable the predicted intracellular fluxes will be. In liver cancer, the predicted intracellular fluxes revealed why glutamate was excreted as a metabolic end-product [83].

Enzymes or even whole metabolic pathways that were not present in some lifecycle stage models usually result in alternative flux distributions. Models can be used to systematically assess these flux changes. For this, functional enzymes or pathways can be removed from the stage-specific models, or vice versa, absent enzymes or pathways can be included in silico.

3.4.3 Predicting Essential Genes beyond the Validation Dataset

The gene essentiality analysis focused on the genes that were tested in experimental datasets, so a model-wide gene essentiality test may reveal novel drug targets (e.g., [11]). The reactions that turn out to be non-essential for growth may be further evaluated in a double-knockout set-up, where all possible reaction pairs are selected and both reactions of the pair are turned off at the same time [84].

3.4.4 Design of Optimized Culture Media

Some cells may be limited in their growth because the medium does not contain all components they need in the right concentrations. Through a reduced cost analysis, it can be identified which metabolite(s) currently limit(s) growth. This can steer medium design. Interestingly, such an analysis may also show which chemical elements are scarce in the medium and can identify which metabolites could potentially provide this element given the organisms' metabolic capabilities. In case of several alternative metabolites, availability and costs may be taken into account for the final choice. Such an analysis has been previously done to allow bulk production of the whooping cough pathogen *Bordetella pertussis* [36].

The above are just a few examples of what is possible. The power of GEMs is that they give access to the genome-encoded metabolic potential of the organism. The added value of lifecycle stage–specific GEMs is that they go beyond the genome sequence to explore the relevant submaps that are expressed at a particular stage in the life cycle.

4 Notes

1. Keep in mind that some of these packages are written in programming languages like MATLAB that require obtaining a license (free trial version available).
2. In this reaction, ATP is hydrolyzed into ADP and inorganic phosphate ($ATP + H_2O \rightarrow ADP + P_i + H^+$).
3. Sometimes, the exact formula is unknown or irrelevant (e.g., does not participate in chemical conversion); then, to represent these parts of the molecule, the "remainder" within the formulae can be labeled as "R." Consider the pair of transfer RNAs, the free tRNA, and its version loaded with the amino acid. Prior to loading, the tRNA could be represented as "RH," meaning that we are not interested in the molecular composition apart from the site that receives the amino acid residue. The loaded tRNA then would be represented by swapping "H" with the amino acid residue (free amino acid minus the -OH which make up water during the aminoacyl–tRNA ligation reaction). For instance: glycine $C_2H_5NO_2$ will be loaded on the respective tRNA RH to form C_2H_4NOR and H_2O.
4. Typically, an assumption of intracellular pH to be close to neutral (pH = 7) is a good starting point.
5. FAIR stands for Findable, Accessible, Interoperable, and Reusable.
6. It remains a challenge to keep biomass constant while in relevant conditions. The example of this 12-h study was done in a growth medium. Therefore, growth can be expected, but they calculated fluxes as if the cell numbers stayed constant. Close inspection of the data shows that not all the points are perfectly on the line. This may be because with a doubling time of BSF *T. brucei* of approximately 6 h, there may have been some growth in their experiments. The authors only report no cell death, but if they would have monitored cell numbers, they may have been able to use the flux calculation for growing cells.
7. The exponential growth phase is often the most relevant and theoretically stationary phase cells may alter their size (dry weight) and biomass composition.

8. Cell culture protocols are different for different (lifecycle stages of) parasites and therefore outside the scope of this chapter.
9. Centrifugation speed depends on the species—the cells should all be spun down but stay intact.
10. In many protocols for other organisms, cells are suspended in (and washed with) water or phosphate-buffered saline (PBS) to exclude proteins from the fetal calf serum (FCS)-containing growth media. We have experienced that this is detrimental to fragile parasites and you lose a lot of biomass. Instead, we do a control with medium alone to correct for carry-over of proteins from the medium.
11. Avoid the liquid from touching the ring of the clamp; if it does touch, make sure to rinse it down in the next step.
12. The BOFdat tool [64] can also automate the calculation of the growth-associated maintenance and non-growth-associated maintenance. However, we recommend performing calculations manually since the BOFdat tool was made with the intention to infer parameters for microbial GEMs from extensive (and precise) physiological data.
13. Sometimes, precise measurements are not available for the organism of interest; in these cases, a good first approximation of the $q_{ATP,\ NGAM}$ and n_{GAM} values can be taken GEM from a closely related organism [85].
14. The P/O is ~2.5 ATP/O if electrons are passed onto proton-pumping Complex I; otherwise (Complex II/Complex I without a proton pump) the value is ~1.5.
15. The physiological concentrations at which the cells were collected will be maintained using this supernatant.
16. For NADH: 10/4 = 2.5 ATP/O, for $FADH_2$ (Q-junction): 6/4 = 1.5 ATP/O.
17. May vary (e.g., 3.3 H^+/ATP in bacteria).
18. The factor 2 accounts for 2 oxygen atoms per O_2.
19. ASCT (acetyl-CoA:succinyl-CoA transferase) enzyme catalyzes the CoA transfer from acetyl-CoA or propionyl-CoA onto succinate. SCS (succinate-CoA ligase), a step of the canonical TCA cycle, catalyzes phosphorylation of ADP to ATP via succinate-CoA hydrolysis into succinate and free CoA.
20. NMR signal acquisition is influenced by instrument-specific settings (e.g., receiver gain, scan number) and reference compound selection (e.g., TSP, formate); validate and adjust these parameters prior to data collection.
21. After spectrum acquisition, the characteristic signals for each metabolite and the quantification standard must be integrated using specialized software. Metabolite concentrations shall

then be calculated by comparison with the internal standard, applying appropriate normalization procedures.

22. ATPM is for "ATP maintenance." Some GEMs also call this reaction "NGAM" to explicitly specify which kind of ATP maintenance this reaction represents.

Acknowledgments

This work was supported by Fundação de Amparo à Pesquisa do Estado de São Paulo (FAPESP) grant 2021/12938-0 (awarded to A.M.S.) and Conselho Nacional de Pesquisas Científicas e Tecnológicas (CNPq) grant 307487/2021-0 (awarded to A.M.S.). P.G. also thanks SURF (https://www.surf.nl) for computational support through the access to the Dutch National Supercomputer Snellius.

References

1. Judge A, Dodd MS (2020) Metabolism. Essays Biochem 64(4):607–647. https://doi.org/10.1042/EBC20190041
2. van Niekerk DD, Penkler G, du Toit F et al (2016) Attacking blood-borne parasites with mathematics. In: Müller S, Cerdan R, Radulescu O (eds) Comprehensive analysis of parasite biology: from metabolism to drug discovery: from metabolism to drug discovery. Wiley-VCH Verlag GmbH & Co. KGaA, pp 513–541
3. Stelling J (2004) Mathematical models in microbial systems biology. Curr Opin Microbiol 7(5):513–518. https://doi.org/10.1016/j.mib.2004.08.004
4. Bakker BM, Michels PA, Opperdoes FR et al (1997) Glycolysis in bloodstream form Trypanosoma brucei can be understood in terms of the kinetics of the glycolytic enzymes. J Biol Chem 272(6):3207–3215. https://doi.org/10.1074/jbc.272.6.3207
5. Kerkhoven EJ, Achcar F, Alibu VP et al (2013) Handling uncertainty in dynamic models: the pentose phosphate pathway in Trypanosoma brucei. PLoS Comput Biol 9(12):e1003371. https://doi.org/10.1371/journal.pcbi.1003371
6. Saavedra E, Marin-Hernandez A, Encalada R et al (2007) Kinetic modeling can describe in vivo glycolysis in Entamoeba histolytica. FEBS J 274(18):4922–4940. https://doi.org/10.1111/j.1742-4658.2007.06012.x
7. Bexkens ML, Martin OMF, van den Heuvel JM et al (2024) The unusual kinetics of lactate dehydrogenase of Schistosoma mansoni and their role in the rapid metabolic switch after penetration of the mammalian host. Int J Parasitol 54(7):367–378. https://doi.org/10.1016/j.ijpara.2024.03.005
8. van Niekerk DD, du Toit F, Green K et al (2023) A detailed kinetic model of glycolysis in Plasmodium falciparum-infected red blood cells for antimalarial drug target identification. J Biol Chem 299(9):105111. https://doi.org/10.1016/j.jbc.2023.105111
9. Montagud A, Navarro E, Fernandez de Cordoba P et al (2010) Reconstruction and analysis of genome-scale metabolic model of a photosynthetic bacterium. BMC Syst Biol 4:156. https://doi.org/10.1186/1752-0509-4-156
10. Chiappino-Pepe A, Pandey V, Billker O (2021) Genome reconstructions of metabolism of plasmodium RBC and liver stages. Curr Opin Microbiol 63:259–266. https://doi.org/10.1016/j.mib.2021.08.006
11. Shiratsubaki IS, Fang X, Souza ROO et al (2020) Genome-scale metabolic models highlight stage-specific differences in essential metabolic pathways in Trypanosoma cruzi. PLoS Negl Trop Dis 14(10):e0008728. https://doi.org/10.1371/journal.pntd.0008728
12. O'Leary NA, Cox E, Holmes JB et al (2024) Exploring and retrieving sequence and metadata for species across the tree of life with NCBI datasets. Sci Data 11(1):732. https://doi.org/10.1038/s41597-024-03571-y

13. Kanehisa M, Furumichi M, Sato Y et al (2025) KEGG: biological systems database as a model of the real world. Nucleic Acids Res 53 (D1):D672–D6D7. https://doi.org/10.1093/nar/gkae909
14. Chang A, Jeske L, Ulbrich S et al (2021) BRENDA, the ELIXIR core data resource in 2021: new developments and updates. Nucleic Acids Res 49(D1):D498–D508. https://doi.org/10.1093/nar/gkaa1025
15. Consortium, T U (2024) UniProt: the universal protein knowledgebase in 2025. Nucleic Acids Res 53(D1):D609–DD17. https://doi.org/10.1093/nar/gkae1010
16. Henry CS, DeJongh M, Best AA et al (2010) High-throughput generation, optimization and analysis of genome-scale metabolic models. Nat Biotechnol 28(9):977–982. https://doi.org/10.1038/nbt.1672
17. Terzer M, Maynard ND, Covert MW et al (2009) Genome-scale metabolic networks. Wiley Interdiscip Rev Syst Biol Med 1(3): 285–297. https://doi.org/10.1002/wsbm.37
18. Bruggeman FJ, Westerhoff HV (2007) The nature of systems biology. Trends Microbiol 15(1):45–50. https://doi.org/10.1016/j.tim.2006.11.003
19. Bordbar A, Monk JM, King ZA et al (2014) Constraint-based models predict metabolic and associated cellular functions. Nat Rev Genet 15(2):107–120. https://doi.org/10.1038/nrg3643
20. Cuevas DA, Edirisinghe J, Henry CS et al (2016) From DNA to FBA: how to build your own genome-scale metabolic model. Front Microbiol 7:907. https://doi.org/10.3389/fmicb.2016.00907
21. Thiele I, Palsson BO (2010) A protocol for generating a high-quality genome-scale metabolic reconstruction. Nat Protoc 5(1):93–121. https://doi.org/10.1038/nprot.2009.203
22. Machado D, Andrejev S, Tramontano M et al (2018) Fast automated reconstruction of genome-scale metabolic models for microbial species and communities. Nucleic Acids Res 46(15):7542–7553. https://doi.org/10.1093/nar/gky537
23. Alstott J, Breakspear M, Hagmann P et al (2009) Modeling the impact of lesions in the human brain. PLoS Comput Biol 5(6): e1000408. https://doi.org/10.1371/journal.pcbi.1000408
24. Mendoza SN, Olivier BG, Molenaar D et al (2019) A systematic assessment of current genome-scale metabolic reconstruction tools. Genome Biol 20(1):158. https://doi.org/10.1186/s13059-019-1769-1
25. Agren R, Liu L, Shoaie S et al (2013) The RAVEN toolbox and its use for generating a genome-scale metabolic model for Penicillium chrysogenum. PLoS Comput Biol 9(3): e1002980. https://doi.org/10.1371/journal.pcbi.1002980
26. Gu C, Kim GB, Kim WJ et al (2019) Current status and applications of genome-scale metabolic models. Genome Biol 20(1):121. https://doi.org/10.1186/s13059-019-1730-3
27. Monk JM, Lloyd CJ, Brunk E et al (2017) iML1515, a knowledgebase that computes Escherichia coli traits. Nat Biotechnol 35(10): 904–908. https://doi.org/10.1038/nbt.3956
28. Marinos G, Kaleta C, Waschina S (2020) Defining the nutritional input for genome-scale metabolic models: a roadmap. PLoS One 15(8):e0236890. https://doi.org/10.1371/journal.pone.0236890
29. Orth JD, Thiele I, Palsson BO (2010) What is flux balance analysis? Nat Biotechnol 28(3): 245–248. https://doi.org/10.1038/nbt.1614
30. Prieto-de Lima TS, Rojas-Jimenez K, Vaglio C (2024) Strategy for optimizing vitamin B12 production in pseudomonas putida KT2440 using metabolic modeling. Meta 14(11). https://doi.org/10.3390/metabo14110636
31. Pelicaen R, Gonze D, De Vuyst L et al (2020) Genome-scale metabolic modeling of Acetobacter pasteurianus 386B reveals its metabolic adaptation to cocoa fermentation conditions. Food Microbiol 92:103597. https://doi.org/10.1016/j.fm.2020.103597
32. Herrmann HA, Dyson BC, Vass L et al (2019) Flux sampling is a powerful tool to study metabolism under changing environmental conditions. NPJ Syst Biol Appl 5:32. https://doi.org/10.1038/s41540-019-0109-0
33. Lieven C, Beber ME, Olivier BG et al (2020) MEMOTE for standardized genome-scale metabolic model testing. Nat Biotechnol 38(3):272–276. https://doi.org/10.1038/s41587-020-0446-y
34. Raman K, Kratochvíl M, Olivier BG et al (2024) FROG Analysis Ensures the Reproducibility of Genome Scale Metabolic Models. bioRxiv:2024.09.24.614797. https://doi.org/10.1101/2024.09.24.614797
35. Orth JD, Conrad TM, Na J et al (2011) A comprehensive genome-scale reconstruction of Escherichia coli metabolism—2011. Mol Syst Biol 7:535. https://doi.org/10.1038/msb.2011.65

36. Branco Dos Santos F, Olivier BG, Boele J et al (2017) Probing the genome-scale metabolic landscape of Bordetella pertussis, the causative agent of whooping cough. Appl Environ Microbiol 83(21). https://doi.org/10.1128/AEM.01528-17
37. Zhang C, Sanchez BJ, Li F et al (2024) Yeast9: a consensus genome-scale metabolic model for S. cerevisiae curated by the community. Mol Syst Biol 20(10):1134–1150. https://doi.org/10.1038/s44320-024-00060-7
38. Robinson JL, Kocabas P, Wang H et al (2020) An atlas of human metabolism. Sci Signal 13(624). https://doi.org/10.1126/scisignal.aaz1482
39. Wang H, Robinson JL, Kocabas P et al (2021) Genome-scale metabolic network reconstruction of model animals as a platform for translational research. Proc Natl Acad Sci USA 118(30). https://doi.org/10.1073/pnas.2102344118
40. Anton M, Almaas E, Benfeitas R et al (2023) Standard-GEM: standardization of open-source genome-scale metabolic models. bioRxiv:2023.03.21.512712. https://doi.org/10.1101/2023.03.21.512712
41. Lahtvee PJ, Seiman A, Arike L et al (2014) Protein turnover forms one of the highest maintenance costs in Lactococcus lactis. Microbiology (Reading) 160(Pt 7):1501–1512. https://doi.org/10.1099/mic.0.078089-0
42. Elsemman IE, Rodriguez Prado A, Grigaitis P et al (2022) Whole-cell modeling in yeast predicts compartment-specific proteome constraints that drive metabolic strategies. Nat Commun 13(1):801. https://doi.org/10.1038/s41467-022-28467-6
43. Michels PAM, Villafraz O, Pineda E et al (2021) Carbohydrate metabolism in trypanosomatids: new insights revealing novel complexity, diversity and species-unique features. Exp Parasitol 224:108102. https://doi.org/10.1016/j.exppara.2021.108102
44. Liu Y, Beyer A, Aebersold R (2016) On the dependency of cellular protein levels on mRNA abundance. Cell 165(3):535–550. https://doi.org/10.1016/j.cell.2016.03.014
45. Clayton CE (2014) Networks of gene expression regulation in Trypanosoma brucei. Mol Biochem Parasitol 195(2):96–106. https://doi.org/10.1016/j.molbiopara.2014.06.005
46. Zikova A, Verner Z, Nenarokova A et al (2017) A paradigm shift: the mitoproteomes of procyclic and bloodstream Trypanosoma brucei are comparably complex. PLoS Pathog 13(12):e1006679. https://doi.org/10.1371/journal.ppat.1006679
47. Shameer S, Logan-Klumpler FJ, Vinson F et al (2015) TrypanoCyc: a community-led biochemical pathways database for Trypanosoma brucei. Nucleic Acids Res 43(Database issue): D637–D644. https://doi.org/10.1093/nar/gku944
48. Doyle MA, MacRae JI, De Souza DP et al (2009) LeishCyc: a biochemical pathways database for Leishmania major. BMC Syst Biol 3: 57. https://doi.org/10.1186/1752-0509-3-57
49. Stouthamer AH, Bettenhaussen C (1973) Utilization of energy for growth and maintenance in continuous and batch cultures of microorganisms. A reevaluation of the method for the determination of ATP production by measuring molar growth yields. Biochim Biophys Acta 301(1):53–70. https://doi.org/10.1016/0304-4173(73)90012-8
50. Haanstra JR, van Tuijl A, van Dam J et al (2012) Proliferating bloodstream-form Trypanosoma brucei use a negligible part of consumed glucose for anabolic processes. Int J Parasitol 42(7):667–673. https://doi.org/10.1016/j.ijpara.2012.04.009
51. Pineda E, Thonnus M, Mazet M et al (2018) Glycerol supports growth of the Trypanosoma brucei bloodstream forms in the absence of glucose: analysis of metabolic adaptations on glycerol-rich conditions. PLoS Pathog 14(11):e1007412. https://doi.org/10.1371/journal.ppat.1007412
52. Teusink B, Wiersma A, Molenaar D et al (2006) Analysis of growth of lactobacillus plantarum WCFS1 on a complex medium using a genome-scale metabolic model. J Biol Chem 281(52):40041–40048. https://doi.org/10.1074/jbc.M606263200
53. Simensen V, Schulz C, Karlsen E et al (2022) Experimental determination of Escherichia coli biomass composition for constraint-based metabolic modeling. PLoS One 17(1):e0262450. https://doi.org/10.1371/journal.pone.0262450
54. Benthin S, Nielsen J, Villadsen J (1991) A simple and reliable method for the determination of cellular RNA content. Biotechnol Tech 5(1):39–42. https://doi.org/10.1007/BF00152753
55. Smith PK, Krohn RI, Hermanson GT et al (1985) Measurement of protein using bicinchoninic acid. Anal Biochem 150(1):76–85. https://doi.org/10.1016/0003-2697(85)90442-7
56. Bradford MM (1976) A rapid and sensitive method for the quantitation of microgram quantities of protein utilizing the principle of

protein-dye binding. Anal Biochem 72:248–254. https://doi.org/10.1016/0003-2697(76)90527-3

57. Van Handel E (1985) Rapid determination of total lipids in mosquitoes. J Am Mosq Control Assoc 1(3):302–304
58. Kawai Y, Moribayashi A (1982) Characteristic lipids of Bordetella pertussis: simple fatty acid composition, hydroxy fatty acids, and an ornithine-containing lipid. J Bacteriol 151(2): 996–1005. https://doi.org/10.1128/jb.151.2.996-1005.1982
59. Ralton JE, Sernee MF, McConville MJ (2021) Evolution and function of carbohydrate reserve biosynthesis in parasitic protists. Trends Parasitol 37(11):988–1001. https://doi.org/10.1016/j.pt.2021.06.005
60. Van Handel E (1985) Rapid determination of glycogen and sugars in mosquitoes. J Am Mosq Control Assoc 1(3):299–301
61. Masuko T, Minami A, Iwasaki N et al (2005) Carbohydrate analysis by a phenol-sulfuric acid method in microplate format. Anal Biochem 339(1):69–72. https://doi.org/10.1016/j.ab.2004.12.001
62. Parrou JL, Francois J (1997) A simplified procedure for a rapid and reliable assay of both glycogen and trehalose in whole yeast cells. Anal Biochem 248(1):186–188. https://doi.org/10.1006/abio.1997.2138
63. Saldanha AJ, Brauer MJ, Botstein D (2004) Nutritional homeostasis in batch and steady-state culture of yeast. Mol Biol Cell 15(9): 4089–4104. https://doi.org/10.1091/mbc.e04-04-0306
64. Lachance JC, Lloyd CJ, Monk JM et al (2019) BOFdat: generating biomass objective functions for genome-scale metabolic models from experimental data. PLoS Comput Biol 15(4): e1006971. https://doi.org/10.1371/journal.pcbi.1006971
65. Nascimento JF, Souza ROO, Alencar MB et al (2023) How much (ATP) does it cost to build a trypanosome? A theoretical study on the quantity of ATP needed to maintain and duplicate a bloodstream-form Trypanosoma brucei cell. PLoS Pathog 19(7):e1011522. https://doi.org/10.1371/journal.ppat.1011522
66. Opperdoes FR, Michels PA (2008) Complex I of Trypanosomatidae: does it exist? Trends Parasitol 24(7):310–317. https://doi.org/10.1016/j.pt.2008.03.013
67. Clarkson AB Jr, Bienen EJ, Pollakis G et al (1989) Respiration of bloodstream forms of the parasite Trypanosoma brucei brucei is dependent on a plant-like alternative oxidase. J Biol Chem 264(30):17770–17776
68. Gnaiger Erich, M T G: (2020) Mitochondrial physiology. Accessed 2025
69. Hinkle PC (2005) P/O ratios of mitochondrial oxidative phosphorylation. Biochim Biophys Acta 1706(1–2):1–11. https://doi.org/10.1016/j.bbabio.2004.09.004
70. Alencar MB, Girard R, Crispim M et al (2025) The role of l-serine and l-threonine in the energy metabolism and nutritional stress response of Trypanosoma cruzi. mSphere 10(3):e0098324. https://doi.org/10.1128/msphere.00983-24
71. Alencar MB, Marsiccobetre S, Mengarda AC et al (2025) Energy metabolism in Trypanosoma cruzi: the validated and putative bioenergetic and carbon sources. mBio:e0221524. https://doi.org/10.1128/mbio.02215-24
72. Faria JDN, Eufrasio AG, Fagundes M et al (2025) Inhibition of L-threonine dehydrogenase from Trypanosoma cruzi reduces glycine and acetate production and interferes with parasite growth and viability. J Biol Chem 301(2): 108080. https://doi.org/10.1016/j.jbc.2024.108080
73. Souza ROO, Damasceno FS, Marsiccobetre S et al (2021) Fatty acid oxidation participates in resistance to nutrient-depleted environments in the insect stages of Trypanosoma cruzi. PLoS Pathog 17(4):e1009495. https://doi.org/10.1371/journal.ppat.1009495
74. Becker SA, Palsson BO (2008) Context-specific metabolic networks are consistent with experiments. PLoS Comput Biol 4(5): e1000082. https://doi.org/10.1371/journal.pcbi.1000082
75. Agren R, Mardinoglu A, Asplund A et al (2014) Identification of anticancer drugs for hepatocellular carcinoma through personalized genome-scale metabolic modeling. Mol Syst Biol 10(3):721. https://doi.org/10.1002/msb.145122
76. Gustafsson J, Anton M, Roshanzamir F et al (2023) Generation and analysis of context-specific genome-scale metabolic models derived from single-cell RNA-Seq data. Proc Natl Acad Sci USA 120(6):e2217868120. https://doi.org/10.1073/pnas.2217868120
77. Molversmyr H, Oyas O, Rotnes F et al (2023) Extracting functionally accurate context-specific models of Atlantic salmon metabolism. NPJ Syst Biol Appl 9(1):19. https://doi.org/10.1038/s41540-023-00280-x
78. Renz A, Widerspick L, Drager A (2021) First genome-scale metabolic model of Dolosigranulum pigrum confirms multiple Auxotrophies. Meta 11(4). https://doi.org/10.3390/metabo11040232

79. Muller M, Mentel M, van Hellemond JJ et al (2012) Biochemistry and evolution of anaerobic energy metabolism in eukaryotes. Microbiol Mol Biol Rev 76(2):444–495. https://doi.org/10.1128/MMBR.05024-11
80. Kovarova J, Novotna M, Faria J et al (2022) CRISPR/Cas9-based precision tagging of essential genes in bloodstream form African trypanosomes. Mol Biochem Parasitol 249: 111476. https://doi.org/10.1016/j.molbiopara.2022.111476
81. Picchi-Constante GFA, Hiraiwa PM, Marek M et al (2022) Efficient CRISPR-Cas9-mediated genome editing for characterization of essential genes in Trypanosoma cruzi. STAR Protoc 3(2):101324. https://doi.org/10.1016/j.xpro.2022.101324
82. Alsford S, Turner DJ, Obado SO et al (2011) High-throughput phenotyping using parallel sequencing of RNA interference targets in the African trypanosome. Genome Res 21(6): 915–924. https://doi.org/10.1101/gr.115089.110
83. Nilsson A, Haanstra JR, Engqvist M et al (2020) Quantitative analysis of amino acid metabolism in liver cancer links glutamate excretion to nucleotide synthesis. Proc Natl Acad Sci USA 117(19):10294–10304. https://doi.org/10.1073/pnas.1919250117
84. Goldstein YA, Bockmayr A (2015) Double and multiple knockout simulations for genome-scale metabolic network reconstructions. Algorithms Mol Biol 10(1):1. https://doi.org/10.1186/s13015-014-0028-y
85. Grigaitis P, Grundel DAJ, van Pelt-KleinJan E et al (2022) A computational toolbox to investigate the metabolic potential and resource allocation in fission yeast. mSystems 7(4): e0042322. https://doi.org/10.1128/msystems.00423-22

Chapter 20

Enhanced Detection of Homology Using Artificial Intelligence in Euglenids

Mark C. Field

Abstract

Identification of similarity between protein sequences is an important component for the assignment of function. With ever-growing databases of genome sequence, this becomes an increasing challenge, and especially in the detection of relationships between distantly related sequences, which is frequently an issue with euglenids. The introduction of artificial intelligence tools to the prediction of protein structure has been, without exaggeration, revolutionary. In particular, AlphaFold3 (AF3), the latest iteration of the AI predictor from DeepMind, a Google subsidiary, offers a potent combination of speed, accuracy, and ease-of-use, all free of charge. Here I will describe a basic workflow for the detection of low similarity between proteins, that is otherwise cryptic, using AF3, discuss how to interpret the predictions, and highlight examples of bizarre predictions or hallucinations.

Key words AI, Protein structure, Sequence evolution, AlphaFold, Homology

1 Introduction

Assigning function to a specific protein sequence usually requires multiple lines of evidence, one of which is sequence similarity. This can be in the form of overall sequence relatedness, which implies orthology or paralogy, or of more limited scope, such as the presence of a conserved domain or fold. With an explosive growth in low-cost nucleic acid sequencing providing an ever-increasing landscape for evolutionary analysis, both the opportunities and challenges for such studies have become greater. Additional challenges arise in terms of evolutionary distance of gene sequences from canonical model organisms, upon which accurate annotation is often based. In the case of euglenids and their kinetoplastid relatives this is particularly important—it is now clear that these organisms lie very close the root of the eukaryotic tree [1] and that this provides both the promise of deep evolutionary insight, coupled with the challenge of identification of homology. Given that

Paul A. M. Michels et al. (eds.), *Euglenozoa: Methods and Protocols, Volume 1*, Methods in Molecular Biology, vol. 3013, https://doi.org/10.1007/978-1-0716-5142-1_20,

structural information is considerably more conserved than simple sequence [2, 3], the availability of such data is of great value in the identification of relationships.

Understanding the principles underpinning protein structure has been a long-standing goal and has included efforts to define restrictions to amide bond configuration in the Ramachandran plot [4] and physicochemical properties as used in the Kyte–Doolittle hydropathy score [5]. However, accurate models have required a template, as used by Phyre2, but which of course have meant that the vast majority of proteins have been refractory to structural predictions [6]. The emergence of DeepMind's AlphaFold3 (AF3) has provided a true leap forward in accuracy, speed, and user-friendliness and brings the use of structural comparisons to augment identification of relationships between proteins truly to all [7–9].

2 Materials

2.1 A Basic Bioinformatics Environment

Computational platform requirements can be very modest as the heavier computation is web-based. A mid-range Mac/PC and an internet connection are obviously essential, but beyond this requirements are simple. A large screen is a significant advantage, as laptop screens are restrictive for sequence comparisons and structural visualizations. Most of the software that is recommended is either available for download (with academic licenses usually free) or can be used via a web browser. Table 1 details a small number of applications that the author finds indispensable, together with some of the major global data repositories. You will also require a free Google account for running AF3.

3 Methods

3.1 Basic Running of Alphafold3

1. Log in to https://alphafoldserver.com/welcome. Here you will find the basic start page. Paste your sequence into the box annotated as "input" and select sequence type as "protein."
2. At the free tier you can run 30 searches in a 24-h period, with each search limited to 5000 tokens. A token is one amino acid residue or one nucleotide, or an atom of another kind of molecule. There are therefore higher costs to adding post-translational modifications such as N-glycans, if these are particularly important to your prediction, but in the vast majority of analyses, this is likely not the case.
3. The input box is somewhat unforgiving and any numbers or non-amino acid characters will result in the job failing. You can

Table 1
Software, databases, and tools

Sequence manipulation
JalView (*https://www.jalview.org/*)
An excellent "Swiss Army knife" of functionality, multiplatform, continually updated, slightly dated design. JalView also supports viewing of protein structure files. Online support is also offered. JalView is cross platform.
BBEdit (*https://www.barebones.com/products/bbedit/index.html*)
A commercial text editor/programming environment, which is actively developed and maintained. Full functionality requires a paid license, but a free version with some restricted functionality remains excellent and is sufficient for most purposes. MacOS only.
Notepad++ (*https://notepad-plus-plus.org/*)
Notepad++ is free and open-source, supporting several languages. It is highly customizable and suitable for various programming needs. Windows only.
Visual studio code (*https://code.visualstudio.com/*)
Visual studio Code from Microsoft is a free and extensible code editor for building web, desktop, and mobile applications, and allows plain text manipulation. Cross platform, so particularly useful if collaborating with folks on multiple OSs.
https://www.reverse-complement.com/cleanup.html
An indispensable tool for removing non-residue characters from sequence files. Can also perform some simple sequence manipulations.
Structure display
ChimeraX (*https://www.cgl.ucsf.edu/chimerax/*)
An excellent "Swiss Army knife" of functionality, multiplatform, continually updated and like JalView, slightly dated design. The newer versions now include the ability to implement a-fold analysis, either by retrieval of pre-predicted structures or de novo analysis using Google playbook (which requires a Google login). Chimera can visualize protein structures in a huge range of formats and display options.
PyMOL (*https://pymol.org/*)
A commercial program, but which has an impressive array of features for processing images for presentation and examination.
Software resources
European bioinformatics institute (*https://www.ebi.ac.uk/*)
Sequence analysis, recalculated protein structure predictions, domain annotation, and many more options.
Expasy at Swiss Institute of Bioinformatics (*https://www.expasy.org/*)
Offers a suite of software, together with the SwissProt database, a highly curated sequence repository.

The examples are used by the author regularly and have been arrived at by trial and error after evaluating many packages. These are all stable, curated, and unlikely to cease functionality any time soon. The list is far from exhaustive, and there are a great many other high-quality tools available, at some level which different workers prefer is one of personal choice. With one exception, all of these tools can be used free of charge clean your sequence at https://www.reverse-complement.com/cleanup.html if needed.

4. You may add more than one sequence to the input field, if you believe that you are working with a heterooligomer, but the total sequence entered must not exceed the 5000 token limit. Furthermore, you may also add more than one copy of a given sequence, again being mindful of the token limitation, if you are aware that the prediction should be of a homo-oligomer.
5. Once you trigger your job by clicking the "Continue and preview job" button, you will be presented with a new window that will provide the opportunity to name the analysis. By default, each job is named by date and time. It is recommended that you rename the job with something more specific for later retrieval.
6. Once your job is completed, it will become available in the history panel. A vertical ellipsis at right provides several options, most relevant are "Open results" and "Download." It is recommended that you download each analysis so that you may examine the predictions offline as well as archive the data.
7. The results page contains a lot of information. First, the top model prediction is displayed in an interactive window and the structure is color-coded according to confidence (Fig. 1). Coloring uses the pLDDT score, which is essentially a measure of confidence in the position of atoms on a 0–100 scale. Next to this is the predicted alignment error, a grid plot that illustrates where high confidence positions are situated for the individual residues plotted against each other. This is particularly useful in detecting if there are domains present and detected as these tend to have lower PAE scores, and can indicate if the protein has just interactions in a contiguous manner or if there is higher order structure produced.
8. The final parameters available on the results page are ipTM and pTM. These are the calculated scores for the overall accuracy, with a pTM above 0.5 indicating a likely accurate prediction. The ipTM value is similar but is only calculated for the modelling of subunits in a complex [10, 11].
9. Finally download the dataset and decompress it. This is essential for accessing additional models to the top prediction.

3.2 Interpretation

1. The following are recommendations and depending on the user's expertise may well be supplanted by more detailed approaches.
2. The folder that you have downloaded contains 26 files; fold_year_month_day_ hour_min_full_data_0.json to fold_year_month_day_hour_min_full_data_4.json, fold_year_month_day_hour_min_model_0.cif to fold_year_month_day_hour_min_model_4.cif, fold_year_month_day_hour_min_

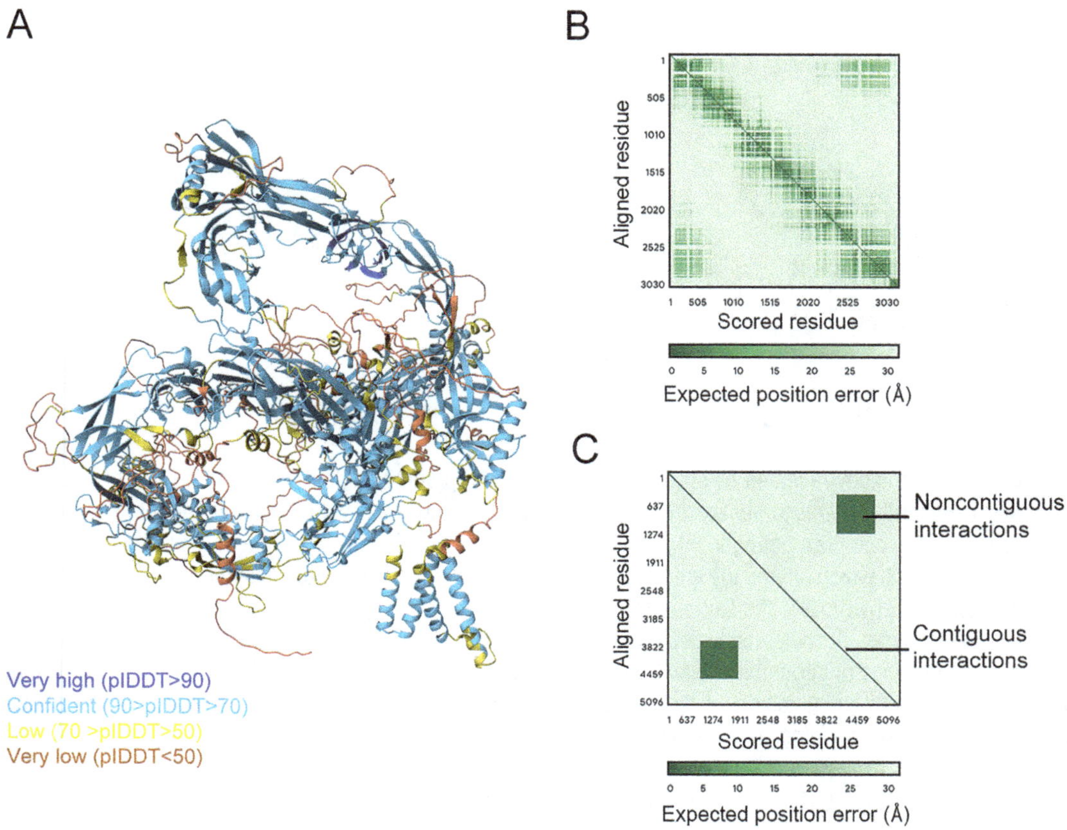

Fig. 1 Data output from Google AlphaFold 3 server. Panel (**a**) AF3 prediction for *Trypanosoma brucei* LAP333, revealing confident prediction scores for much of the structure. Where low confidence is displayed (pIDDT<50) is almost exclusively for linker regions and at the termini of domains (70 > pIDDT>50). Panel (**b**) PAE plot for TbLAP333 which is the estimate of the error in relative separation of two amino acids in the predicted structure. Higher values indicate higher predicted error which equates to lower confidence in the predicted structure. Panel (**c**) The PAE data can be considered also as evidence for close residue interactions, that is, within domain, which would sit upon the diagonal, or for more distant structural connections, off diagonal within the PAE plot

summary_confidences_0.json to fold_year_month_day_hour_min_summary_confidences_4.json and terms_of_use.md.

3. These correspond to five predictions of the structure. The fold_year_month_day_ hour_min_full_data_0.json etcetera files contain the raw data while fold_year_month_day_hour_-min_summary_confidences_0.json is a summary of the confidences of the prediction, including the pTM/ipTM scores, percent disorders and other parameters. The .json files (javascript object notation) are human-readable and can be edited with a text editor (Table 1).

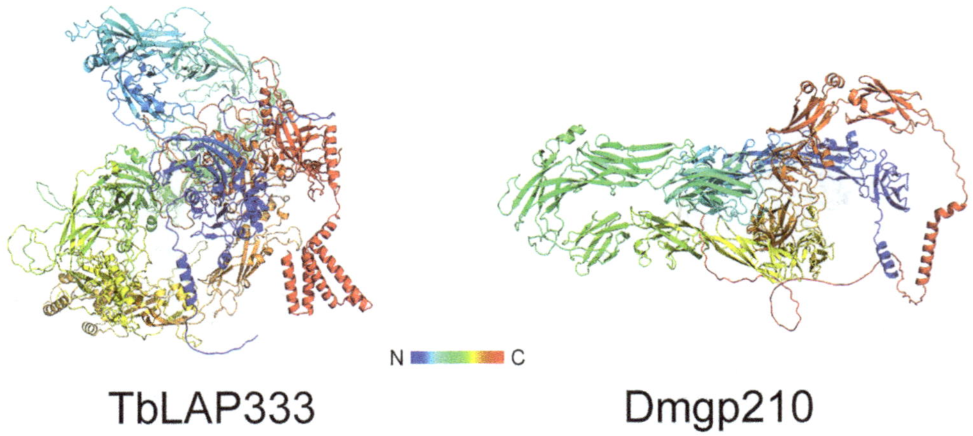

Fig. 2 Cryptic homology revealed by structure prediction. AF3 predictions for *Trypanosoma brucei* TbLAP333 (left) and *Drosophila melanogaster* Dmgp210 (right). Gp210 is a nuclear pore complex component and crucial to assembly of the complex, acting to interface with the nuclear envelope membrane. The equivalent to gp210 in yeast is Pom152, which shares the architecture of gp210 but is quite divergent in sequence, and neither will identify each other by BLAST; the same is the case for LAP333. The trypanosome protein was identified from interactome analysis of nuclear envelope proteins [13], and localized as a nuclear envelope protein: AF3 predicts the presence of Ig domains, which suggests that LAP333, gp210 and Pom153 are homologs, despite varying considerably in molecular weight. Protein structures were rendered with PyMOL and are shown colored with rainbow shading (N-terminus blue to C-terminus red)

4. The .cif files contain the structural model data (.cif = crystallographic information file), with file 0 the most confident and file 4 the least. These files can be viewed in Chimera and/or PyMOL.
5. Standard colorization options can help to interpret the predictions. In Fig. 2 are shown two nuclear pore complex (NPC) protein predictions, from *Trypanosoma brucei* (TbLAP333) and *Drosophila melanogaster* (Dmgp210). There is no detectable sequence similarity between these two proteins, but AF3 reveals a clear architectural relatedness in that both proteins are predominantly composed of Ig-like domains [12, 13], and together with additional evidence is sufficient to suggest a common function in anchoring the NPC to the nuclear envelope, as well as retention of architecture across eukaryotes.

3.3 Alternatives and Extending AlphaFold Access

1. Many software packages are now incorporating AF as a component, perhaps a reflection of the rapid adoption of this technology.
2. ChimeraX is a software package for the visualization of molecular structural data and is an example of software with AF connectivity. It can be downloaded from https://www.cgl.ucsf.edu/chimerax [14]. Download the appropriate version for your platform, unpack, and install.

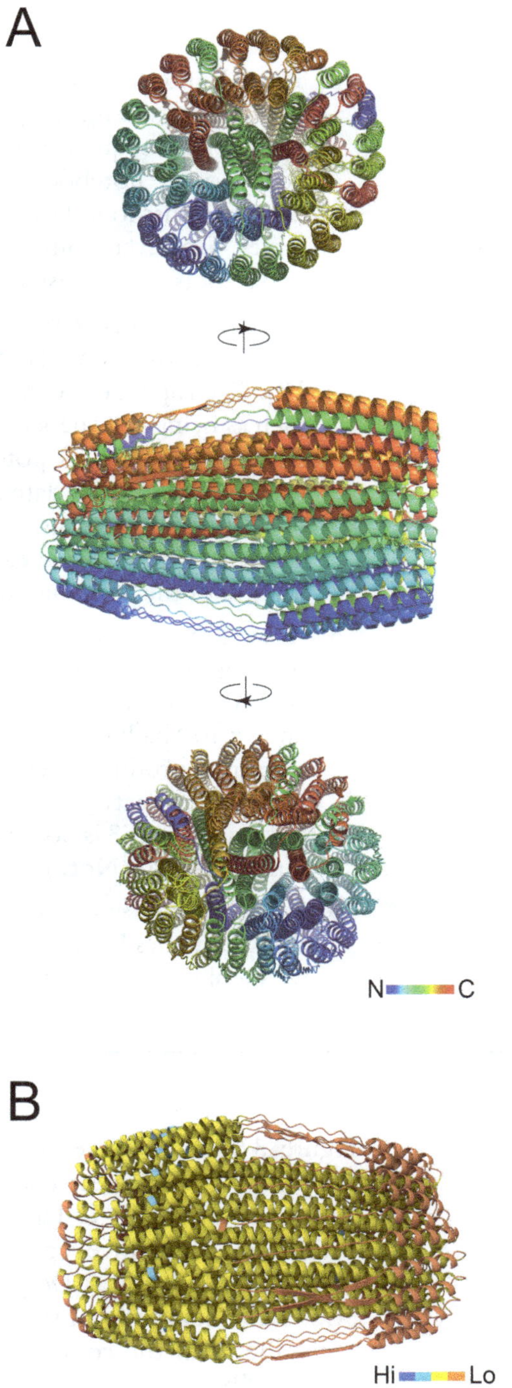

Fig. 3 Example of hallucinatory structure prediction. AF3 prediction of the homodimer of the central repeat region of *Trypanosoma brucei* NUP-1. The sequence contains 17 perfect repeats, which are coiled coil with interspersed unstructured linker regions. The predicted structure has collapsed these repeats into a nested structure, which is inconsistent with known in cellulo behavior [20]. Protein structures were rendered with PyMOL and each monomer is colored with rainbow shading (N-terminus blue to C-terminus red). In panel b, structural confidence is shown colored by pIDDT value: dark blue >90, light blue $90 >$ to >70, yellow >50, and orange <50. Note that essentially all of the prediction falls into the lower two confidence categories

3. The AF interface is under Tools/Predictions/AlphaFold. At the time of writing the present version of ChimeraX does not allow longer sequences to be analyzed and relies on a ColabFold Google notebook [15] that interfaces with AF2, which the author has found to be slower and less reliable than the AF3 server. However, some users may find the presence of additional tools for the visualization of structure an advantage.
4. A further alternative to calculating an AF3 prediction ab initio is to use the AlphaFold database and/or FoldSeek [16–18]. The AlphaFold database has some obvious disadvantages, such as being fully dependent on data that may not be updated frequently as well as potentially using predictions not made with the most up to date approaches; for example, at the time of writing predictions, use AF2. FoldSeek allows structural datasets to be searched easily and rapidly by innovative simplification of the structural data format used to search [17].
5. As with many predictive software, it is important to have some concept of what an expected/accurate result would be. AI software, such as ChatGPT and other LLM-based software, is known to "hallucinate," a polite way of saying that these systems can produce bizarre results, with little or no basis in reality. Polydactyly in images of humans or felines is not uncommon. AF3 is not dissimilar as very unrealistic predictions can emerge (*see* **Note**).
6. An online course in the use of AF3 is offered by the European Bioinformatics Institute (EBI) and can be accessed at https://www.ebi.ac.uk/training/online/courses/alphafold/ [19].

4 Notes

Figure 3 provides an example of such a hallucination, and is a prediction of the repeat core region of the trypanosome nuclear lamina protein NUP-1. The submission to AF3 is two copies of a 1750 residue repetitive region and which AF3 predicts to fold back on itself multiple times. Despite a clear aesthetic appeal, there is no evidence to support that such a configuration is adopted in cellulo but rather that the conformation is more likely to be open and forming fibers.

Acknowledgments

I am heavily indebted to Erin Butterfield (University of Dundee) for her careful and detailed work in structure prediction and considerable patience in educating the author in appropriate and best practice with AlphaFold.

References

1. Williamson K, Eme L, Baños H et al (2025) A robustly rooted tree of eukaryotes reveals their excavate ancestry. Nature 640(8060):974–981
2. Krissinel E (2007) On the relationship between sequence and structure similarities in proteomics. Bioinformatics 23(6):717–723. https://doi.org/10.1093/bioinformatics/btm006. Epub 2007 Jan 22. PMID: 17242029
3. Chothia C, Lesk AM (1986) The relation between the divergence of sequence and structure in proteins. EMBO J 5(4):823–826
4. Ramachandran GN, Ramakrishnan C, Sasisekharan V (1963) Stereochemistry of polypeptide chain configurations. J Mol Biol 7:95–99
5. Kyte J, Doolittle RF (1982) A simple method for displaying the hydropathic character of a protein. J Mol Biol 157(1):105–132
6. Powell HR, Islam SA, David A et al (2025) Phyre2.2: a community resource for template-based protein structure prediction. J Mol Biol 23:168960
7. Abramson J, Adler J, Dunger J et al (2024) Accurate structure prediction of biomolecular interactions with AlphaFold 3. Nature 630(8016):493–500
8. Wee J, Wei GW. Benchmarking AlphaFold3's protein-protein complex accuracy and machine learning prediction reliability for binding free energy changes upon mutation. ArXiv [Preprint]. 2024 6:arXiv:2406.03979v1
9. Mifsud JCO, Lytras S, Oliver MR et al (2024) Mapping glycoprotein structure reveals Flaviviridae evolutionary history. Nature 633(8030):695–703
10. Zhang Y, Skolnick J (2004) Scoring function for automated assessment of protein structure template quality. Proteins 57(4):702–710
11. Xu J, Zhang Y (2010) How significant is a protein structure similarity with TM-score = 0.5? Bioinformatics 26(7):889–895
12. Upla P, Kim SJ, Sampathkumar P et al (2017) Molecular architecture of the major membrane ring component of the nuclear pore complex. Structure 25(3):434–445
13. Butterfield ER, Obado SO, Scutts SR et al (2024) A lineage-specific protein network at the trypanosome nuclear envelope. Nucleus 15(1):2310452
14. Meng EC, Goddard TD, Pettersen EF et al (2023) UCSF ChimeraX: tools for structure building and analysis. Protein Sci 32(11): e4792
15. Mirdita M, Schütze K, Moriwaki Y et al (2022) ColabFold: making protein folding accessible to all. Nat Methods 19(6):679–682
16. https://alphafold.com/
17. van Kempen M, Kim SS, Tumescheit C et al (2024) Fast and accurate protein structure search with FoldSeek. Nat Biotechnol 42(2): 243–246
18. Elfmann C, Stülke J (2025) Cutting-edge tools for structural biology: bringing AlphaFold to the people. Trends Microbiol 33:S0966-842X(25)00110-6
19. https://www.ebi.ac.uk/training/online/courses/alphafold/
20. Padilla-Mejia NE, Koreny L, Holden J et al (2021) A hub-and-spoke nuclear lamina architecture in trypanosomes. J Cell Sci 134(12): jcs251264

INDEX

Paul A.M. Michels et al. (eds.), *Euglenozoa: Methods and Protocols, Volume 1*, Methods in Molecular Biology, vol. 3013, https://doi.org/10.1007/978-1-0716-5142-1,

www.ingramcontent.com/pod-product-compliance
Lightning Source LLC
Chambersburg PA
CBHW081223070426
42612CB00024B/67
* 9 7 8 1 0 7 1 6 5 1 4 1 4 *